Nuclear Politics

Nuclear Politics

ENERGY AND THE STATE IN THE
UNITED STATES, SWEDEN, AND FRANCE

James M. Jasper

PRINCETON UNIVERSITY PRESS
PRINCETON, NEW JERSEY

Copyright © 1990 by Princeton University Press
Published by Princeton University Press, 41 William Street,
Princeton, New Jersey 08540
In the United Kingdom: Princeton University Press, Oxford

All Rights Reserved

Library of Congress Cataloging-in-Publication Data

Jasper, James M., 1957–
Nuclear politics : energy and the state in the United States,
Sweden, and France / James M. Jasper.
Includes bibliographical references.
1. Nuclear energy—Government policy—United States. 2. Nuclear
energy—Government policy—Sweden. 3. Nuclear energy—Government
policy—France. I. Title.
HD9698.U52J37   1990   333.792'4—dc20   89-49566

ISBN 0-691-07841-6 (alk. paper)

This book has been composed in Linotron Caledonia

Princeton University Press books are printed
on acid-free paper, and meet the guidelines for
permanence and durability of the Committee on
Production Guidelines for Book Longevity of
the Council on Library Resources

Printed in the United States of America by
Princeton University Press, Princeton, New Jersey
10 9 8 7 6 5 4 3 2 1

For Jane Howard-Jasper

Contents

List of Figures and Tables   ix
Preface   xi
Abbreviations   xvii

PART ONE: *Explaining Nuclear Policies*

CHAPTER 1
Introduction   3

CHAPTER 2
Partisan Cleavages and Policy Styles   21

PART TWO: *Creating Nuclear Systems: The Triumph of Technological Enthusiasm, 1960–1973*

CHAPTER 3
The Triumph of Technological Enthusiasm in the United States   41

CHAPTER 4
Early Victory for Light Water in Sweden   64

CHAPTER 5
The Difficult Transition to Light Water in France   74

CHAPTER 6
Commercial Success in Three Countries   98

PART THREE: *To Build or Conserve: Dilemmas Arising from Public Opposition and the Oil Crisis, 1973–1976*

CHAPTER 7
The Reassertion of the Economic Perspective in the United States   107

CHAPTER 8
Party Politics in Sweden   129

CHAPTER 9
Technological Enthusiasm at the Top in France   148

CHAPTER 10
Elite Discretion in Three Countries   178

PART FOUR: *The Structures Tighten: Policy Divergence and the Loss of Flexibility, 1976–1989*

CHAPTER 11
High Costs and Decentralization of Control in the United States   187

CHAPTER 12
Political Paralysis and Antinuclear "Compromise" in Sweden   218

CHAPTER 13
Political Repression and Low Costs in France   237

CHAPTER 14
Structures and Flexibility in Three Countries   256

PART FIVE: *Conclusions*

CHAPTER 15
What Have We Learned?   267

List of Informants   278

Bibliography   283

Index   311

Figures and Tables

FIGURES

3.1	Commercial Reactors Ordered in the United States since 1953	47
3.2	Size of American Reactors Being Ordered, 1953–1978	48
5.1	Sources of EDF Funds, 1952–1984	93
6.1	Reactor Orders in the United States, Sweden, and France	99
14.1	Acceptability of Nuclear Energy in American, Swedish, and French Public Opinion	261

TABLES

2.1	Views of Selected Groups on How to Solve Environmental Problems, United States, 1980	34
4.1	Commercial Reactors Ordered in Sweden	69
5.1	EDF Nuclear Reactor Orders, to 1985	90
11.1	Construction Delays for American Nuclear Plants	200

Preface

SOCIAL SCIENTISTS love structures. They lend a sense of rigor to an analysis. They are hidden to the casual observer. But they are important in shaping and constraining our actions. Political analysts have recently rediscovered "structures" as a rallying cry, using political and economic structures to explain not only policies and decisions but also to identify contestants and their goals. Some popular variables examined in this manner include the autonomy of state bureaucracies, centralization of authority within the state, public ownership of economic production, financial structures, and openness to international trade. Structural constraints like these are enormously important in shaping state policies.

But the structuralists often overstate their case, ignoring other important causal variables when they concentrate on relatively formal political and economic structures. *Nuclear Politics* documents some limitations of formal structures as explanations of public policies and political choices. Structures are crucial to our explanations, but they are not the whole story. First, they must be "filled in" with accounts of actors' beliefs and goals: structures can often explain why one group attains its goals, but not always why it has the goals it does. Personal biographies, as well as cultural beliefs and practices should also enter our explanations. Second, we must recognize what David Riesman called the counter-cyclical compensation of ideas. The structural approach to state policies suffers because it ignores many features of reality; proposing an alternative that is just as partial would be no solution. Instead, I shall supplement—not eliminate—structural factors with two kinds of explanatory factors often overlooked: cultural meanings and individual biographies. I began research on nuclear energy with a structural orientation, but I soon realized that political and economic structures alone missed important realities. The more I poked into the details of nuclear energy in the United States, Sweden, and France, the less convincing a purely structural account became.

My framework is just as state-centered as that of most structuralists. Politicians, bureaucrats, and other state actors play more important roles than do nonstate actors like business executives and antinuclear activists. But unlike structuralists who define the state as a set of institutions or structures, I follow Eric Nordlinger in defining it as the public officials with discretion to set policies. This focus allows me to sort out the various influences—structural, cultural, and biographical—on their decisions.

The result will not be a spare, elegant model that reduces policymaking to several key variables. Instead, my goal is to generate useful concepts, metaphors, and insights: tools that can be used to understand state policies in many contexts. My project is more concept-building than theory-building.

Nuclear energy politics is particularly suited to an analysis that shows why structural explanations need to be supplemented with additional factors. There have been excellent comparative studies of nuclear politics, providing a strong rather than a weak target. If even these works need to be bolstered by consideration of cultural and biographical factors, then so also do most structural accounts of state policy. A broader approach to the topic is especially useful for examining nuclear energy's "critical" period, the years following the oil crisis of 1973–74, when elites had especially great discretion to recast laws and administrative structures. The structure of this book encourages a comparison between this period and those before and after, when the balance between unyielding structures and elite flexibility was different.

Because the reader will not be satisfied with assurances about scholarly objectivity, let me briefly state my position on nuclear energy. I am mildly antinuclear, in that I think the risks of nuclear fission are currently greater than society should bear. My objections rest more on economic grounds than moral ones, a position that distinguishes me from much of the activist core of the antinuclear movement. But my opinions of specific national efforts vary greatly. The Swedish program is much safer than the American, for example. The American program is both unsafe and uneconomical. The French program is at least cost-effective. What matters to me most, however, is democracy. I am satisfied if, after intense public discussion, a majority of the public makes an informed decision. I believe Sweden came close to this, at least closer than any other country.

That I have an opinion about nuclear energy should not prevent my being relatively objective in analyzing how policies unfolded. I criticize many decisionmakers, and they are most often pronuclear decisionmakers. This verdict reflects not my personal prejudices, but rather the fact that the nuclear proponents were the ones with power, the ones who made most of the critical decisions. Many mistakes were made in the deployment of nuclear reactors, the most common being that they were adopted prematurely, before adequate operating knowledge had accumulated. Pointing out these mistakes, however, is not evidence of antinuclear bias. Many journalists believe that objectivity comes from presenting two extreme positions and then placing one's own account midway between the two. That is nonsense.

In presenting research results the author usually lists people who have aided the research, just as she lists those interviewed. Other than occasional elaboration for a loving spouse or a demanding dissertation adviser, the contributions of interviewees and supporters are not detailed. The implication is that the research report is objective and straightforward; its origins need not be described since they are irrelevant to the results. This thinking is especially faulty in cross-national and qualitative research. The history and context of the research matter greatly. Thus I shall try to elaborate how various people facilitated my research, just as I list at the end of the book how many times I talked to interviewees, and for how long. Extended conversations over several months are not the same as one sixty-minute interview.

This book began as a dissertation. It took six months in 1983 for my research topic to emerge from a mush of ideas about cross-national comparisons and explanations of policy outcomes. My dissertation adviser, Hal Wilensky, was extraordinary in his diligent reading and line-by-line criticism, helping me clarify my research focus through drafts of grant proposals. My experience with his research project on taxing and spending taught me a lot about comparative research techniques. Conversations with Neil Smelser, Todd La Porte, Gene Rochlin, Aaron Wildavsky, Peter Steen, Jean-Paul Schapira, and especially Alain Touraine also added insights that I hope have stayed with me and appear in the written outcome.

Hal Wilensky's textual criticism paid off, literally. A patchwork of grants enabled me to begin research in 1984 and continue it in 1985. Thanks go to the University of California, which provided a Chancellor's Patent Fund Grant and a Regents' Traveling Fellowship; the Institute of International Studies at Berkeley, which provided two years of research apprenticeships; the Scandinavian-American Foundation, which provided a grant for travel to Sweden; the Fulbright apparatus and the French government, which provided a one-year scholarship in France; and the MacArthur Foundation, which added a year of support after I returned from Europe.

I began my research in California in mid-1984, conducting a case study of the controversy over the Diablo Canyon reactor (Jasper 1985). My first interview, with a rancher in the stunning countryside near San Luis Obispo, convinced me I would enjoy this mode of research. Evie Walters helped me in this research, as she has in so many ways in the years since.

I had extraordinary luck in France in 1984 and 1985. Not only did the dollar hit ten francs, but two organizations graciously gave me places to feel at home in Paris. I was officially attached to Alain Touraine's Centre d'Analyse et d'Intervention Sociologiques, and I learned a great deal from Touraine, Zsuzsa Hegedüs, and their students. Touraine's perspective permeates this study in many ways. In addition, Hélène Meynaud introduced me to the Groupe de Recherche, Energie, Technologie et Société of Electricité de France (EDF), where she works. Not only were Hélène, Jocelyne Smadja, Georges Morlat, Eric Stemmelen, and others very helpful, but the GRETS energy library had nearly everything I wanted on French nuclear energy. Almost all my French informants were helpful, but several added hospitality as well, including Philippe Roqueplo, Bénédicte Vallet, and Jean-Paul Schapira.

My French research concentrated on Electricité de France, since that remarkable organization is almost synonymous with nuclear energy in France. No one declined to be interviewed, even though many of the people I contacted were roughly equivalent to vice presidents in American corporations (chefs du services et des départementes). Contrary to the myth of the elusive French bureaucrat, all were helpful and, I believe, frank. When I interviewed them, nuclear energy was no longer a political issue: EDF had vanquished the antinuclear movement. Officials were even willing to talk about how they had done it. They were also at the right level of the corporate hierarchy; they had several secretaries they could set to work gathering documents while I was there, and they were responsible for large enough chunks of the company that I could profitably spend two or three hours finding out about them. This level of interviewee, just under the top rank, was perfect. Messrs. Clavel and Haond especially stand out.

In addition to personal interviews, EDF allowed me to sit in obscure offices and read the documents related to various siting disputes. I read hundreds of comments written by farmers in distant regions opposed to reactors, antinuclear poems written by members of the counterculture, and form letters written and signed by thousands. Even more revealing, in many ways, were the arrogant, dismissive responses written by elite engineers; there could hardly be a better way to understand French technocracy. My visits to reactor sites were also illuminating. At one of them fuel had not yet been loaded, so I could wander inside; at another a vast field of razor wire had just been planted.

In Sweden I interviewed more politicians than utility managers, since the locus of decision making had been in the Riksdag (parliament) rather than private companies. The cynicism that allowed both government officials and antinuclear activists in France to speak frankly about the state's crushing of dissent was replaced by an earnestness that caused Swedes to be candid but tense. They were candid because the Swedish elite often sees its policymaking practices as a model for others and because it is always polite to foreigners, but they were tense because they thought the nuclear issue was sure to be revived again. Of all my Swedish informants, I especially want to thank Peter Steen and Evert Vedung, two people who helped even more than the others, with innumerable conversations, suggestions, and personal hospitality.

Outside my research, many people made me feel welcome in both France and Sweden. Natalie, Daniel, and Barthélémy Touffu were forever gracious and patient; Hélène Meynaud and Denis Duclos shared their intense warmth; Alf and Eva Lundqvist disproved the stereotype of the cold Swede; and Sylvester Kupczak provided an unusual but productive living situation in Cormeilles-en-Parisis.

Good fortune continued when I returned to the United States in late 1985. Robert Mitchell secured an office for me at Resources for the Future in Washington, D.C., a perfect situation in several ways. I had ready access to the libraries at RFF, the Brookings Institution, the Environmental Law Institute, and the Nuclear Information and Resource Service; I could pester John Ahearne—former NRC commissioner, former member of the Schlesinger energy team, and extremely knowledgeable about nuclear energy in the United States—with my naive questions; and Robert Mitchell and Luther Carter provided stimulating conversations as well as access to their own vast collections of papers and articles on nuclear energy.

The kind of information I gathered and used differed slightly for the three cases. I relied the most on written documents for the United States—since so many exist—and the least for Sweden. But another factor shaped my information gathering: decisions about nuclear energy did not occur in exactly the same institutions in each country. Individual utilities were more important in the United States; the Riksdag more important in Sweden; the intersection of state and EDF in France. Since I could not interview dozens of American utilities, I relied more on the mass of research already done.

News accounts were relevant for all three countries, although I quickly learned how inaccurate reporting can be. Different papers gave conflicting dates and names; their interpretations of events and of politics diverged even more. Among top newspapers, the *New York Times* coverage of European nuclear politics is especially embarrassing. In spite of limitations, I have used the following regularly and extensively: *New York Times*, *Wall Street Journal*, *Economist*, *Le Monde*, and *Dagens Nyheter* among newspapers; and the *Bulletin of the Atomic Scientists*, *Science*, *Nucleonics Week*, *Nuclear News*, *Groundswell*, *Revue générale nucléaire*, and *Gazette nucléaire* among periodicals. *Forbes'* occasional articles on nuclear energy have been accurate and sensitive to political subtleties.

Hal Wilensky, Charles Perrow, and Harvey Feigenbaum provided ferociously close readings and very constructive commentaries on the manuscript, and Valentine Moghadam, Dorothy Nelkin, Neil Smelser, Robert Alford, and Todd La Porte added useful insights on drafts. Conversations with a variety of colleagues helped me sharpen the analytic framework, especially groups at the Russell Sage Foundation and Harvard's Council for European Studies. All the interviewees quoted or cited in the book had a chance to comment, and several wrote extensive suggestions. At New York University, Bess Buckley, Cynthia Beal, Cynthia Gordon, and Jane Poulsen helped in the final stages of preparing this book, and Dean Ann Burton provided some financial assistance.

Finally, my parents have been extremely supportive of my work, especially while I was in Washington. In spite of their very different styles, they have both provided the same positive qualities: confidence in me, a sense of humor, vicarious anxiety, and not a few dinners out. Living with my mother during my year in Washington was far more than convenient; it was a rare chance I shall cherish forever. I proudly dedicate this book to her.

All translations from French and Swedish are my own.

Abbreviations

TECHNICAL AND INTERNATIONAL

BWR	Boiling Water Reactor
ECCS	Emergency Core Cooling System
EEC	European Economic Community
GWe	Gigawatt (electric) = 1000 MWe
IAEA	International Atomic Energy Agency
IEA	International Energy Agency
kWe	kilowatt (electric)
kWh	kilowatt hour
LWR	Light Water Reactor
MWe	Megawatt (electric) = 1000 kWe
NEA	Nuclear Energy Agency (of OECD)
NSSS	Nuclear Steam Supply System
OECD	Organisation for Economic Co-operation and Development
PIUS	Process Inherent Ultimate Safety (Swedish reactor design)
PWR	Pressurized Water Reactor
SECURE	Safe and Environmentally Clean Urban Reactor (based on PIUS)
UNGG	Natural Uranium Gas Graphite (Reactor)

UNITED STATES

ACRS	Advisory Committee on Reactor Safeguards
AEC	Atomic Energy Commission
DOE	Department of Energy
EPA	Environmental Protection Agency
ERDA	Energy Research and Development Administration
FEA	Federal Energy Administration
FERC	Federal Energy Regulatory Commission
GE	General Electric
INPO	Institute of Nuclear Power Operations
JCAE	Joint Committee on Atomic Energy
NIMBY	Not In My Backyard (Used to characterize nuclear opponents)
NRC	Nuclear Regulatory Commission
OTA	Office of Technology Assessment
PIES	Project Independence Evaluation System
PUC	Public Utility Commission (state-level regulator)

TMI	Three Mile Island
TVA	Tennessee Valley Authority
WIPP	Waste Isolation Pilot Project (New Mexico)
WPPSS	Washington Public Power Supply System

Sweden

AB	aktiebolag: incorporated, as in a firm
CDL	Centrala Driftledningen (association of electricity producers)
FKA	Forsmarks Kraftgrupp AB
KBS	Kärnbränslesäkerhet: Nuclear Fuel Safety Project
OKG	Oskarshamns Kraftgrupp
SKBF	Statens Kärnbränsle Förlag: Swedish Nuclear Fuel Supply Company
SKI	Statens Kärnkraftsinspektion: Nuclear Inspectorate
SOU	Statens Offentliga Utredningar: Official State Reports
SSI	Statens Strålskyddsinstitut: National Institute of Radiation Protection

France

AEE	Agence pour les Economies d'Energie
AESOP	Association pour l'Etude des Structures de l'Opinion Publique
AFME	Agence Française pour la Maîtrise de l'Energie
CEA	Commissariat à l'Energie Atomique
CFDT	Confédération Française Démocratique du Travail
CFTC	Confédération Française des Travailleurs Chrétiens
CGE	Compagnie Générale d'Electricité
CGP	Commissariat Géneral du Plan
CGT	Confédération Générale du Travail
CNRS	Centre National de la Recherche Scientifique
Cogema	Compagnie Générale des Matières Nucléaires
CRS	Compagnies Républicaines de Sécurité
DUP	Déclaration d'Utilité Publique
EDF	Electricité de France
ENA	Ecole Nationale d'Administration
FFSPN	Fédération Française des Sociétés de Protection de la Nature
GSIEN	Groupement de Scientifiques pour l'Information sur l'Energie Nucléaire
IEJE	Institut Economique et Juridique de l'Energie
INSEE	Institut National de la Statistique et des Etudes Economiques
IPSN	Institut de Protection et de Sûreté Nucléaire

PCF	Parti Communiste Français
PEON	Commission Consultative pour la Production d'Electricité d'Origine Nucléaire
PS	Parti Socialiste
PSU	Parti Socialiste Unifié
RPR	Rassemblement pour la République
SCPRI	Service Central de Protection contre des Rayonnements Ionisants
SCSIN	Service Central de Sûreté des Installations Nucléaires
UDF	Union pour la Démocratie Française
X	Ecole Polytechnique

PART ONE

Explaining Nuclear Policies

CHAPTER 1

Introduction

WHEN the oil crisis struck in the fall of 1973, the United States, Sweden, and France each had plans to increase reliance on nuclear energy as a source of electricity. Each country had been developing nuclear energy since the 1940s, and each had an ambitious nuclear industry capable of producing reactors and optimistic about exporting them. Each had significant numbers of reactors already in operation, as well as many more planned or under construction, and each had domestic uranium supplies. By 1973 each country was committed to the American light water technology that was thought to be the least expensive and most efficient reactor design. The three countries also had similar regulatory structures governing the licensing and operation of nuclear plants and designed primarily to encourage the construction of new reactors.

By 1973 all three countries faced scattered but growing antinuclear forces, and the three political and regulatory systems dealt with this protest in similar ways. Each system allowed formal participation by the public but blocked that participation from affecting the design or siting of nuclear reactors. France and the United States reorganized their regulatory systems in the early 1970s to further smooth the licensing process. Everyone except the protesters themselves expected antinuclear protest to weaken soon as the public became more knowledgeable about and accustomed to nuclear energy. Almost no one doubted that the three countries' nuclear programs would continue to expand rapidly.

When the price of oil quadrupled in the final months of 1973, it seemed a final guarantee of nuclear energy's bright future. If the relative costs of nuclear energy and oil before the oil crisis favored a large commitment to nuclear reactors, then after the oil crisis the figures should have been even more in favor of nuclear power. The French and American governments were especially quick to proclaim the need for accelerated nuclear construction, with President Nixon's Project Independence in November 1973 and Prime Minister Messmer's Energy Plan in March 1974. Plans were laid to deploy similar light water technology widely enough to produce most of the electricity in all three countries by the year 2000.

Nuclear plans that had been almost identical in 1973, however, diverged markedly within several years. Only France continued with a massive deployment of nuclear energy, pursuing a program scheduled to produce almost 80 percent of that country's electricity when completed in the 1990s. The United States did not even complete all the reactors that had been ordered or that were under construction in 1973. No new reactor orders were placed and completed in the following ten years, while over one hundred existing ones were canceled. At its peak around 1989 nuclear generation produced only 18 percent of American electricity. Between these two extremes, Sweden added ten reactors to the two operating in 1973. They produce 50 percent of Swedish electricity, but the Swedish government has committed itself to shutting down all twelve reactors by the year 2010. The three nuclear commitments could hardly have diverged more: the triumph of one reactor program, the collapse of a second, and the control and curtailment of a third.

A STATE-CENTERED EXPLANATION

Why did the United States, Sweden, and France begin to diverge in their commitment to nuclear energy? The question is important first because large-scale technologies have widespread social effects when they are functioning properly and especially when they are not. We should understand how societies come to use technologies as risky as nuclear energy. The case of nuclear energy also reveals how these three democracies resolve visible and bitter social conflicts over technology. In the 1960s all three had chosen a strong reliance on nuclear energy, but in different ways all rethought that commitment in the mid-1970s. Because democracy demands occasional reconsiderations of policy, we should study what causes them.

I will show that the divergent nuclear commitments are best explained by concentrating on the dynamics of state policymaking. The nuclear case reveals a lot about how, by whom, and for what reasons public policies are made. It especially demonstrates how much the state can be autonomous from groups in civil society. I will also show that we cannot explain all policy decisions on the basis of formal political and economic structures, but must add analyses of elite discretion, structural change, social psychology, and cultural meanings. Finally, I intend to show that there are political and cultural boundaries to economic rationality and calculation, so that the latter are allowed to operate only in certain circumstances. The most precise methods for making policy recommendations eventually run into the realities of political resistance and conflict.

To begin with the third theme, one popular explanation of nuclear policy divergence is that economic rationality dictated different policies in the three countries, based largely on the different natural resources available. The United States has large quantities of coal and oil, while France has little of either. Sweden also has little fossil fuel, but more hydroelectric potential than France. To begin with, this economic rationality explanation in its simple form does not work for the period before the oil crisis, when all three countries pursued similar nuclear plans in spite of contrasting resource endowments. If the explanation is amended to argue that the oil crisis of 1973 imposed economic rationality on energy policies for the first time, it is more plausible. Energy modeling and sophisticated forecasting techniques were rarely employed before 1974 and 1975, and the very idea of an "energy policy" comparing the costs and benefits of different energy sources became widespread only after the oil crisis. Even so, different groups and organizations made conflicting claims about what policies would be most economically rational, and these conflicts were resolved through the use of political power. Even when there was one policy that seemed economically rational, it was often not accepted for political reasons. In other words, economic rationality influenced policy only when the political system allowed it to. It was usually only one policy position among several.

Choices in nuclear energy policy were made through political processes. Some positions and arguments referred to economic rationality, but not all of them. And those that won were not always the most economically cogent. Part of the reason is that convincing economic data have not always been available. Until the late 1970s the precise costs of nuclear energy were unknown, so sound cost comparisons were impossible. The calculations that were used in debates reflected the arbitrary assumptions of the speakers. Speculation was often presented as rational economic calculation. An account of nuclear choices has a place for preferences based on economic criteria, but we must examine how these were influenced or thwarted by politics and policies. What is more, the chosen policies strongly influenced the very cost and safety data that should have been the basis for economically rational decisions.

Political conflicts within the state are crucial to an explanation of French, Swedish, and American commitments to nuclear energy. Part of the explanation must deal with the high politics of elected officials, voting, interest groups, and media coverage; but a larger part must cover the infighting among different bureaucracies and politicians that occurred behind the scenes. Decisions made outside the state did matter—especially in the United States, where utilities, reactor manufacturers, and banks played important roles. But even here, these decisions were shaped decisively by public policies, so that our explanation leads back to the state. If the empirical question of this study is why nuclear commitments diverged, the theoretical one concerns how state policies are made.

How do states generate policy decisions? Most recent accounts of political action and outcomes take a structural approach to political systems and a utilitarian approach to actors within those systems. Actors are assumed to pursue wealth, power, and prestige (in a word, utility), while constrained by their available resources, degree of political savvy and mobilization, and relationship to the state. If an observer can ascertain a group's position in the economic and political structures, she can describe its interests and thereby predict its actions. Political analysis then consists of looking at the constraints imposed by other actors through the structures themselves, and of using the actors' structural positions to explain the relative success of contending groups. Key questions are who has the power to block a certain proposal, who has the power to fulfill her own project, and what projects are compatible with existing structural constraints. Recent analysts have usefully argued that state bureaucrats, having a clear place in the political structure, have their own interests and pursue them. In both public choice theory and political sociology, this state autonomy perspective is a good antidote to theories that saw the political sphere as derivative from the economic.[1]

It is important to "bring the state back in," but in the process many theorists reduce the state to political and economic structures. They may soften their language by speaking of an "institutional approach," but the result is still that human beings are replaced by formal organizations and legal arrangements. In contrast, Eric Nordlinger (1981, 3) defines the

[1] Public choice theory is the application of economic assumptions and models to political phenomena such as voting, party behavior, coalitions, and policymaking. It concentrates on individual behavior, assuming that individuals rationally maximize whatever bundle of satisfactions is most important to them. For example, an elected politician may be analyzed as maximizing her ability to attract votes and hence stay in office (Downs 1957, 1960; Tullock 1967; Palda 1973; and Crain and Tollison 1976), while a bureaucrat is seen as trying to accumulate resources and discretion (Niskanen 1971; Brennan and Buchanan 1977). The public choice approach was honored in 1986, when one of its founders, James Buchanan, was given the Nobel Prize for Economics.

At the same time, many sociologists and political scientists claim to have rediscovered the "state" as an actor having distinct interests of its own and enough autonomy to pursue them. A volume entitled *Bringing the State Back In* (Evans et al. 1985) makes explicit what various authors (e.g., Katzenstein et al. 1978; Krasner 1978; Stepan 1978; Skocpol 1979) have undertaken, "examination of the organization and interests of the state, specification of the organization and interests of socioeconomic groups, and inquiries into the complementary as well as conflicting relationships of state and societal actors" (Skocpol 1985, 20). Debates over this "new" state-centered approach include Krasner (1984), Lentner (1984), and Almond (1988) and his commentators.

Alford and Friedland (1985) describe the "state-centered" tradition of political research as "managerial" to distinguish it from pluralists who concentrated on voters and interest groups in civil society, as well as from class theorists who saw political actions and outcomes as following from class position. My work criticizes this managerial tradition, but from within. Both the managerial and class traditions (which are not always easily separated) are structuralist in the sense I use it in this study. I am interested in "filling in" those structures.

state as "public officials taken all together." This approach allows us to compare the officials' preferences with policy outcomes and to sort out the effects of political and economic structures, cultural meanings, and individual biographies. Another advantage is a focus on conflict within the state—which is the bulk of the action in nuclear politics. As Nordlinger (p. 15) says of the factors that shape officials' preferences: "Simply to mention some of them—the officials' career interests, organizational loyalties, and professional knowledge—makes it abundantly apparent that state preferences are rarely unified preferences. They are usually the product of all sorts of conflict, competition, and pulling and hauling." What factors determine the outcome of that pulling and hauling?

POLITICAL AND ECONOMIC STRUCTURES

Previous works on nuclear energy policies have focused on political and economic structures.[2] They have described the goals of the contending groups and organizations, traced their actions in the political system, and described the ways in which political and economic structures facilitated or blocked each group's influence. In the simplest form, they argued that nuclear programs were curtailed in countries where the political structure allowed the antinuclear movement access to the state, whether through courts, regulatory agencies, or elected officials. Sweden, the United States, the Netherlands, and West Germany are commonly said to follow this pattern. Nuclear programs were unscathed where nuclear and electric industries had access, and where government agencies and officials were insulated from public pressures. France, Japan, and the Soviet Union are used as examples.[3]

[2] I use the term *structure* to refer to the institutional resources and constraints that groups and individuals use in trying to attain their goals. Structures are not limited to negative constraints on actions; they can also *enable* actors in a positive way. In fact, they do both simultaneously, since a pattern of action excludes certain possibilities and allows others. One person's enabling structure may be another's constraining structure. Structures must have some continuity over time, even though they do change. Decisions or events that occur just once are distinct from structures. So are the skills, mental grids, and goals that actors carry in their heads.

In addition, I discuss only political and economic structures rather than social structure (which might be construed as simply any patterns of behavior that persist over time). It is the former that political analysts have used extensively, and rightly so, since they have more explanatory bite and clarity. Social structure defined as patterns of behavior threatens to become tautological: we hardly wish to explain a particular action or decision by the fact that it is one instance of a recurring pattern.

My concern with the openness of political and economic structures is inspired by several social theorists, among them Anthony Giddens (1979, 1984), Alain Touraine (1971, 1977), and Roberto Unger (1987a, 1987b).

[3] Almost all the best works on comparative nuclear energy policy follow this structuralist

This simple account is inadequate empirically since in virtually every advanced country—and certainly in France, Sweden, and the United States—the nuclear industry had far greater access to government agencies and politicians than the antinuclear movement did. As a result, antinuclear movements had little effect on nuclear energy policies in any country, and it is impossible to explain our three countries' policy divergence by means of the antinuclear movement. In those countries that did curtail their nuclear programs, especially the United States and Sweden, the curtailment came largely at the initiative of economic elites or state officials, and the antinuclear movement had only indirect effects. In each of our three countries there was disagreement within the state structures over what nuclear energy policies to pursue, and it was these bureaucratic battles that shaped nuclear energy policies.

Taken separately, many structural factors have influenced nuclear power deployment. Important ones include electoral thresholds for parties to enter parliament; the ability of corporations in mixed economies to withhold investment; the relationship between utilities and regulators; mechanisms for financing nuclear reactors; and competition between reactor manufacturers. But taken all together, structural factors like these do not adequately explain state policies or nuclear outcomes. *Structures* are important, while a structural *approach* (allowing nothing but structures) is insufficient to explain concrete policies.

Some limitations of a purely structural explanation of political outcomes become especially apparent with intrastate conflicts like those over nuclear energy.[4] First, structural approaches tend to recognize formal, legal, codified power more readily than the informal power that comes, for example, from skills at persuading people or attracting favorable public opinion. Informal power is harder to see but very important. A shared language, ritual, or worldview gives a group better access to a government agency, no matter what the formal or legal situation. A group that "speaks a different language" from the agency's can have little influence despite all the formal access in the world. Groups have cultural resources, such as rhetorical or political know-how, that are not reducible to the usual structural resources of money and power.

---

pattern (Kitschelt 1982, 1986; Campbell 1988; Camilleri 1984; and Nelkin and Pollak 1981). When nonstructural factors are mentioned, they are rarely integrated into the main explanatory framework. Bupp and Derian's (1978) concern for the psychology of nuclear "intoxication" is exceptional.

[4] The limitations of the structural approach are imposed by traditions of research and ways of formulating problems rather than by necessary, logical conditions. Many structural analyses rely on biographies, cultural systems of meaning, informal power, and subtle changes in structure, even while they deny them at the explicit theoretical level. I have elsewhere discussed the ways in which problem formulation, theory, methods, and findings cling together in distinct styles of comparative research (Jasper 1987a).

Second, intrastate controversies are often decided by the discretion of top officials rather than by the relative power of the feuding bureaucracies. To the extent that these decisions depend on biographical and psychological factors, the structural approach typically provides an incomplete explanation. Some degree of discretion is found in almost all policy decisions, but that degree varies greatly. Only by comparing goals and projects with structures and outcomes can we judge how much discretion actors have in a given situation. For example policymakers in all three countries had more discretion in nuclear energy policy after the oil crisis in the mid-1970s than they had by the end of the 1970s.

Third, if elected officials are important, so are electoral contests and coalitions. Political parties compete with each other, and they use policy decisions as a way of doing this. Policy outcomes are influenced by how a party distinguishes itself from rivals (typically, but not always, by its ideology), which party is in power, and who are its allies. Changes of government almost always bring the possibility of new policies.

Fourth, structures change continually. In pursuing their goals groups transform structures as well as simply use them. The balance of power among various agencies and elected bodies changes over time in ways a purely structural perspective is slow to see. Informal power changes when bureaucracies develop new skills or better political habits, and both formal and informal power can be changed by decisions from the top. Knowing the goals, perspectives, and ambitions of policymakers helps us understand how and why structures change and develop as they do. The political projects of competing individuals and groups become embodied in the structures as those structures change. A perception of crisis sometimes heightens the possibility for change, whether the crisis is at the level of the entire state, a policy subsystem, or a single bureaucracy.

Finally, the policy positions adopted by government agencies do not always follow clearly from any organizational "stake." This is especially true for bureaucracies that are relatively autonomous or dominated by clear professional ethics, such as councils of economic advisers. Not only individuals but organizations themselves occasionally act contrary to their own material and organizational interests. The structural approach is better at explaining the likelihood of an organization's attaining its goals than at explaining why it has the goals it does.[5]

These weaknesses indicate that a purely structural approach to policy explanation must be "filled in" with additional, especially cultural, fac-

---

[5] The utilitarian streak in most structural explanations is parallel to that in economics, which explicitly brackets the question of where desires come from in order to concentrate on how people try to fill them.

tors.[6] These can help us recognize sources of informal power, understand the perspectives of officials and ideologies of political parties, see why officials change structures in the ways they do, and discern nonmaterial motivations for action. However, to say that explanations based largely on structural factors are inadequate is not to say that particular structural factors are unimportant. Cultural factors complement structural ones; they are not alternatives. Structural factors are the backbone of this book. Market competition, financing mechanisms, bureaucratic autonomy, federal versus unitary systems of government, and other factors all play important roles in my explanation of nuclear energy policies. But the structural backbone needs a lot of other bones, and some flesh.

CULTURAL MEANINGS

Cultural factors in social action—such as worldviews, rhetoric, and problem-solving styles—have a bad reputation among many empirical social scientists. When structural-functionalism dominated the teaching of social science, culture came to be equated with values: vague entities that usually played the role of unmoved mover, a kind of untestable set of beliefs that were the residual category in many explanations. In the Parsonian system values "explained" the ends of social action, although they themselves had no history and hence could not be explained sociologically. Instead of being anchored in concrete social groups, values characterized a society as a whole. Worse, values were often measured or recognized by the same behaviors they were meant to explain (Barry 1978, chap. 4).[7] Marxism too, when it revived in the late 1960s and 1970s, often slighted cultural factors by clinging to a modified base-superstructure model in which economic structures were determinant "in the last instance."

In recent years cultural analysis has returned to social theory, becoming to the 1980s what Marxist analysis was to the previous decade, a set of guidelines and concepts that provide a new way of looking at social structure and action. In the early 1970s young social scientists read Althusser and Poulantzas; in the early 1980s Foucault, Bourdieu, and Habermas. Inspired by anthropology, linguistics, and the philosophy of science, social theory has revived the idea that mental frameworks influence actions, so that actions must be *interpreted*. To explain an action we must

---

[6] Firth (1951, 27) described the role of culture in similar terms: "If society is taken to be an aggregate of social relations, then culture is the content of those relations."

[7] Exceptions to this pattern include the symbolic interactionist tradition as well as sociologists like Erving Goffman and Harold Garfinkel, who studied the construction of meanings in localized settings but whose methods were not easily transferred to macrosociological research programs.

be able to identify what sort of action it is, and to do this involves knowing what the actors believe they are doing. Taking the lead from philosophy and social theory, empirical social science has rediscovered that humans are symbol makers as well as toolmakers.[8]

The language used for cultural factors is varied. Some terms refer to structured beliefs about the world: *Weltanschauung*, worldview, paradigm, mental grid, mental map, cognitive grid or map, and schema, to name a few. Some seem to refer to elements in these totalities: exemplar, prototype, myth, symbol, metaphor, rhetoric, stereotype, bias, and heuristic. Some words indicate more of an unfolding framework: a script or narrative. Then there are rituals, the enacted embodiment of these symbols and meanings. *Nuclear Politics* uses two main cultural factors. *Ideologies* refer to explicit, stated tenets of political parties or movements. *Policy styles*, or *worldviews*, are distinct clusters of images, symbols, rhetoric, and techniques that an individual or group can use in thinking about public problems, developing solutions, and persuading others. Chapter 2 elaborates on these concepts.

All these cultural factors have two sides. They refer on the one hand to publicly observable definitions or actions: we can read books, listen to rhetoric, watch rituals. French structuralism has given us the idea that signs get their meanings through their differences from other signs in the same system (the color tan is defined by comparing it to beige or ivory), not from the intentions of the people who use them. Cultural meanings form a structure with some autonomy from those who use it. On the other hand, culture also consists of unobservable meanings that individuals carry around with them. Symbols have meaning for one person because they have meaning for her friends. Rituals typically embody beliefs their participants hold. The duality of culture is to be public and social at the same time it is internal and individual. Many cultural analysts try to show that they study things shared rather than individual, external rather

---

[8] Among the many works describing and contributing to this change in social thought are those of Habermas (1979, 1984, 1987), Giddens (1976, 1979, 1984), Dallmayr (1981, 1984), Turner (1980), Bernstein (1976), and Hookway and Pettit (1978). One source of the shift is anthropological debates over how to understand alien cultures and rationalities, on which see Winch (1958, 1964) as well as the debates in Wilson (1970), Ryan (1973), and Hollis and Lukes (1982). More generally, anthropology never lost an awareness of the symbolic aspects of human action (Douglas 1970, 1978, 1982, 1986 and Geertz 1973, 1983). Another source is the recent recognition that even the natural sciences are characterized by shifts in worldviews rather than simply by the collection of ever better data (Kuhn 1962, 1977; Feyerabend 1977; Lakatos and Musgrave 1970; Laudan 1977; Hesse 1974, 1980; and Hacking 1981). Third, analysis of subcultures and of cultural institutions and practices themselves revealed the importance of interpreting the mental sets of actors (Hall et al. 1980 and Hall and Jefferson 1976). The philosophy of social explanation reflects trends similar to those in social theory, for example, Taylor (1985) and Turner (1980).

than internal (Wuthnow 1987), but the contrast is false. Cultural symbols are both.[9]

How should these cultural factors change how we study politics? First, we can see additional motivations and goals of political action. Traditional analyses have assumed that groups are motivated to act to further their own material interests, their power over others and over their own lives, or their group's social status. But if ideas are important to people, they will fight to protect them and to prove their symbols and beliefs are correct, even when no power, wealth, or status is at stake. People care about their mental worlds. It is as disturbing to lose one's reality as it is to lose one's wealth. These "ideal interests" count as much as material and status interests, but the structuralist tradition has ignored them.[10]

Cultural factors also help explain political tools and tactics. They acknowledge a broader range of actions as rational than the structuralist tradition does. The latter claims to be able to calculate the optimal (or in some cases satisficing) course of action on the basis of a social actor's structural position and interests. The result is a determinate model like those of the natural sciences, and if an actor deviates from the predictions she may be considered irrational (Pettit 1978). The cultural approach recognizes rationality in a broader range of choices. First, political tactics, which structuralists tend to see as neutral means to goals, can be viewed instead as having moral and symbolic worth in and of themselves. If all political action has a symbolic dimension, actions will sometimes be taken that seem an inefficient way of attaining a group's stated goals. For example, decisionmaking by consensus is awkward and time-consuming, and it discourages swift action; in the structural, utilitarian perspective it is inefficient. But antinuclear groups often adopted it because of the symbolic satisfaction provided. Second, each group—even each individual—will have a different repertoire of strategies, habits, and symbols that will

---

[9] Culture can be seen as a kind of structure, although not to be confused with political and economic structures. Duality characterizes all structures: they both constrain and constitute individual actions; to individuals they seem unchangeable except in minor ways; yet the actions of many individuals add up to changes in the structures. They are "both medium and outcome" of social practices (Giddens 1984, 25). Individuals can use, twist, and play the existing structures, thereby changing them; but they cannot stray too far from the accepted roles and meanings without losing their connection to other people. This tension between innovation at the individual level and stability at the public, common level perfectly characterizes cultural meanings.

[10] Although some cultural analysts wish to abandon the image of culture as providing goals of action (largely because this threatens to be tautological, like Parsonian values), others see this as central. Aaron Wildavsky (1987) claims that political preferences arise from the answers that cultures give to two questions: Who am I? and What shall I do? These questions are the heart of culture, and political goals arise from the answers that culture gives.

influence both its goals and the actions taken to achieve them (Tilly 1978, chap. 5). Structural position does not imply one optimal path of action; the word *repertory* implies freedom to choose between different possibilities within it. This is why actions are viewed as "constructed." Throughout this book contending groups are assumed to rationally pursue their goals, even though their actions are not fully calculable in advance. Different groups have distinct clusters of images, symbols, and rhetorics at their disposal; chapter 2 describes these as "policy styles."

Cultural factors like ideologies and policy styles arise from specific, concrete social practices and settings. But humans inhabit many different settings; we are socialized into groups as diverse as families, churches, nations, work organizations, and professions. Which of these sources of cultural meanings will dominate varies greatly, and their effects will occasionally even be in conflict. This kind of indeterminacy is a central weakness of most cultural analysis—and the reason many social scientists avoid it. But this avoidance has exacerbated the problem; because many cultural phenomena cannot be measured with the rigor needed for statistical analysis, they have not been studied with much rigor at all. They have been left in a ghetto of mushy theorizing and ad hoc use. But the effect of culture can be described clearly, and in many cases cultural phenomena such as worldviews can at least be placed in typologies. Meanings are complex, but their complexity can be reduced to units which social science can use. The first step toward increased rigor is always to specify the concrete group that holds or uses a certain worldview, strategy of action, or rhetoric, whether that group is as extensive as a nation-state or as limited as a political cell of ten people. Systems of meaning and habits of thought—whether we call them worldviews, rhetorics, or policy styles—should always be linked to identifiable social practices.

Finally, cultural considerations can change our image of what a public policy is. Instead of a single choice, with a range of effects on economic and organizational interests and the distribution of power, it becomes a composite of material and symbolic decisions, statements, and commitments. Like nuclear energy itself, nuclear policies mean different things to different groups: a pronuclear policy can be a commitment to large scale business, technological change, progress, economic growth, national strength, regulatory bureaucracy, police surveillance, nondemocratic decisionmaking, or many other things. And at the symbolic level all could be accurate at once. Nuclear energy became a symbolic *lightning rod*, attracting various charged meanings from different groups and individuals. If people's beliefs about the world matter, then we cannot dismiss one component of a policy as "merely symbolic." For example many antinuclear groups thought nuclear energy would lead to increased police surveillance, while their governments dismissed these fears as empirically invalid. Even if they were invalid, the governments' decisions to ignore their fears was a real assault on these groups. These groups believed that the world worked in a certain way (for example, that nuclear energy required police surveillance), and they expected the state to make commitments in line with their worldviews. When states did not, groups concluded that either their beliefs were wrong or that the state was not protecting their interests. Either possibility injured these groups' sense of security.[11]

Worldviews, ideologies, rhetorics, and other cultural factors are intimately connected to political and economic structures. At an ideal typical extreme, each niche in a structure would have its characteristic culture; structure and culture would reinforce each other, and they would not be separable. In the modern industrial world, structures are so complex, and people move around in them so often that this simple correspondence rarely works completely. Instead cultures overlap: an engineer is torn between professional ethics and loyalty to her employer; a politician is torn between ideological commitment and her own advancement. Cultural meanings and structural interests occasionally conflict, providing the observer with a chance to sort out their relative effects.

Individual Biographies

In their emphasis on mental states, cultural factors are related to social psychology. While culture is a system of meanings and tools of action shared by a particular group, psychology examines how the components of culture become embodied and then function in individuals. While cultural factors play a major part in this book, individual biographies are also occasionally used to assess the motives of key actors such as heads of state. Their discretion is often crucial in explaining public policies. Political decisions are made by individuals who have a wide variety of fears, enthusiasms, weaknesses, and biases.

Explanations of the perspectives of individual politicians and officials need not be arbitrary. Biographical information is used to relate individuals to concrete practices, subcultures, and training in their pasts, on the assumption that traces remain. What university did someone attend, and what outlook is it independently known to inculcate? What professional training did a person have? What political parties has she belonged to, with what ideologies? In what organizations has she worked, and what interests and perspectives did they have? The bulk of an individual's

---

[11] Murray Edelman (1964, 1977) has emphasized the symbolic role of policies, although he has also argued (1964) that symbolic elements can reassure the masses while the real effects of policies help elites. It is a mistake to think that the psychic benefits of symbolic politics are illusory.

worldview derives from cultural and structural factors like these, although there is also an idiosyncratic residue not explained through the lens of sociology. Political and economic structures, cultures, and individual biographies must be combined to give adequate explanations of policy outcomes.

CHANGE AND CRISIS

A strongly structural approach to state policies is often slow to see changes in political and economic structures; its focus is on stable structures that shape individual actions, not the reverse. It recognizes changes in structures most readily when they result from dramatic, visible crises. Stephen Skowronek (1982, 10) says: "Crisis situations tend to become the watersheds in a state's institutional development. Actions taken to meet the challenge often lead to the establishment of new institutional forms, powers, and precedents." The result is a concentration on periods when old structures are replaced by new ones, such as revolutions, wars, or economic depressions. Structural factors explain extensive historical transformations of structures—like industrialization or the rise of the nation-state—better than smaller, less visible changes within structures. Yet smaller changes may be vital in explaining policies.[12]

Many kinds of crisis occur within a political system. The broadest crises involve paralysis of the entire system, and these may enhance the possibility for changing the entire system. But subsystems of policymaking organizations can also face crises, increasing the possibility for reorganizing those subsystems. The oil crisis was such an event for energy policy organizations throughout the world, leading to many changes in administrative and decisionmaking structures. Individual agencies, or even departments within agencies, can also lose legitimacy and enter a crisis, raising the chances of change within those agencies. Crises of various kinds increase the chance of structural change; they do not guarantee it. To understand why crisis sometimes causes change and other times does not, we must examine the goals of particular groups and individuals as well as conflicts within the state.

Structures change even without crises, especially in small ways. New laws are passed. Existing laws are interpreted in new ways. New agencies are formed. Existing agencies are given new leaders. Constant interaction occurs between resilient political structures and the officials operating within them. A purely structural orientation biases the case in favor

[12] At the extreme, structural approaches sometimes define crises as "processes in which the structure of a system is called into question" (Offe 1984, 36). I wish to avoid this circularity, defining crises as problems that systems face, which may or may not lead to questioning of those systems.

of the influence of structures on individuals, but a fair assessment of the interaction demands equal weight for the goals and perspectives of individuals and groups. Structural, cultural, and biographical factors are needed to explain transformations over time, although there will always remain a residue of unpredictable chance and discretion.

COMPARATIVE ANALYSIS

A comparative study should be explicit about why its cases were chosen rather than others, and about what is being compared and what not. The United States, Sweden, and France, differ from most countries in that they are advanced industrial democracies, and they differ from other advanced democracies in that they had, in the early 1970s, ambitious and established nuclear industries, domestic uranium supplies, and plans for rapid nuclear expansion. Britain's nuclear industry was as old, but less ambitious. Germany's and Japan's were ambitious but not yet as established. Canada had a significant industry, but it faced the hurdle of convincing other countries that its alternative technology was better than the American light water reactors that most other national industries manufactured. (Sweden's reactor manufacturer, although small, was well established thanks to innovation and a significant construction program.) What is more, our three countries faced more visible and vocal antinuclear movements than most other industrial democracies. Similarities among the three also distinguish them from other industrial democracies. These include centralized nuclear regulators, insulated from public pressures, and almost unanimous state support for nuclear expansion in the early 1970s. We must look to other variables to explain our countries' policy divergence.

What were the relevant differences in our three political systems in the early 1970s that could help explain later policy divergence? It is misleading simply to list various differences, since that portrays them as static factors that only later became relevant. Instead, several insignificant differences early in the 1970s came to have large effects later in the 1970s, because of the interaction between the small original differences, the sharply contrasting policy choices after the oil crisis, and the resulting nuclear programs themselves. Little by little differences appeared and widened, but it is difficult to point to original structural differences that had large effects. Not all the ways in which the countries differed in 1973 became relevant to policy divergence, and some similarities later developed into differences.

Several political differences in 1973 later became relevant, even though they had had no effect on nuclear policies before that time. First, the "dominant cleavages" between political parties or coalitions—the grounds on which they habitually attacked each other—differed in the three countries. American Republicans and Democrats fight over free markets versus government intervention; the French Left and Right debate the interests of capital versus those of labor; the Swedish Left and Right dispute both capital-labor issues and the hegemony of the Social Democrats. When energy policy became a "public problem" after 1973, it came to be debated along these same lines (even though France and the United States lacked significant antinuclear parties). Hence the policy debates themselves diverged in France, Sweden, and the United States. Second, the role of engineers and economists as policymakers and policy advisers in government differed in ways discussed in part three. When the oil crisis hit, these groups began to disagree, with engineers facing the crisis as a problem of supply, and the economists one of price. The relative influence of each profession and the policy styles they used affected policy formulations and outcomes. Third, American nuclear regulation "from the outside," unwilling to interfere with management practices, contrasts with more cooperative regulation from the inside in France and Sweden. This difference was irrelevant in the period when nuclear energy was noncontroversial and without obvious problems. But as the American industry developed internal and external problems during the 1970s, American styles of regulation blocked solutions. Finally, as financing for reactor construction became more difficult during the 1970s, ownership of the reactors and patterns of financing began to influence policies. This was less a matter of state versus private ownership than of state officials' willingness to support nuclear energy.

These differences should be compared as a group rather than separately as individual factors, since they influenced each other in complex ways throughout the 1970s. And no aspect of political structure was as influential as the discretionary choices policymakers had to make in the two years after the oil crisis. Each policy decision changed the very structure for setting nuclear policies and brought formerly irrelevant factors into salience; dependent variables suddenly become causal ones; what should have been causal variables (public attitudes, reactor costs) became dependent ones. With each policy decision the countries moved in their own directions, but throughout much of this period, bold policy decisions could have reunited their policy paths. Yet as the years passed, bolder and bolder decisions would have been required, thus becoming far less likely. By the 1980s, they were impossible. Discretion—especially in France and the United States—grew smaller. What must be compared for our three countries is not individual differences in political systems, but entire *trajectories* and processes for solving problems. These trajectories combine independent and dependent variables (in fact the distinction dissolves into complex feedback processes), but the three trajectories are quite characteristic of the three political systems.

We can see the policy trajectories gradually diverging, as the three countries faced the same international phenomena: the same light water technology in the late 1960s, the quadrupling of oil prices in 1973, rising antinuclear movements in the early 1970s, electricity demand that rose less steeply in the late 1970s than predicted, and nuclear accidents at Three Mile Island and Chernobyl. The three countries reacted to each of these in slightly different ways, ways characteristic of the three political systems. But by the end, their reactions were much further apart than they had been early on. Contrasting nuclear policies had influenced their own structural contexts, which had hardened and taken away some of the policymakers' discretion.

The structure of *Nuclear Politics* reflects the tension between two approaches: presenting each country's story in depth and comparing those stories systematically. A separate chapter at the end of each part provides systematic comparisons, freeing the chapters on individual countries to deal with the historical complexity of each policymaking system.

Why three countries instead of two or four or six? Two cases give too little variation, leaving explanations too free to favor certain variables. I have elsewhere (Jasper 1987b) criticized the common tendency to compare French and American nuclear efforts, which have come to differ in so many ways that one can explain French success and American failure in almost any way one wishes. Sweden, which shares many characteristics with France, discourages such facile comparisons. Studying a fourth country would have been possible with more time, funding, and language capacities—but which country? Britain lacked a large antinuclear movement. West Germany was both too different and too similar: it differed in having an antinuclear movement that mixed weapons issues with energy issues much more than those in the other three countries; it was similar in having a federal system with strong courts, as the United States did. It would have provided little explanatory power.

The two main techniques used in this study were elite interviewing and archival research. One hundred state officials, utility employees, antinuclear activists, and other observers and participants were interviewed, since only through personal interviews can one probe the policy styles and motivations behind the policy positions and debates. Various kinds of written documents supplemented the interviews, from testimony at congressional hearings in the United States to comments written by farmers protesting nuclear plants in France. The newspaper and magazine accounts as well as the social scientific analyses in all three countries are endless, and I hardly claim to have exhausted them.

Chapter 2 describes two kinds of cultural meaning that figure prominently in this book. The major one consists of three policy styles I call technological enthusiasm, the cost-benefit approach, and ecological moralism. Each is a cultural repertory with which groups try to avoid the negotiation and struggle of regular politics by referring to nonpolitical, transcendent principles for selecting policies. Unfortunately the repertories had few common references and engendered strong misunderstandings between participants in nuclear debates. The second set of meanings is that generated by the competition between political parties, which have to distinguish themselves along clear ideological and organizational grounds. Policy debates are twisted to reflect these dominant partisan cleavages. The rest of the book uses these mental grids to fill in political and economic structures and to explain policy outcomes.

Part two describes how the United States, Sweden, and France each came to accept the commercial use of light water reactors in the 1960s. No matter what their political structures, all three created strong, centralized agencies for promoting nuclear energy. Policies and decisions arose through "insider politics"—the close cooperation between reactor manufacturers, electricity producers, and various government agencies—and in each country technological enthusiasts had to vanquish the doubts expressed by "cost-benefiters" (those with the cost-benefit policy style). This victory occurred early and easily in Sweden and the United States, but late and with difficulty in France. By the early 1970s, however, policies and deployment seemed to be converging; problems were thought to have been ironed out and the future of nuclear energy seemed bright in all three countries.

Part three is concerned with the emergence of nuclear energy policy from the arena of bureaucratic politics into that of public, partisan debate. Nuclear energy became a visible public issue in all three countries—politicians began to take stands, the news media to cover the issue, and the public to develop opinions—because of small but growing antinuclear movements and the oil crisis that shocked all policymakers in late 1973. It was largely in this arena of high politics that policy trajectories began to diverge. There was nothing inevitable about any of the policy paths; they were the result of struggles and decisions by politicians and bureaucrats who faced many options. A key choice was between improving efficiency and decreasing electricity demand (the cost-benefit solution) and building more generating capacity to avoid future oil shocks (the technological enthusiasts' solution).

Although in the first months after the oil crisis all three countries reaffirmed their commitments to nuclear energy, in the following two years their policy actions diverged markedly. France took the technological enthusiast's route and launched an enormous construction program, reorganizing heavy industry and opening new financing mechanisms in the process. The United States, by delaying any strong action, allowed mechanisms for careful cost comparisons to be put in place. These encouraged the recognition of alternative energy sources and helped the United States follow the cost-benefit program of discouraging new demand. Sweden also hesitated long enough to follow the cost-benefit solution, less out of awkward fumbling as in the United States than out of a political principle of keeping all options open and avoiding controversial decisions. The contrasting steps that our countries took, however, did not yet look like clear policy paths, since all three maintained a formal commitment to nuclear deployment.

Part four traces the debates and policy decisions as they continued in the late 1970s and 1980s. During this time the discretion policymakers had in making decisions shrank gradually, because the policy paths taken after the oil crisis began to influence their own structural contexts. In France, the state's strong commitment to nuclear energy lowered construction costs, prevented recognition of alternative energy sources, and eventually even forced the public to favor nuclear energy. In the United States, the federal government's casual commitment allowed costs to rise, alternatives to be sought, and the public to become antinuclear. It even allowed political authority to devolve to local and state levels so that future nuclear plans might be more easily blocked. Sweden's decision to phase out nuclear energy by 2010 has forced it to search for alternative energy sources, but the possibility that the decision will be reconsidered has kept the political and economic structures from becoming rigid. Although the United States has had discussions of new, safer reactor designs as a partial solution to global warming and acid rain, Sweden has a better chance of revising its future plans.

CHAPTER 2

Partisan Cleavages and Policy Styles

THERE ARE many systems of meaning that men and women bring to their political actions, and these can supplement the structural account of politics. Two are especially useful to me in explaining how political systems affect nuclear policy. One set arises from competition between political parties in electoral systems. Members of any given party see themselves as a certain kind of person standing for a certain kind of policy and principle. Party members see and use any political action or policy as a way of reaffirming their party's distinct identity and competing for votes. Thus it is important to understand how the political parties in France, Sweden, and the United States distinguish themselves, in other words to identify the *dominant partisan cleavages* in each country. Although typically ideological, they may be institutional as well: the government versus the opposition as well as free marketeers versus government interventionists.

A second set of cultural meanings are *policy styles*: clusters of related symbols, language, images, and algorithms (both implicit and explicit) that can be used to recognize, analyze, and solve social problems and to convince others of one's own solutions.[1] Policy styles are distinct from the specific goals derived from them. They combine rhetoric—the tools and language used in arguments—with worldviews, the beliefs that make those tools and language plausible.[2] Several types of arguments and rhetoric appear again and again in policy debates, since they refer to basic human conditions and thus are plausible to wide audiences. Three policy styles common in nuclear policy debates center around references to costs and prices, technical change, and morality.

To the extent that policy styles are simple rhetorical tools, any individual or group can choose among them to appeal to its current audience. It can use them strategically, sometimes even cynically. Hence antinuclear groups might refer to costs or technological change as well as to morality.

---

[1] The phrase is inspired by Ludwik Fleck's (1979) term *thought style*, which he used in the sense of a conceptual scheme, based on beliefs about the world, used to generate action, and tied to concrete social groups. I am interested in the thought styles used to generate policy recommendations.

[2] Because rhetoric is the art of persuasion, it is useful for analyzing policy arguments and positions. Kenneth Burke (1941, 1950) has analyzed many examples of rhetoric in politics. More recent uses include Bormann (1972) and Gusfield (1981). I have compared moralistic and instrumental rhetorics (Jasper 1987c).

But since policy styles also involve beliefs about the world, a given group or individual typically finds one style more compelling than others. Because people's symbols and beliefs arises from concrete social practices, skills, and experiences, people differ in which symbols they find most plausible. They see the social and physical worlds as open to change in different places. For this reason, I shall often characterize groups according to their typical policy styles: "cost-benefiters," "technological enthusiasts," and "moralists," even though a group may use other rhetorics on occasion. Because it is often impossible to separate strategic rhetoric from underlying worldview, I shall generally use the terms *policy style*, *worldview*, and *rhetoric* interchangeably.

Policy styles and partisan cleavages are public, observable statements that structure the way individuals think, talk, and act. By providing people with something to believe, say, and do, they help constitute political arguments and actions. They are themselves concrete practices employed in specific settings, but they are also beliefs and meanings that individuals hold. (Rhetorics often work to convince the speaker as well as the audience.) Cultural meanings are partly formed by political and economic structures, but once formed they gain autonomy and persist on their own. They even help to reshape those structures. It is tempting to label the cultural meanings "structures," since they have clear forms, but to do so runs the risk of making all social practices into structures. I prefer to call them culture, what Ann Swidler (1986) calls a "tool kit" of language, beliefs, and skills. Yet we must always try to ground culture in concrete groups, organizations, and practices.

Perhaps most importantly for the study of nuclear energy, policy styles and partisan cleavages provide assumptions about how the world will work in the future. In the absence of abundant and convincing information, people make decisions based on their assumptions about the world. Such assumptions played a large role in nuclear policymaking in the 1960s and early 1970s, since many key decisions were made about nuclear energy when little was known about it. General belief in the benign effects of technology was often enough to make policymakers minimize the risks of emerging nuclear technology, while a general suspicion of technology could lead to opposition.

Dominant Partisan Cleavages

People are attracted to a party because of shared values or policy goals, but once a party is formed an additional dimension appears. A party affirms its symbolic commitment to certain goals even when it cannot attain them, and it is quick to use these goals as a way of distinguishing itself from other parties. Thus discussions of any policy will be twisted slightly to reflect the lines along which each party distinguishes itself and attacks the others. The lines of cleavage are usually, but not always, ideological, and they differ from one national party system to another.

E. E. Schattschneider (1960, 68) said that "every major conflict overwhelms, subordinates and blots out a multitude of lesser ones." Similarly Victor Turner (1974, 38) described the tendency for any conflict to escalate in order to line up with a social system's "major dichotomous cleavage." A partial reason is that different factions see any disturbance to the system as a chance to gain advantages over competitors. In political systems any conflict is a chance for political parties to attack their rivals, and nuclear energy debate was often used in this way. The cleavages between parties may or may not reflect broader social and economic cleavages in society (Sartori 1969, 87ff.), but they are often simply an arena for parties to slug it out. In addition to competitive advantage, however, partisan cleavages allow people a chance to reaffirm their political identities and party support. They also reflect genuine beliefs about how the world works.

In the United States, political parties lack strong ties to organized interest groups, so Republicans and Democrats typically fight over ideological questions, especially whether markets should be left to their own dynamics or whether government should intervene and regulate market activities.[3] Thus questions of nuclear energy policy often became questions of how much government should interfere in the decisions of private utility companies. Democrats could attack Republicans for placing free markets above citizen welfare; Republicans could attack Democrats for ruining the economy with tight governmental regulations.

The empirical evidence for this cleavage is extensive. Aage R. Clausen (1973, chap. 5) found that Congressional voting regarding the government-market cleavage consistently revealed the strongest differences between Democrats and Republicans as well as the strongest consensus within each party. In their study of the ideologies of bureaucrats and politicians, Joel D. Aberbach et al. (1981, 122–25) found that, out of six industrial democracies, the United States displayed by far the most disagreement over this issue. When he studied how several American groups understood economic incentives to stop polluters, Steven Kelman (1981b) found that politicians and those involved in political debate tended to see issues in terms of free markets versus government inter-

[3] For descriptions and evidence of this ideological division, see Clausen (1973), Clausen and Van Horn (1977), Schneider (1979), and Shaffer (1980). For its historical roots in the public's distrust of big business, see Vogel (1986). Ferguson and Rogers (1986, chap. 1) discuss the deep ambivalence most Americans feel on the issue, showing suspicion of business and favoring government interventions in specific cases, even while responding favorably to the symbols and ideology of free markets.

vention. Instead of describing incentives in terms of economic efficiency, environmentalists, industry representatives, and congressional staffers viewed them in ideological terms. All three used the market-government contrast, the political staffers most strongly. (The environmentalists also introduced questions of right and wrong and justice—the policy style described below as ecological moralism.)

French political parties identify themselves clearly along a Right-Left axis, and voters take clear positions along the same axis, so that the dominant partisan cleavage is also strongly ideological.[4] But unlike the American cleavage, the French counterpart is concerned with how policies affect Labor or Capital: issues like income distribution, the safety of workers, and profits. Whereas American parties would debate the merits of government intervention, French parties are more likely to debate whether a policy will help the working class or increase the profits of capitalists. Nuclear energy was often seen as an issue of industrial structure and profits and of the safety and health of workers. In neither country did partisan debate place pronuclear parties on one side and antinuclear parties on the other, since no major parties were antinuclear; nevertheless many aspects of nuclear energy were vociferously argued. French and American parties are ideological in part because they must attract voters directly, whereas Swedish parties are closely tied to class-based organizations that act as intermediaries.

Swedish parties have a Labor-Capital split similar to that in France. But the dominant partisan cleavage in Sweden is institutional as well as ideological, since one large party, the Social Democrats, faces four (since 1988, five) small parties and has dominated the government for most of the last six decades. The Social Democrats take the lead in developing policies, and the other parties attack their suggestions. This was partly true even when the Social Democrats were out of office. Thus, in the early 1970s, nuclear energy was identified with Social Democratic emphasis on urbanization and economic growth; the main opposition party (the Center Party) could attack it as a symbol of Social Democratic tyranny. In contrast to France and the United States, aspects of nuclear energy were not twisted into Right-Left issues. The dominance of the Social Democrats was an especially salient cleavage in the mid-1970s, when they had governed for forty straight years.

When energy policy moved from the bureaucratic arena to the arena of partisan politics because of the antinuclear movement and the oil crisis, the dominant partisan cleavages became important in all three countries. They influenced how the policy issue was formulated as well as how

---

[4] Michael Lewis-Beck (1984) describes the strong Left-Right identification of voters, while Charlot (1975, 240) analyzes the Left-Right rhetoric of presidential candidates.

it was solved. Instead of a simple question of cost and risk, nuclear energy became a lightning rod for all sorts of partisan concerns. Party competition is a concrete social practice that influences people's cultural meanings and mental grids. Politicians' assumptions about nuclear energy are often linked to their party ideologies and partisan cleavages.

THREE POLICY STYLES

In addition to partisan positions, three policy styles are especially common in technical policymaking. *Cost-benefiters* typically propose, in market settings, to change relative prices (or let the markets change themselves) and, in non-market settings, to perform cost-benefit analyses that mimic market prices. *Technological enthusiasts* usually want to change the physical structure of the world. *Moralists* generally wish to change people themselves and their tastes and desires, or simply to prohibit objectionable practices.

Cost-benefiters, typically economists, seek to distribute resources efficiently to maximize society's production. To change prices is to change resource allocations, because humans maximize their utility in calculable ways, so that a price rise for one good or activity will decrease its consumption. Cost-benefiters' rhetorical tools for making specific policy proposals include the construction of demand and supply curves and price elasticities, as well as various kinds of cost-benefit analyses designed to create the same price information for nonmarket goods.

Technological enthusiasts are skeptical about the rationality of uncoordinated individual choices, feeling more comfortable with physical structures whose behavior can be predicted with the certainty of the natural sciences. Policies of research, development, and construction to change the physical ability of humans to do what they wish is the most certain way they see to solve most social problems.

Moralists claim that principles of right and wrong should dictate policies, since both prices and technology can be used toward improper ends. In particular we shall be concerned with ecological moralists, whose criteria of right and wrong derive from concern for the balance of nature.[5]

These three approaches appear again and again in the struggles over nuclear energy. Because they entail distinct symbols and forms of ratio-

[5] To describe the three positions in economic jargon, technological enthusiasts wish to expand society's production possibility frontier or (for the individual) ease the budget constraint; cost-benefiters wish to change the slope of the budget constraint by changing relative prices or to discover what the true budget constraint is by means of cost-benefit analysis; moralists wish to change the slope of the indifference curves or to remove one activity in the tradeoff altogether.

nality, it is often difficult for one group to take the others seriously. They often cannot communicate and understand each other, so each seems irrational to the others. Each wants to avoid the negotiation and compromise of normal politics by substituting another way of making policy decisions: cost-benefit analysis for the economists, technological development for the enthusiasts, and moralistic rules for the ecologists. When political debate is unavoidable, their rhetorics provide ammunition for convincing the unconvinced. No policy style succeeded in gaining a permanent and complete hegemony over nuclear energy politics in the United States, Sweden, or France, but the power that each group gained and the alliances it was able to make with others help explain the policy divergence in those countries.[6]

I constructed this typology of three policy styles inductively, while trying to make sense of debates over nuclear energy policy. I have used them because they work, because they help me explain choices of state policy. There may be other distinct styles, and there are certainly refinements to be made in my typology, which I have refrained from making for fear of pushing the evidence beyond what it can sustain. But I believe these policy styles will be useful in thinking about other technological policy debates.[7]

THE RHETORIC OF COST AND PRICE

In two hundred years economics has developed a most elaborate policy style involving changes in the prices of goods and services. The root impulse is, "Trust markets, overrule them at your peril."[8] Much of the rhetoric of economists involves letting markets find their own proper price levels. When this is impossible, cost-benefit techniques are available to

---

[6] Compare my use of policy styles with efforts to discern *national* policy styles and use them to explain policy outcomes (e.g., Freeman 1986; Richardson 1982). National policy styles or cultures cannot be separated from national political structures in a clear way, unless one attempts to ground them in public opinion differences (Hill 1985). Yet public opinion often has little effect on policy decisions, as my study reveals. Instead of looking for national styles, we can identify contrasting styles within a country and then see how they interact with each other and with political structures to determine policy outcomes.

[7] The policy styles should also characterize many conflicts over infrastructure, like water supply, transportation systems, and agriculture. In policy areas that concern the remaking of nature into a form humans can use, technological enthusiasts want to build bigger, better systems; cost-benefiters to charge higher prices; moralists to exhort people to save resources and to ban wasteful practices. I am currently using this three-fold typology to understand water supply debates in the New York region and debates over animal experimentation.

[8] *The Economist*, 3 January 1987, p. 66. McCloskey (1986) analyzes economics as a rhetorical style.

approximate the price structure of a free market.[9] Prices can then be set to reflect real costs of production, and resources can be allocated to maximize total production. Prices typically diverge from those of a hypothetical free market through the intervention of powerful producers, consumers, or the state; and some goods and services are never bought and sold on markets (for example many state activities). In these situations market outcomes must be approximated through cost-benefit analyses. If a government proclaims a goal that is inefficient economically, the economic perspective can often still recommend how to accomplish it through price changes. For example an economist may or may not believe that conservation is an economically efficient energy policy, but she can help achieve it by recommending certain increases in the price of energy.

The cost-benefit policy style is to recommend price changes since most of its tools involve predicting the causes and effects of price changes. That is what economists know about. To use Kuhn's phrase, the "exemplars" that economists learn in school involve setting up social problems in terms of relative price changes. Supply and demand curves, price and income elasticities, indifference curves, budget constraints, and many other constructions are designed to examine the behavior of prices. Although economists speak as though demand curves actually exist in the real world, the curves are models constructed by economics for predicting changes in data also constructed by economics. The economist's world consists of these constructions, however, and she makes policy recommendations to change parts of this world. Technological enthusiasts and ecological moralists see quite a different world.

Unlike the other two policy styles, cost-benefit rhetoric is clearly grounded in a single profession, economics. But it employs such a clear and convincing set of symbols and tools that it has spread far beyond professional economists. While only economists have the tools for a rigorous analysis of price behavior, other policymakers often use the rhetoric of cost-benefit analysis. One does not have to be a professional economist to recognize a scarcity of resources and the need for a method of allocation. But whereas economists are generally agnostic about the total amount of state money that should be allocated to a public problem, others often use cost-benefit rhetoric as a way to limit the total budget. Under Ronald Reagan many policymakers used cost-benefit analyses to reject certain expenditures as inefficient, without replacing them with

---

[9] I use the terms *cost-benefit* and *economic* interchangeably to refer to the rhetoric and perspective based on price changes, even though they may be distinct subsets of the rhetoric. Cost-benefit analysis itself has many forms: Lave (1981, chap. 2) describes risk-risk (both direct and indirect), risk-benefit, cost-effectiveness, regulatory budget, and benefit-cost analyses. My discussion describes the entire family under the rubric of cost-benefit analysis, since their similarities are more relevant to us than their differences.

more efficient spending. Whereas many are sincerely attracted to price analysis and rhetoric as a way of dealing with scarce resources, others cynically use it to avoid real political negotiation and compromise.

What is the broader rhetorical appeal of cost-benefit analysis? It holds out the promise of making decisions scientifically by quantifying all aspects of various possible policies, tallying these, and choosing the one that maximizes benefits or minimizes costs. In its pure form, it quantifies on the basis of market prices or approximations thereof, implicitly assuming there are objective valuations. But in many cases such values do not exist—sometimes public goods have no conceivable market; sometimes the value of future actions is very uncertain—and the political disagreements that cost-benefit analysis was designed to avoid are merely displaced. They reappear as conflict over what values to assign to the potential costs and benefits.

Cost-benefit analysis has had a long and controversial role in policy analysis, especially in France and the United States, perhaps the two countries that have done the most to develop it. Its clearest policy triumph may be Executive Order 12291, signed by President Reagan in February 1981, which subjects the actions of government agencies to the cost-accounting of the Office of Management and Budget. This was a significant change for the Nuclear Regulatory Commission, originally charged with a moralist goal of pursuing safety regardless of economic cost. But in June 1982 it duly chopped certain safety regulations deemed not cost-effective by the budget office. Executive Order 12291 is probably a case of noneconomists cynically using the cost-benefit policy style to carry out policies developed on other grounds (namely a crusade against big government).[10]

In its pure form the economic perspective is agnostic about the likely costs and benefits of nuclear electricity generation, so that to complete its calculations and make specific recommendations it must turn to one of the more complete worldviews associated with the other two policy styles. Each of them has strong ideas about the likely costs and risks of nuclear technology since they are both closely tied to broad worldviews and thus to basic assumptions about technological development. In the 1960s most holders of the economic perspective adopted the technological assumptions of the enthusiasts, but in the late 1970s many came to accept some of the assumptions of the ecological moralists.

[10] Karp (1985) discusses Executive Order 12291, emphasizing the conflict between cost-benefit accounting, as it was ideologically applied by the Reagan administration, and democratic participation. Smith (1984) discusses the effects of the order on environmental policy.

TECHNOLOGICAL ENTHUSIASM

For most people, the rhetoric of price is less stirring than the rhetoric of conquering nature. If the economic policy style is best suited for finding the most cost-effective of available alternatives and for influencing individual behavior by changing relative prices, then the technological enthusiast wants to expand available alternatives and lower prices altogether through technological development. Her rhetoric implies that an engineering solution, or technological fix, is available for most social problems. The physical environment that conditions and enables social life is the easiest point to change, as well as the most certain. Humans are forever behaving in unforeseen ways that derail policies, but the natural world behaves according to predictable laws.

Many skills are available for finding technological solutions to problems. Research and development funding assures a long-run supply of new ideas, techniques, and control over nature. In the short term, engineers are remarkably resourceful at solving smaller challenges, and they generally have confidence in their ability to build things that have never been built before. The motto of the U.S. Navy engineers, the Seabees, sums up this attitude: "The hard things we do right away; the impossible ones take a little longer." The panoply of tools and know-how at the engineer's disposal makes the claim plausible. During the last one hundred years the combination of engineering experience and theoretical science has provided evidence that most problems involving physical infrastructure can eventually be solved. The Manhattan and Apollo projects are two famous and—to some—inspiring exemplars.

Technological enthusiasm has broader public appeal than the cost-benefit or moralist rhetorics, since its underlying assumptions about the world are widely shared in the advanced industrial countries. Recent research has dubbed these beliefs and values the "dominant progrowth paradigm." Drawing on Ronald Inglehart's (1977) idea that widespread higher education and affluence since World War II allowed younger generations to pursue satisfactions beyond mere material comfort, researchers have distinguished an emerging "new environmental paradigm" from the traditional "dominant progrowth paradigm."[11] In the latter, humans are seen as having the ability and the right to dominate and change the natural environment for their own ends; progress, especially in the form of economic growth, is seen as perhaps the only sure goal for human society; this growth is closely attached to the domination of nature and is

---

[11] On the contrast between the dominant progrowth paradigm and the new environmental paradigm, see Dunlap and Van Liere (1978), Catton and Dunlap (1980), Cotgrove (1982), and Milbrath (1984).

based on the development of science and technology. These beliefs make the rhetoric of technological enthusiasm especially plausible.

Just as noneconomists use cost-benefit rhetoric because they know economists have rigorous tools to back it up, technological enthusiasts know that engineers and scientists have the skills for changing the world's physical shape. Nonprofessionals have supreme confidence that technological solutions are possible, though they lack the specific training to see *how* they are possible. They want to let the scientists and engineers have the resources and freedom to find solutions. Scientists and engineers usually (but not always) share this orientation, but they also have specific formulations about how to proceed. Naturally, they propose to solve problems by changing what they know how to change.

Because technological enthusiasm dominates Western societies, it is not difficult to explain why some groups and individuals embrace it. What needs explaining is why other groups do not. Family background helps explain socialization into dominant beliefs and symbols. So do, in some cases, leisure activities that involve scientific or technical tinkering and that breed a fascination for technological development. Some groups rely on technological enthusiasm because it legitimates their political power, in that they can hide their own discretion behind a claim of technological determinism. In the case of nuclear energy, uneasiness and guilt about the atom bomb undoubtedly contributed to many scientists' enthusiasm for the atom as an instrument of peace (Lilienthal 1963, 109). But in general the nonprofessionals have simply picked up protechnology values and beliefs that are dominant in Western societies.[12]

Occasionally professional and nonprofessional technological enthusiasts disagree, since professionals are sometimes more realistic about the pitfalls of technological progress than the nonprofessionals. The practice of engineering leads to a kind of artisanal pride as well as to technological enthusiasm: engineers are confident they can build anything, but they want to take the time to build it properly. These two tendencies often conflict. Many nuclear engineers resigned because they felt they were not being allowed to develop adequate safety mechanisms, whether by industrial promoters or regulators. The latter two groups often pushed a technology out the door before the engineers thought it was ready—a major source of nuclear energy's problems in the United States. Both kinds of technological enthusiast have confidence in technological development, but for the nonprofessionals that confidence is sometimes blind

[12] For evidence that the split between new environmental paradigm and dominant social paradigm is similar in France, Sweden, and the United States, see Mitchell (1984) and Zetterberg (1980) as well as those works cited above.

faith. I shall refer to some professionals as technological *realists* to distinguish them from the hard-core enthusiasts.[13]

For both kinds of technological enthusiast, crisp policies of research and construction replace messy political negotiation and struggle. Technological development can avoid the scarcities that cause people to fight over resources. In this policy worldview, political problems are not seen as social, but as concerning material well-being—as a problem of the society-nature interface rather than the group-to-group one. The further back one looks in the history of nuclear research, the more one finds this confidence that technology could solve political problems. A California Institute of Technology physicist, R. M. Langer (quoted in Hilgartner et al. 1982, 18) made explicit the link between technological progress and political conflict in a 1949 *Collier's* article: "Privilege and class distinction and the other sources of social uneasiness and bitterness will become relics because things that make up *the good life* will be so abundant and inexpensive. War itself will become obsolete because of the disappearance of those economic stresses that immemorially have caused it."

This rhetoric of technological enthusiasm, with its clear policy implications, eventually persuaded both President Eisenhower, who launched his "Atoms for Peace" program in 1953, and the Atomic Energy Commission, whose chairman Lewis Strauss (1954) uttered the notorious expectation that "our children will enjoy electrical energy too cheap to meter." The enthusiasm was reinforced and exploited by commercial interests, resulting in pamphlets like Westinghouse's "Infinite Energy" in 1967. Not until the mid-1970s did many technological enthusiasts have doubts about the light water nuclear reactor, but their enthusiasm, rather than dying down, was easily transferred to fusion and breeder technologies.[14]

ECOLOGICAL MORALISM

Moralism is a common policy style that tries to change people themselves, or at least to ban activities defined as immoral. *Ecological moralism* is a form common in debates over technology and the environment. Moralist policy styles, which in their rhetoric and policy proposals make clear distinctions between right and wrong, have been especially common in American politics. They have usually taken their definitions of right and wrong from religious beliefs, lifestyle preferences, or some combination of the two. The temperance movement (Gusfield 1963), the

---

[13] Even among professionals who are technological enthusiasts, disagreements are common, for example, geologists versus engineers (Meehan 1984).

[14] For other cases of technological enthusiasm in the early history of nuclear energy, see Lilienthal (1963), Hilgartner et al. (1982, chaps. 2, 4), and Ford (1982, part 1).

antiabortion crusade (Luker 1984), and the antipornography campaign (Zurcher and Kirkpatrick 1976) provide examples of advocates who would completely ban activities they considered wrong. More recently, the civil rights movement used a moralist policy style, drawing on the Quaker tradition of nonviolence and using the pulpits of black churches (Morris 1984). This style was adopted by the women's movement, the environmental movement, and many social movements of the sixties and seventies throughout the developed West. Many core beliefs of the antinuclear movements in France, Sweden, and the United States came from ecological moralism.

Ecological moralism differs from other moralist styles in its source of moral beliefs and framework for distinguishing right and wrong. At the heart of the position are the beliefs that there are fundamental laws of nature, that modern industrial societies are ignoring them, and that catastrophe (or the retribution of nature) will result (Cotgrove 1982). Antinuclear discourse is filled with stylistic traits of the moralist: references to the laws of nature, a concern for future generations, the potential for catastrophe, the rights of "the people," outrage at the putting of profits before lives. Their actions are often presented explicitly as moral examples. Touraine et al. (1982, 76) present the moralist style, first quoting and then describing ecologist and antinuclear activist Philippe Lebreton: " 'The classic name for it, 200, 300, 500 years ago was saintliness, quite simply. A saint was not passive, quite the opposite. He experienced a sort of upsurge of instinct, but he was able to turn it back upon himself.' This was the perfect description of exemplary action impelled by pure conviction, as opposed to political militancy, the language of the nature of things against the language of social relations."

If professional training helps explain technological enthusiasm and the cost-benefit policy style, what social practices encourage ecological moralism? This position can partly be explained as an absence: unlike economic and technological policy styles, moralist styles have few roots in professional training, except for those of occasional clerics like those involved in the early civil rights movement. But even without professional tools that seem relevant to policy development, members of the public can extrapolate from the world they do know. One process they know is child rearing, and many of the moralists in the movements mentioned above have been mothers. Among other processes, child rearing involves imparting a sense of right and wrong and forbidding the latter. A parent does not calculate the costs and benefits of various actions of her child or re-engineer the house to make it child proof. She doesn't glue vases to the tables, but teaches children not to break them.[15] However, moralist policy styles are in fact *less* tolerant than many child-rearing methods,

---

[15] Sara Ruddick (1989) describes various effects of mothering on worldviews.

since adults are assumed to know right from wrong already. Moralist styles may be like old-fashioned child-rearing, which refers to suprahuman laws, whether of God or of nature, and to dire consequences that threaten if these laws are ignored (as opposed to interpersonal reasons such as not hurting other people) (Bernstein 1975). This perspective is similar to what Weber (1958, 120ff.) called the "ethic of ultimate ends," which defines certain actions as right and others as wrong without weighing the full range of intermediary and indirect effects. Its "tools" are not precise algorithms like those of economists, but rather the rhetoric of good and evil and the retribution of God or Nature, which are powerful tools of persuasion (Burke 1941, 1954, 1969).

Another source of ecological moralism is the lack of direct interests and participation in the industrial sector that depends on economic growth and technological expansion. Stephen Cotgrove and Andrew Duff (1980) have argued that people in the service sector are more likely to have environmentalist attitudes. This is also true of younger people, women, and those with higher education, all of whom are less likely to be involved in processing raw materials into consumer goods. These groups are more likely to develop the postmaterial beliefs of the new environmental paradigm, which questions both the feasibility and the desirability of perpetual rapid growth, hoping to replace it with a balance between humans and the natural environment in which humans are seen as part of nature rather than superior to it. The harmful side effects of growth and technology (sometimes even science) are emphasized, and the fairness and efficiency of unregulated markets are often questioned. These are the key beliefs of ecological moralism, and they directly contradict the dominant progrowth paradigm undergirding technological enthusiasm.[16]

Table 2.1 nicely reveals the split between these two basic paradigms, showing what changes survey respondents thought were necessary for solving society's environmental problems (Milbrath 1984, 124). The most striking divergence is between environmentalists on the one hand and business and labor leaders on the other: 70 percent of business leaders favored greater scientific and technological development, while only 14 percent of environmentalists did. In a question comparing preserving nature for its own sake or using it to produce the goods we use, Milbrath (1984, 120) found that 78 percent of environmentalists favored a society that emphasizes preserving nature for its own sake, as opposed to 24 percent of business leaders. At the other end of the scale, 62 percent of business leaders favored a society that emphasizes using nature to produce goods. Only 8 percent of environmentalists favored such a policy.

[16] Jasper (1985), Ladd et al. (1983), and Scaminaci and Dunlap (1986) show that antinuclear activists typically held the beliefs of the new environmental paradigm.

Note that labor leaders are much closer to business leaders than to environmentalists in their responses in table 2.1, underlining the importance of economic sector as an explanatory variable. The dominant progrowth paradigm is compatible with both technological enthusiasm and a cost-benefit policy style, indicating that the split between these two may be less fundamental than the gulf between them and moral ecology. At the same time, the economic perspective aims at distributing scarce goods and resources; it does not necessarily imply economic expansion and technological change. Thus technological enthusiasm is closer to the heart of the dominant progrowth paradigm, and the opposition between it and ecological moralism is particularly strong.

Mary Douglas and Aaron Wildavsky (1982) have provided an additional explanation of the moralist rhetoric of the environmental and antinuclear movements. They argue that the hearts of these movements consist of small voluntary groups under continuous pressure to prevent their members from quitting. Throughout history groups of this kind—Douglas and Wildavsky call them sects—have used tactics and rhetoric to remind members of the security of the group and the dangers outside it. The distinction between the sect and the rest of society is strongly moral (p. 124): "The sect needs enemies. It encourages thinking in either/or terms because of the political focus on that dividing line between saints and sinners. Compromise is ruled out because there is only one important

TABLE 2.1
Views of Selected Groups on How to Solve Environmental Problems, United States, 1980

	Greater Scientific and Technological Development			vs.	Basic Change in Nature of Society			
	1	2	3	4	5	6	7	Mean
General Public (N = 1513)	18	10	9	11	13	13	25	4.5
Environmentalists (225)	5	3	6	14	14	19	39	5.4
Business Leaders (223)	40	23	7	7	8	6	7	2.5
Labor Leaders (85)	40	13	1	8	8	10	19	3.4
Elected Officials (78)	15	13	13	11	25	11	11	4.1
Appointed Officials (153)	16	18	12	13	20	11	11	3.8
Media Gatekeepers (105)	20	16	14	14	14	9	14	3.7

Source: Adapted from Lester Milbrach, *Environmentalists*, 124, 150, by permission of the State University of New York Press. Copyright 1984.

Note: To the question "What kind of change is most needed to solve our environmental problems?" respondents could chose a number on a scale from 1 to 7, ranging from an emphasis on technical development (1) to one on basic political and economic change (7). Figures below columns 1 to 7 indicate the percentage response for each choice.

distinction, between those who are loyal and the betrayers. Purity becomes a dominant motif." Douglas and Wildavsky accurately describe environmental politics as moralistic, but sects—by any reasonable definition—represent only a small segment of the environmental and antinuclear movements. The main source of ecological moralism is the new environmental paradigm and activities outside industrial production.[17]

Two victories for the moralist policy style in regulation and technology are the Delaney Clause of the Food, Drug, and Cosmetic Act, and the 1970 Clean Air Act amendments. The Delaney Clause stipulates that any substance be banned if evidence shows it causes cancer in humans or animals. The Clean Air Act amendments demand air clean enough not to harm any humans, presumably even those most vulnerable to pollution. Air pollution and carcinogens are defined as wrong, and there is to be no further consideration of costs and benefits. Nor is there special provision for technological changes to ameliorate the problems. As Lave (1981, 13) points out, this kind of no-risk policy "is like attempting to legislate morality: the rhetorical appeal is evident, but regulation can hope to affect only a tiny proportion of the relevant risk, and at rapidly increasing cost."

In place of policy styles that would make decisions on the basis of weighing costs and benefits, or that would research and build better technologies, the moralist defines certain activities as wrong and not to be pursued no matter what the costs or potential benefits. The economic perspective is willing to patch over such issues: small amounts of carcinogens are tolerable; in certain cases it may be cost-efficient to pay someone to stop doing something objectionable. To the moralist, government should not condone immoral conduct, even in minor cases and even when the costs of stopping it are high. Compromise on questions of morality is itself immoral.

INTERACTIONS BETWEEN THE STYLES

Each policy style has its own characteristic opinions of the other styles. Cost-benefiters find the others somewhat naive and dreamy in their interest in the future rather than the factual details of the present, but they see the moralists as more dangerous in their rejection of cost-benefit

[17] Douglas and Wildavsky (1982) explain nuclear conflict as the result of opposed attitudes to technological risk. Risk attitudes, despite the growing literature on the subject, were not very helpful to me in explaining nuclear policies. For concrete groups and individuals it is very difficult to discern a single dimension that might be termed risk aversion or risk acceptance. Attitudes toward particular technologies have many dimensions, based on ownership patterns, political support for the technology, aesthetics, moral objections, and other factors in addition to fears and risk aversion. On risk and on Douglas and Wildavsky, see Jasper (1987d).

analysis, while the technologists merely want to expand it to account for their envisioned developments. The technologist sees the cost-benefiter as somewhat shortsighted, lacking the ability for imaginative, long-term planning, but she sees the ecological moralist as a downright Luddite opposed to and blocking civilization's greatest good, technological progress. The moralist sees both the others as soulless, but the technologist more so since she wants to replace the soul with something else, while the cost-benefiter merely misses that aspect of human nature.

Different ideas about how the world works, different notions of what rational policies and political action are—even different languages for describing policy proposals and their rationales—assure there will be little mutual understanding and appreciation between advocates of the three policy styles. But certain alliances are possible. The most likely is between cost-benefiters and technological enthusiasts, who typically share the beliefs of the dominant progrowth paradigm as well as, in many cases, a feeling of professional authority. Cost-benefiters are usually willing to accept the assumptions about future developments that the enthusiasts provide, since they lack clear information on such changes themselves. In some cases, however, cost-benefiters can drop the technologist's assumptions and adopt those of the moral ecologist, especially when events make the costs of technological change prominent. Least likely is an alliance between technological enthusiasts and moral ecologists, whose beliefs about the likely effects of technological change are diametrically opposed. Changing people and changing physical environments are very different solutions.

All three policy styles have wide appeal since they are based on beliefs and symbols widespread in modern societies. Technological progress plays upon deep hopes for a better life; the need to allocate scarce resources among desirable uses is widely recognized; right and wrong are appealing concepts in themselves. Few people are opposed a priori to improving technologies, changing prices, or even changing human beings, but they put different priorities on which should be tried first in solving social problems. Each rhetoric has its core group for which it has the highest level of plausibility, but it has varying levels for other groups as well. Despite conflict and misunderstanding between the three policy styles in the nuclear energy debates, we must hope and try to mediate between them.

Although I have labeled one of the policy styles *moralist*, all three carry a confidence and sense of moral purpose. Each group believes its own project could save the world, which the others are threatening. This moral fervor is clearest on the part of the moralists and the technologists; those early radiation pioneers were true martyrs, heroes who died for the betterment of mankind and the pursuit of truth. But only the ecological moralists concentrate on explicitly moral arguments.

The typology of policy styles could be further refined, with more systematic data than I collected. The "professionals"—like engineers and economists—could be compared with the nonprofessionals in the cost-benefit and technological enthusiast styles. Within each style, contrasting attitudes might be found that parallel the three styles. For example, some individuals might have a moral fervor for the cost-benefit style, while others use it instrumentally. Finally, some use their style as a decisionmaking *process*, while others are more concerned with the *outcomes* generated. But these subtle distinctions are not necessary for explaining our three countries' policy divergence.

USING CULTURAL MEANINGS IN ANALYSIS

What are the limits of using partisan cleavages and policy styles in an explanation of nuclear policy divergence? First, some groups and individuals do not display any combination of the policy styles or partisan ideologies. These are generally people who do not care to cloak their own material and organizational interests behind claims concerning the good of society or the way the world works. But they are rare and rarely successful. Because influencing policy most often requires persuading people who do not share one's interests, one usually appeals to broader issues. It is difficult for the analyst to distinguish positional interests from sincere policy styles and beliefs about the world, but cases of conflict between the two will often help us.

A second caution: analysts must avoid identifying policy styles by means of specific policy positions, and then explaining those positions as a result of the styles. This is the circularity of values and behaviors for which structural-functionalism was often criticized (Barry 1978, 88). It results in merely redescribing political structures rather than adding a dimension to them. For all groups and individuals who enter our explanation of nuclear energy policies, we should find the origins of their policy styles as well as the effects of these on their political successes and failures. Policy styles are useful as explanatory variables because they summarize various cultural tools: language, symbols, skills, myths, and other meanings.

Succeeding chapters will use these partisan cleavages and policy styles to characterize key individuals and groups who played a role in nuclear energy policies, always trying to give an account of how the people came to their cultural meanings. These policy styles are the most important way we shall use a cultural analysis of political systems to fill in a structural one. How different groups wanted to solve general problems of energy supply depended on their worldviews, characteristic skills, and policy styles, and their success in achieving these goals depended on political structures. Both cultures and structures are needed to avoid empty formalism and incomplete explanations on the one hand and idealist generalizations about culture on the other.

PART TWO

Creating Nuclear Systems: The Triumph of Technological Enthusiasm, 1960–1973

CHAPTER 3

The Triumph of Technological Enthusiasm in the United States

> It is a tragedy in the classical sense. A technology with heroic potential laid low by an enthusiasm—arrogance if you will—that blinded its handlers. Their failings come not from too much weakness but from the wrong kind of strength.
> —Peter Bradford (1982a, 7)

THE UNITED STATES, Sweden, and France all use the light water reactor (LWR) in their nuclear energy programs. Originally developed under Hyman Rickover to power American submarines, this reactor uses a heavily processed, "enriched" uranium as fuel, and regular water as coolant. The LWR was the design most commonly adopted throughout the world from among dozens of types that have been attempted and developed (everything short of sawdust fuel and beer coolant, one engineer said). The LWR in turn has two main forms, the pressurized water reactor, in which pressurized water cools the fissioning uranium and causes an additional loop of water to boil and drive turbines, and the boiling water reactor, in which the water that flows over the uranium also drives the turbines. Westinghouse and the French manufacturer Framatome produce pressurized water reactors, while GE and the Swedish manufacturer ASEA-ATOM produce boiling water reactors.

Partly because they were developed here, light water reactors achieved commercial success in the United States earlier than in France or Sweden. No other reactor type was so fully developed here, so there were no serious alternatives to be vanquished. But the commercialization of the LWR was hardly inevitable even in the United States. Technological enthusiasts at the Atomic Energy Commission (AEC) had struggled for a decade to force the LWR on utilities before the first commercially competitive LWR was sold in 1963. The enthusiasts succeeded only when their optimism spread to the two giant electrical manufacturers, Westinghouse and General Electric, a victory made possible by the rivalry between the two companies.

The commercialization of the LWR occurred in three stages. First came pressure and encouragement from the AEC but reluctance on the part of utilities. Then, when reactor manufacturers gained the confidence to sell loss-leading nuclear plants beginning in 1963, they touched off a "great bandwagon market" for nuclear generating capacity that lasted, in two major waves, until 1974. This latter period can be divided at the point when the AEC began to withdraw its research support from LWR technology in the late 1960s.

ENTHUSIASM AT THE AEC AND THE JOINT COMMITTEE

Throughout the 1950s extravagant technological enthusiasm dominated the AEC but ran aground on the economic calculations of virtually all electricity producers. Fueled in part by guilt over the development of the atomic bomb (Lilienthal 1963), scientists had embraced nuclear energy in the 1940s, and by the mid 1950s they had convinced many policymakers that it could save the world. In 1953 and 1954 President Eisenhower launched his "Atoms for Peace" program, and the AEC optimistically predicted that 89,000 MWe of capacity would be in place by 1975.[1] To encourage this rapid development, the Atomic Energy Act of 1954 permitted private ownership of nuclear reactors and provided for the lease of nuclear fuel from the government. The AEC's Power Reactor Demonstration Program was launched in 1955 to encourage and fund private reactors to produce electricity, and more than a dozen reactors were built under the program from then until the mid-1960s. Several of these were built with little or no government funding.[2]

The special Senate-House Joint Committee on Atomic Energy (JCAE) was given extraordinary powers to oversee and encourage the development of nuclear power. It was the only joint committee able to introduce legislation, and it was the only congressional committee to oversee the AEC and nuclear power. Such centralization of control seemed justified by the military uses of nuclear fission, and it contrasts sharply with the usual dispersion of policy authority in American politics. Since committee assignments are partially voluntary, the committee attracted members who were extreme enthusiasts for the emerging technology. As a result, according to Harold Green (1982, 60), "The Joint Committee was almost always more aggressive and expansionist than the Atomic Energy Commission and the Executive Branch, and it constantly pressed for

[1] See Pringle and Spigelman (1981, 226). The "Atoms for Peace" idea was raised in the president's speech of 8 December 1953 to the United Nations General Assembly and involved international control of arms and fissile materials, the sharing of nuclear technology with allies, and the promotion of peaceful atomic energy. The speech bore fruit in 1954 with the Atomic Energy Act and the creation of the IAEA (Hall 1965; Pilat et al. 1985).

[2] As described in Del Sesto (1979, 56, 73), Commonwealth Edison of Chicago applied to build its Dresden plant with several other companies; Consolidated Edison of New York applied to build Indian Point; and General Electric applied to build a reactor at its laboratories in Vallecitos, California.

larger and more ambitious atomic energy projects." This is the first instance of a common pattern: it is those closer to the technical details who have a more tempered realism, while those whose knowledge is less specific and hands-on are extreme technological enthusiasts.

Because they themselves were so confident in nuclear technology, the AEC and the JCAE felt frustrated with so little industry participation. This frustration led to the Gore-Holifield Bill of 1956, a proposal for a major program of government reactor construction. The bill should also have spurred more participation by industry, simply by raising the specter of "a TVA for nuclear energy," anathema to private utilities. One effect of the bill was to shift the nuclear debate toward the lines of the dominant partisan cleavage: free markets versus government intervention.[3] Not only was the bill defeated, but it did little to jolt private utilities into greater participation. Nevertheless enthusiasm remained high among government policymakers.

Such was AEC confidence that it released the Brookhaven report in 1957 without any concern that the report's description of enormous potential accidents might cause public distress. "WASH-740" described a worst-case scenario in which three thousand people died immediately and more than forty thousand were seriously injured. There was a potential for property damage of $7 billion, over an area of 150,000 square miles.[4] The effect of the report was to scare off potential participants, especially insurance companies. The AEC and the JCAE then redoubled their promotion efforts. The Price-Anderson Act was passed in 1957, almost completely underwriting liability for nuclear accidents; utilities and industry were off the hook.[5] The Power Reactor Demonstration Program

---

[3] At the same time Congress was debating the Dixon-Yates proposal to create a private company to provide electricity to AEC facilities, undermining the public utility, the Tennessee Valley Authority. This debate also helped frame the nuclear issue in terms of the dominant ideological cleavage. See Clarke (1985) and Wildavsky (1962).

[4] "Theoretical Possibilities and Consequences of Major Accidents in Large Nuclear Power Plants" (U.S. AEC 1958). In later years this kind of report either would not have been written or would have been suppressed. Its 1964 update, which described a much worse possible scenario, was not released until 1973 under the threat of a Freedom of Information Act suit (Clarfield and Wiecek 1984, 350–51). Daniel Ford (1982, 68ff.) describes the ways in which the AEC pressured the Brookhaven scientists who wrote the second report to tone it down.

[5] Public Law 85-256, "An Amendment to the Atomic Energy Act of 1954," in *United States Statutes at Large 1957*, vol. 77 (Washington, D.C.: Government Printing Office, 1958), p. 576ff. The act required nuclear operators to carry the maximum amount of insurance available, but it also provided government indemnity above that, up to a cutoff of $560 million (despite WASH-740's estimate of $7 billion in possible damages). The private insurance available turned out to be a mere $65 million (Fuller 1975, 26). Congress extended the act with few changes in both 1967 and 1975. A 1977 court ruling found the act unconstitutional (*Duke Power Company v. Carolina Environmental Study Group*), but the Su-

was expanded to finance all research and development as well as provide free uranium for at least seven years (Del Sesto 1979, 59). Finally, in 1964 it became legal for utilities to own their own fuel, through the "Private Ownership of Special Nuclear Fuels Act." Senator Clinton Anderson (1965, 13) summed up all these efforts as an attempt to "force-feed atomic development."

Few utilities took the AEC's bait. The costs were still far too high. The reactor at Shippingport, the first commercial reactor in the country, went on line in 1957 producing electricity at 6.4 cents per kilowatt, compared to 0.6 cents for a coal-fired plant (Clarfield and Wiecek 1984, 273). Five years later an AEC report was still predicting an imminent fall in nuclear costs, still without much evidence (U.S. AEC, 1962). Technological enthusiasm at the AEC easily resisted disconfirming evidence, but cost-benefit analysis at the utilities in turn resisted the claims of the enthusiasts.

Almost all the nuclear construction in the 1950s took place under the prodding and the active participation of the federal government, but the government was not willing to go so far as outright ownership of plants. The reason was largely Republican Party ideology—as represented by the administration, the Republicans on the JCAE, and Lewis Strauss, the former Wall Street financier and free-market ideologue who headed the AEC from 1953 to 1958.[6] But without government construction of plants, the level of activity remained very low. The incentives available to the AEC were insufficient to overcome the economic uncertainties for the utilities.

What was holding up commercial nuclear energy in the 1950s and early 1960s? The simplest answer involves political structure. The AEC, as enthusiastic as it was, lacked the authority to build commercial reactors itself. But we can fill in this structural account with two factors from the realm of policy styles and ideologies. First, technological enthusiasts had not yet been able to convince cost-benefiters to accept optimistic assumptions for their calculations. Had utility cost-benefit analyses been more optimistic, the incentives offered by the AEC would have been suf-

---

preme Court overturned this finding. Even in 1986 the cutoff was only $630 million, with $160 million of that available privately (*New York Times*, 14 March 1986), but in 1988 the act was renewed with a much higher liability ceiling; see chapter 11. On the Price-Anderson Act, its effects, and the general question of insurance for nuclear energy, see Wood (1982), U.S. NRC (1983), or the brief discussion by Murphy (1976, 122–27). Legal articles include Cavers (1964), Collier (1966), England (1971), and Green (1973). For comparisons with other countries, see OECD NEA and IAEA (1985).

[6] Strauss was also an amateur physics buff (Metzger 1972, 30), leaving him open to technological enthusiasm. He was fascinated by the potentials of scientific development without having detailed knowledge of how that development actually took place. The title of an article he wrote while at the AEC expresses his technological position perfectly (Strauss 1955): "My Faith in the Atomic Future." See also Strauss (1962).

ficient to spur a commercial nuclear program (as happened in the mid-1960s). Second, the structural powers of the AEC were limited by the free-market ideology of the Republican administration. A broader commercial role for the AEC was certainly conceivable; this is precisely what Gore-Holifield proposed. The AEC was already immensely powerful, able to direct huge amounts of money into the private sector; it simply wasn't allowed to own commercial reactors. A slim majority in the House prevented it. The dominant ideological cleavage between Republicans and Democrats was already affecting nuclear policies.

ENTHUSIASM AT GE AND WESTINGHOUSE

With the government unwilling to buy reactors itself, their price had to fall greatly for them to be competitive with fossil fuels. This breakthrough came suddenly in December 1963, when Jersey Central Power and Light signed a contract with GE for a 515 MWe boiling water reactor to be installed at Oyster Creek, New Jersey. The contract was a "turnkey," in that the price was fixed and the utility was simply given the key after construction and testing. Jersey Central claimed that electricity from this plant would be cheaper than that from a coal-fired plant, the first time this rationale had been seriously given for the construction of a nuclear plant.

Behind this contract lay immense technological enthusiasm on the part of GE. It lost a significant amount of money but assumed that rapidly falling costs would compensate for the sale of one or more "loss leaders." Clarfield and Wiecek (1984, 276–77) summarize the assumptions behind this optimism:

1. The future economic performance of essentially untried high-technology ventures could be predicted accurately, even where these predictions were for plants three times the size of any that might provide a base of operating experience, and even where auxiliary supporting technologies were just as untried. 2. The scaling-up process would achieve significant economies of scale. 3. "Learning experience" would further reduce costs as engineers perfected designs and procedures on the basis of operating experience.

GE and Westinghouse signed eleven additional turnkey contracts in the three years after Oyster Creek (Rolph 1979, 56), while claims about costs were continually lowered on the basis of no operating experience at all (Bupp and Derian 1978, 49). The losses from the turnkeys totaled almost $1 billion for the two companies (U.S. Congress, JCAE 1968, 50–51). Only extreme confidence in future technological progress could allow this to happen.

The origins of this technological enthusiasm are not hard to find. GE probably initiated the competitive spiral of enthusiasm. Its reactor division was dominated by scientists who were technological optimists (to the extent that there was even a large breeder reactor program), while Westinghouse, known as an engineer's company dominated by detail people, was slower and more cautious about nuclear technology.[7] But in 1963 both companies got new chief officers with no technological experience: a former light bulb salesman at GE and a financial analyst at Westinghouse (Pringle and Spigelman 1981, 266). With no realism to temper their technological enthusiasm, both men had unlimited confidence in the rapid development of nuclear technology. Their faith was reinforced by the rivalry between the two companies, each anxious to outpace the other. GE, for example, adopted the advertising slogan "Progress is our most important product," and the same sentiment pervaded Westinghouse.

Technological enthusiasm lined up with corporate interests, and both were reinforced by the desires of two corporate heads to establish themselves in new positions. This competition between two active reactor producers is the major reason nuclear reactors were commercialized so early in the United States. Strong faith in the rapid advancement of reactor technology meant that both companies were willing to take significant losses on the first turnkey arrangements. These contracts in turn sparked the great bandwagon market for nuclear reactors. Nuclear industries in other countries were also dominated by technological enthusiasm, but technological realists moderated the actual pace of reactor deployment.

The Great Bandwagon Market

After two years of cheap turnkey contracts and extravagant claims about the future, utilities were ready to take the bait, and in 1965 they began to order plants without cost limits arranged in advance. Coal prices that were no longer dropping and increased capacity for the regional pooling of electricity also made nuclear energy more attractive.[8] For three years

---

[7] The contrast in styles at GE and Westinghouse can be traced to their founding in the late nineteenth century; GE was Thomas Edison's company, while George Westinghouse built his company by buying other people's inventions (Munson 1985, chap. 4).

[8] Arturo Gandara (1977, 52–59) doubts the crucial importance of the turnkey plants in stampeding utilities into nuclear energy, largely on the grounds that there was a lag of almost two years between Oyster Creek in December 1963 and the flow of orders of November 1965. But GE didn't sell its second and third turnkey plants until February 1965 and September 1965. Corporate decisions are not made instantly, and utilities waited to be sure that Oyster Creek was not a fluke. The other factors Gandara provides for the bandwagon market don't seem as important as the turnkey prices, but they certainly reinforce them: coal prices, regional pooling of power, the increasing size of generating plants, environmental concerns, private fuel ownership, and marketing efforts by reactor producers.

utilities stampeded into nuclear energy, so that by the end of 1968 they had ordered 74 plants with a total capacity of more than 63,000 MWe (figure 3.1). Babcock and Wilcox, Combustion Engineering, and the General Atomic Corporation joined GE and Westinghouse in producing reactors. Technological optimism now pervaded both the nuclear and the electric industries. For example, when GE sold a reactor to the Tennessee Valley Authority in 1966, it offered to pay $1,500 per hour of peak demand when the reactor wasn't working properly. The TVA exceeded this display of bravado with its own, by turning the offer down (Pringle and Spigelman 1981, 269). Those who had been trying for years to coerce the electric industry into accepting nuclear energy were ecstatic. In 1963 Congressman Chet Holifield of the JCAE proclaimed "nuclear energy is on the move," and several years later Alvin Weinberg, head of the Oak Ridge National Laboratory, spoke of a "nuclear energy revolution" as finally having arrived because "nuclear reactors now appear to be the cheapest of all sources of energy."[9]

Who was buying the reactors? More than sixty public and private electric utilities adopted nuclear reactors, some of them in partnerships with other companies. In general they were the largest utilities, the elite among the hundreds of large and thousands of small electricity producers in the United States.[10] Most nuclear utilities own only one or two reac-

FIGURE 3.1 Commercial Reactors Ordered in the United States since 1953

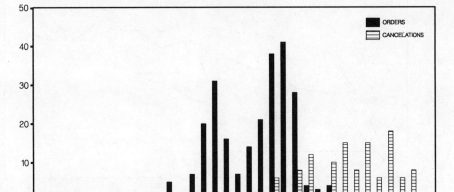

Source: U. S. Department of Energy (1983), Atomic Industrial Forum

[9] Holifield is cited in Del Sesto (1979, 82) and Weinberg in Bupp (1979, 136). For more of the same, see Weinberg and Young (1967) and Hogerton (1968).

[10] Around thirty-five hundred companies produce and distribute electricity in the United States, ranging from tiny rural cooperatives to enormous corporations like Pacific Gas and

tors; only a dozen utilities own more than two, and only three or four own more than six. But seen from the other side, most reactors are owned by utilities that own more than one. Nuclear generating capacity is more concentrated than it appears at first.

In spite of optimism and the surge of orders, little solid information was available about LWR costs throughout the period of commercialization from 1965 to 1974. In the late 1960s data on existing plants began to replace sheer estimates, but these data could not be extrapolated to future plants. The size of the plants was growing so rapidly that at no time from 1962 to 1972 were there any plants in operation as large as the smallest of those being ordered (figure 3.2). By 1968 reactors were being sold that were six times as large as any then in operation (Bupp 1979, 140–42). Nuclear Regulatory Commissioner Peter Bradford (1982c, 10) later commented that "an entire generation of large plants was designed and built with no relevant operating experience—almost as if the airline industry had gone from piper cubs to jumbo jets in about fifteen years." As a result, cost estimation did not improve over the period; in fact, cost overruns beyond original estimates were greater for plants built in the 1970s than for plants built in the 1960s (Bupp and Komanoff 1983).

FIGURE 3.2 Size of American Reactors Being Ordered, 1953–1978

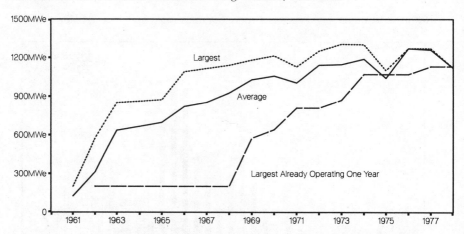

Sources: Bodde (1976, V-24), U.S. Department of Energy (1983), Atomic Industrial Forum.
Note: 1961 figure includes reactors ordered 1953–1961.

Electric Company that serve millions of customers. Of the 2.2 billion kWh electricity produced in 1979, 78 percent came from 237 private utilities; 10 percent from federal power companies (of which TVA and Bonneville Power represent four fifths); and the rest from rural cooperatives and state and municipal utilities (Joskow and Schmalensee 1983, chap. 2).

When larger plants began operating in the late 1960s, the disappointing cost results in no way dimmed enthusiasm for nuclear energy. The community of nuclear enthusiasts redoubled their faith and their proselytizing when faced with disconfirming evidence.[11] Not only were the poor results from smaller plants, but they were written off as "learning experience." In the early 1970s, as in the mid-1960s, wonderful predictions were made of falling costs—just as soon as the kinks could be worked out of this new technology (Bupp and Derian 1978, 82). After a slight dip in orders in 1968 to 1970, a second wave of the great bandwagon market[12] appeared from 1971 to 1974, even larger than the first. It peaked in 1972 and 1973, with seventy-nine reactors ordered in those years. By then the utilities should have known better, but their strong faith in technological breakthroughs discouraged doubts.

In situations where there are no clear data, everyone must rely on intuitions filtered heavily through worldviews. Technological enthusiasts at the AEC fed optimistic information to technological enthusiasts in the nuclear and electricity industries, who recycled it back. Those responsible for cost-benefit analyses at the utilities had little choice but to accept the assumptions about future developments that the technological enthusiasts fed them. Bupp and Derian (1978, 76) describe this process as "a circular flow of mutually reinforcing assertions that apparently intoxicated both parties and inhibited normal commercial skepticism about advertisements which purported to be analyses. . . . Even though more than 100 light water reactors were under construction or in operation in the United States by the end of 1975, their capital cost was almost anyone's guess." This coalition between technological enthusiasts and cost-benefiters was a blind, mutually reinforcing, and generally unstoppable force behind nuclear development. It appeared in similar forms in France and Sweden. One of the main groups in the American economic system with a persistent concern for accurate cost analyses—indeed, whose existence and success depended on them—was the insurance industry. But it had been pushed out of the picture when the government assumed responsibility for accident liability through the Price-Anderson Act.

Why were American utilities willing to plug such optimistic assump-

---

[11] On this "power of preconceptions," see Festinger et al. (1956) and Wilensky (1967, chap. 2).

[12] The phrase "great bandwagon market" comes from Philip Sporn, former president of American Electric Power and a critic of technological enthusiasm in the nuclear industry. The phrase was popularized by Bupp and Derian, in a brilliant analysis (1978) of how technological optimism "intoxicated" the electricity industries of the world. Although Bupp and Derian use the phrase to refer to the wave of orders of 1965–1967, it characterizes the entire period up to the oil crisis of 1974.

tions about the future costs of nuclear energy into their cost calculations despite the lack of firm evidence? Why did they so easily adopt the assumptions of technological enthusiasm? In spite of diverse managerial traditions and structures, almost all electric utilities have engineers in prominent positions, including those of president and chief executive officer.[13] Utilities with high growth rates—the ones most likely to pursue nuclear energy—tend to rely most heavily on engineers (Roberts and Bluhm 1981, 297). In some utilities, such as Pacific Gas and Electric, lawyers too are often found in management positions. But before the oil crisis economists had little influence. Most utility decision makers hold the values of the dominant progrowth paradigm and the policy style of technological enthusiasm (Maher 1977).

For most of the post-war period, electricity production has been seen as the heart of economic growth, in France and Sweden as well as the United States. Until the late 1970s a fairly rigid link was assumed between GNP and energy production, so that the former was thought to grow only to the extent the latter did. In all three countries electricity demand had grown at annual rates of around 7 percent with the regularity of a natural law. Within the electric industry this growth was seen to rest on a long series of technological breakthroughs, and it was assumed that future progress was naturally forthcoming. Ellen Maher (1977, 168) describes the technological enthusiasm of utility executives as a "persisting confidence in a forthcoming solution [that] appears to be more ideological than scientific, the product of a deeply ingrained value commitment." She quotes (168) an executive from Commonwealth Edison: "If [nuclear energy] proves to have hazards which we are unable to manage, with existing methods, we can and must learn to handle the hazards with other methods—not abandon the technology. . . . nuclear waste management methods are new but they are not beyond our knowledge or capacity. If they should turn out to be inadequate, they too can be corrected." Utilities normally perform careful cost calculations, but with such a new technology as nuclear fission, adequate data were not available. Utilities had enough technological enthusiasm to adopt very optimistic assumptions in place of firm data. The same mindset dominates French and Swedish utilities.

Light water commercialization had less to do with the many government policies designed to achieve it than with a change in assumptions

[13] Little sociological work has been conducted on electric utilities. I rely here on Roberts and Bluhm's study (1981) of six large utilities, Maher's study (1977) of executives at Commonwealth Edison of Chicago and the Northern Indiana Public Service Company, as well as a research project on California's utilities in which I participated (Bradshaw et al. 1984). Hymen (1985) and Munson (1985) are useful histories of the electric industry. Baughman et al. (1979) and Gordon (1982) discuss utility financing and operations.

by the reactor producers and then by the utilities. GE and Westinghouse became technological enthusiasts largely because the competition between the two companies gave them many incentives to believe in nuclear technology. The electricity industry had long contained both technological enthusiasts and cost-benefiters, and the latter had resisted nuclear energy until the turnkey plants. After Oyster Creek, the cost-benefiters (outnumbered to begin with) began to accept the assumptions of the technological enthusiasts and feed them into their models. The utilities had been won over.

THE AEC ABANDONS ITS OFFSPRING

As soon as the AEC had fulfilled its dream of commercializing light water technology, it turned its attention—and research funds—to the breeder and advanced gas reactors. A 1967 AEC report (quoted in deLeon 1979, 197) announced that "in view of the striking progress that had been made in the development of light and boiling-water reactors and their construction on a wide commercial scale (i.e., by General Electric and Westinghouse), the Commission would in future concentrate its resources on fast breeders." Partisan political battles prevented AEC interference with industry once a technology was developed, and technological enthusiasm implied that continued improvements would follow even in industry's hands. Technological optimism also dazzled the AEC with the prospect of further wonders to be uncovered with more advanced technology, while blinding it to the mundane but vital tasks involved in the diffusion and routine operation of the LWR.

The AEC's mission was contradictory. It was permissible for the AEC (and the JCAE) to promote nuclear energy, even to the point of bullying and cajoling utilities into accepting it. But it could not own reactors itself or even continue to oversee and cooperate closely with utilities in the operations and management of their reactors. Every kind of external incentive could be offered, but there could be no interference with the internal decisionmaking processes of private companies. In 1964 Milton Shaw became the AEC's director of reactor development, with a mandate to develop a strong government program similar to Hyman Rickover's project on the submarine reactor. His appointment was a victory for the Democrats (favoring stronger government intervention) over the (free-market) Republicans on the JCAE. But instead of bringing the government back into nuclear energy through continued development of the LWR, he concentrated his efforts on breeder reactors. A Democratic government program had to operate within a Republican ideology of noninterference with business, so it took the form of research and development of the breeder rather than dissemination and continued support for the

LWR. The AEC chose to regulate the LWR from the outside instead of cooperating with the utilities and helping them learn better operations procedures.

The commercialization of the LWR changed the role of the AEC drastically. In the tradition of the Manhattan Project and Rickover's submarine propulsion project, the government had previously played the central role, aided by the cooperation of a handful of large corporations. Even government ownership of commercial nuclear plants was not absent from the agenda, as the Gore-Holifield bill shows. Throughout this period the role of the government was very similar to that found in France and Sweden, including the close ties between military and civilian nuclear research. This strong, centralized role is unusual for the United States government, but it was created relatively easily and used effectively in the nuclear realm through the late 1960s.

In the mid-1960s the administrative structure for a strong government program was still in place, reinforced perhaps by Milton Shaw's appointment. But ideological struggles prevented this structure from being used to support the existing commercial LWRs. The AEC was allowed to intervene strongly in developing new nuclear technologies but not to interfere with private companies in managing existing reactors. Technological enthusiasm made the AEC's abandonment of the LWR seem a safe policy.

Enthusiasts, Realists, and Nuclear Safety

When the technological enthusiasts on the JCAE and running the AEC began to withdraw active support for the light water reactors, many AEC researchers familiar with the details of LWR safety reacted strongly. These researchers were realists who saw the amount of work yet to be done to make light water technology safe and who were insulated from commercial pressures to market LWRs. The enthusiasts and the realists represent two contradictory roles of the AEC: as promoter of nuclear technology, to encourage commercialization of the LWR; and as regulator of the technology, to protect public safety. Much of the conflict was between the safety-oriented Advisory Committee on Reactor Safeguards (ACRS) and the rest of the AEC.[14]

Technological enthusiasts heading the AEC believed most accidents were too unlikely to worry about. In the late 1960s the AEC's budget for

[14] Ironically, part of the AEC's inability to intervene in utility operations resulted from an attempt in 1961 to separate safety and regulation from promotional functions. The reorganization failed to resolve the contradiction, since it left safety research in the development wing, but it made communication between safety researchers and regulatory staff very difficult (Gillette 1972, 971–72). As a result, questions and concerns of the safety researchers were often not incorporated in regulatory policy and decisions.

LWR safety research actually fell in real terms, and the AEC did not even spend all the funds allocated to this area. The sums had not been large to begin with.[15] Most commentators have blamed this attitude on Milton Shaw, whose Division of Reactor Development and Technology covered both breeder research and LWR safety research. Rolph (1979, 93) claims that "in February of 1971 Shaw, arguing budget shortages, canceled a key study being done at Oak Ridge National Labs on fuel rod behavior during a loss-of-coolant accident. Clearly, in spite of the attention being given them, he still did not consider low probability, high consequence events to be high priority research topics." Many in the JCAE and the AEC believed that industry should conduct *all* safety research.

At the same time, both the ACRS and the national research labs had large numbers of technological realists: people who believed in the benefits of nuclear energy, but who saw the hard work needed to take advantage of them. As early as 1965 the ACRS campaigned for increased safety research, and in 1966 it sent a letter to Chairman Glenn Seaborg recommending that a "vigorous research program . . . be initiated promptly" (Rolph 1979, 88). Safety research was crucial because the ACRS was being asked to approve reactors far larger than any in operation (figure 3.2; see Okrent 1981, chap. 9); experimental research was the only conceivable substitute for operating experience. In 1971 testimony the ACRS described the situation as even worse (Rolph 1979, 94): "There had been no progress on any of the recommended research topics, several programs had been canceled or delayed, and the R&D budget had suffered continued erosion." The committee concluded that the industry could not (and perhaps should not) be held responsible for the research required to set regulatory standards.

The national laboratories that were conducting safety research, such as Oak Ridge and Idaho, reacted strongly to the AEC's change of focus from LWR safety to the breeder. As described by Robert Gillette (1972), the conflict arose partly from the national labs' organizational interests in continued funding but even more from contrasting assessments of the risk of accidents in LWRs. On one side Shaw told Gillette (p. 772) that "the closeness of researchers to their work may make the possibility of nuclear accidents 'seem more real than it is' and thus may inflate the urgency of safety research in their eyes." On the other side, an Idaho researcher said (p. 974), "It also became clear that Shaw and others just didn't believe a serious accident of this kind could happen, and that it was really worth working on." The disagreement was between technological enthusiasts

---

[15] Rolph (1979) provides a detailed account of AEC operations during the 1960s, with chapter 6 discussing the safety research programs. See also Okrent (1981) and Robert Gillette's (1972) pathbreaking articles on safety issues at the AEC.

who were making policy in Washington and technological realists who were working directly with reactors.

This dissatisfaction became public through the careful midwifery of the Union of Concerned Scientists during generic hearings on the issue of Emergency Core Cooling Systems (ECCS). Begun in January 1972, the ECCS hearings were intended to settle the issue of the adequacy of existing technology for pumping water rapidly into the reactor core if the normal cooling system broke down. With generic hearings, the AEC hoped to keep the ECCS issue out of hearings for individual nuclear plants. Instead of lasting six weeks, the generic hearings lasted more than a year, and instead of solving ECCS uncertainties they highlighted intense disagreement between safety experts as well as a cover-up by the AEC. The hearings became a prime opportunity for the Union of Concerned Scientists and other intervenors to demonstrate problems in AEC operations. The number of scientists willing to leak damning documents and even to testify in spite of the threat of being fired (several were) shows how divided the AEC and its researchers had become, prompting one lawyer to remark, "The AEC leaks like a sieve" (Primack and von Hippel 1974, 220). Still, the AEC hoped that the ECCS hearings would pave the way for quicker licensing of nuclear plants.

REFORMING THE AEC

As though the debates over safety issues were not enough, the AEC had problems processing all the applications it received in the great bandwagon wave of orders. Until 1967 the AEC processed construction permits in an average of eleven months; after that the time increased steadily to twenty-two months in 1971. For operating licenses, processing had averaged eighteen months; after 1967 it climbed to forty-two months in 1974.[16] Before the oil crisis, the main reason for increased licensing periods was simply that the very success of light water commercialization was overwhelming the system. But delays had no effect on orders, which rose steadily from 1969 to 1973.

In 1971 the AEC was rocked by its defeat in *Calvert Cliffs Coordinating Committee v. AEC*. In July, U.S. Appeals Court Judge J. Skelly Wright ruled that the AEC was subject to the National Environmental Policy Act and had to develop serious environmental impact analyses for nuclear plants. He also criticized the agency for arrogance, saying, "We believe that the Commission's crabbed interpretation of NEPA makes a mockery of the act" (quoted in Taylor 1984, 86). Since the Calvert Cliffs reactor was only forty miles from Washington, policymakers watched the case

[16] These figures come from the Atomic Industrial Forum and Campbell (1988, 41).

closely, and their concern with the AEC increased. The ruling was due partly to the personal opinions of the judge, and it might have been overturned if the AEC had appealed it to the Supreme Court.[17] Often seen as a victory for the antinuclear movement, the decision did not stop a single nuclear plant.[18] But it did undermine the AEC's legitimacy and encourage plans for reorganization.

Perhaps unaware that its legitimacy was fragile, in 1971 the AEC launched a campaign to curtail public hearings. With little evidence and encouraged by industry, the AEC blamed licensing delays on public intervenors. It proposed two amendments to the Atomic Energy Act for the JCAE to consider. The first made it optional rather than mandatory for the ACRS to review license applications; the second essentially limited public intervention to hearings concerning reactor sites. But these proposals merely highlighted the AEC's arrogance toward the public and alarmed traditional environmental groups that President Nixon wished to appease. The amendments never reached the floor of either house (Lewis 1972, 259–69).

Internal disagreement, increased delays, and *Calvert Cliffs* left the AEC ripe for schemes to restore its legitimacy. From 1971 to 1973 several important steps were taken. In 1971 Richard Nixon appointed James Schlesinger, an economist with experience in academia and at the Office of Management and Budget, to head the AEC. This outsider to the nuclear establishment would replace, or at least supplement, the blind zeal of the technological enthusiast with the skeptical eye of the cost-benefit perspective. In October Schlesinger (1971, 2–3) gave a strong speech to the annual meeting of the Atomic Industrial Forum and the American Nuclear Society: "You have reason to blush regarding some of these aspects of quality assurance. Gentlemen, these engineering details are not peripheral; they are the heart of our problem. . . . you should not expect

[17] Although Wright was writing for the majority, the tone of the decision reflects his opinions. It is possible that he was sympathetic to moralist arguments about democratic process, having been an activist judge in the South during the civil rights movement. Goldman and Jahnige (1976, chap. 5) discuss the importance of judges' beliefs in legal decisions. Lieberman (1981, 99–107) sees in the *Calvert Cliffs* decision a preference on the part of judges for legalistic, court-like procedures in regulation, which technological enthusiasts at the AEC had felt they could ignore. For further discussion of *Calvert Cliffs*, see Lewis (1972) on the antinuclear organizing and Bronstein (1971), Hoban and Brooks (1987, chap. 4), Landau (1972), Bieber (1973), and Zimmerman (1973) on the legal and procedural implications. Wenner (1982, chap. 3) discusses the case in passing.

[18] The *Calvert Cliffs* case added slightly to licensing periods, since applications then under consideration were delayed roughly one year while the case was being heard. The numbers show no clear effect on operating licenses, but construction permits went from twenty-two months in 1971 to thirty-three in 1973 and back to twenty-six in 1976. This small, temporary effect had no apparent impact on reactor orders.

the AEC to fight the industry's political, social, and commercial battles. These are your tasks—the tasks of a self-reliant industry." If the nuclear industry were now a commercial success, it should not need AEC pampering. It was Schlesinger's decision not to appeal *Calvert Cliffs* to the Supreme Court. Schlesinger (4) was a harbinger of skeptical cost-benefiters who, especially after the oil crisis, asked if we could curb electricity demand rather than always increasing supply: "It is not unreasonable to question whether neon signs or even air conditioning are essential ingredients in the American way of life. More fundamentally, it is not unthinkable to inquire whether energy production should be determined solely in response to market demand." This was one of the first cracks in the alliance of technological enthusiasts and cost-benefiters, and it took an academic outsider to the nuclear establishment to create it. The skepticism of the cost-benefit policy style was beginning to reassert itself.

Personnel changes were not all the AEC faced. Proposals for major organizational reforms had appeared as early as the 1950s, from members of Congress as well as occasional nuclear critics. By 1973 there were several bills before Congress, most of which would have opened up the regulatory process through increased public hearings or through certifying bodies at the state or local level (Ebbin and Kasper 1974, 280). These bills generally died in committee. The one that succeeded was Nixon's limited proposal to separate research and development from regulatory activities. Ford signed this legislation, the Energy Reorganization Act, on 11 October 1974, creating the Energy Research and Development Administration (ERDA) and the Nuclear Regulatory Commission (NRC).[19]

These changes in nuclear regulation in the early 1970s were made for one purpose: to strengthen the nuclear industry and speed its expansion. A stronger and more independent regulator would, it was thought, reassure the public, improve nuclear operations, and speed licensing of new plants. Delays would soon disappear. The NRC annual report for 1975 (quoted in Aviel 1982, 63), its first full year of operation, proudly announced: "More than 70 new and revised regulatory guides were developed and published to provide industry and the public with clear guidance on NRC requirements, and to suggest methods of satisfactorily complying. The use of regulatory guides aids in understanding the requirements for safe reactor design and operation, encourages uniform approaches to regulation, promotes understanding among all parties to the regulatory process and helps eliminate delays."

[19] Dixy Lee Ray, then AEC chair, had tried to reform from within in 1973, revamping the Reactor Development and Technology Division to strengthen safety research. But it was too little too late.

When the oil crisis struck in 1973 and 1974, a system was in place that promised to deploy nuclear reactors promptly and effectively. Early difficulties and bottlenecks had presumably been solved, so that nuclear energy seemed a perfect response to the quadrupling of oil prices.

FORMAL BUT INDULGENT REGULATION

The licensing of new reactors and the regulation of existing ones, by the AEC and then the NRC, have always been a peculiar combination of formal legalism and sympathetic indulgence. Nuclear operators must go through innumerable procedures and hearings, but they are primed beforehand so that they never lose. Strict that the letter of the laws be met, regulators are more indulgent about the actual operations of nuclear plants. In part, this split personality is due to a formal separation between the staff that helps utilities prepare for licensing hearings and the judicial boards that preside at the hearings.

Many observers have perceived an American regulatory style that is "formal," or "hands-off." In his classic comparative study, Andrew Shonfield (1965) noticed an American penchant for a "judicial" style of regulation. This "role of referee rather than that of manager" (330) generally takes the form of negative sanctions rather than positive planning (322–26), and avoids intervening in the management of private firms (330). David Vogel (1986, chap. 5) has similarly characterized American regulation as "rule-oriented" and "legalistic." The effects of this orientation include complex rules; opportunities for public participation; adversarial rather than cooperative relations between regulators and industries; a large role for lawyers rather than technical experts; inspectors expected to follow rules rather than use their discretion; and a large role for courts, since regulatory agencies can be sued for improper administrative procedures. The historical roots of the formalism of American regulation are the same as those of the dominant ideological cleavage between free markets and government intervention: the historical power of big business and the public's resulting distrust of it. Business elites did not respect American civil servants, so they met any regulatory suggestions with contempt. Likewise the public would have mistrusted arrangements in which regulators worked closely with business elites; they preferred a formal, legalistic separation between the two (Vogel 1986).

Procedures for licensing nuclear power plants in the United States—little changed since the 1960s—look quite adversarial at the formal level. All utilities, public or private, must first apply to the NRC (as they earlier had to the AEC) for a construction permit. These applications, typically more than ten volumes long, address three main issues: safety, environmental effects, and antitrust implications. Different segments of the NRC staff simultaneously review the applications and file reports on them, often with the aid of other federal agencies such as the Environmental Protection Agency, the Attorney General's Office (for the antitrust review), and the Army Corps of Engineers. Then the Advisory Committee on Reactor Safeguards (ACRS) examines the safety information and makes a recommendation to the NRC. A three-member Atomic Safety and Licensing Board holds public hearings on the proposed plant and formulates an initial decision. The commissioners finally decide whether to grant the license. When the plant is almost ready to operate, similar procedures are followed for the granting of an operating license, minus the antitrust review. A public hearing is not required by law for the operating license, but opponents of the plant can request them, and they are universal. Additional avenues for public participation have opened over the years.[20]

Despite public, adversarial procedures, there is close informal cooperation between the NRC (AEC) and the license applicants. It is not clear what effect increased public participation has had. Antinuclear intervenors have faced a common hurdle: public, adversarial hearings for a construction permit are held only after many months of preparation and cooperation between the NRC (AEC) staff and the utility. In line with the AEC's abandonment of LWR development, the aim of this cooperation was rapid licensing, not building better reactors. According to Elizabeth Rolph (1979, 75), the commissioner most concerned with regulatory issues was James Ramey, and even he "tended to concentrate on regulatory procedures, particularly as they applied to staff review and licensing, and showed little interest in the structure or support services required for an authoritative, independent regulatory program. The other commissioners rarely concerned themselves with regulatory questions at all."

So even though regulators take a hands-off approach, refusing to interfere in utility operations or management; and an adversarial one, using formal, public hearings; neither of these precludes their "capture," defined as "a predisposition by regulators to make decisions and take actions consistent with the preferences of the regulated industry" (Mitnick 1980, 207). The often described "iron triangle" of AEC, JCAE, and large

[20] Descriptions of the regulatory apparatus can be found in OECD NEA (1980, 1983b, 1984) and Golay et al. (1977). Okrent (1981) details the history of nuclear regulation, especially from the perspective of the ACRS. For a detailed examination of one licensing process, for Seabrook, see Stever (1980). The JCAE typically imposed new requirements as the result of scandals or dubious decisions by the AEC: it required hearings in 1957; set up Atomic Safety and Licensing Boards in 1962; and established Atomic Safety Licensing Appeal Boards in 1969.

corporations had immense powers until the 1970s.[21] The industry's preference is certainly to avoid managerial interference, and it can tolerate adversarial proceedings that inevitably seemed to yield outcomes in its favor. Regulatory staffs often give utilities advice and cooperation before the hearings. But that advice concerns how to survive hearings, not how to manage nuclear energy. Similarly, utilities are often more concerned with passing inspections than getting help with their problems from inspectors.[22] If the adversarial, legalistic proceedings seem unduly harsh, the concrete demands made on utility operations are probably unduly indulgent. Either way, utilities do not see regulators as a source of help for the challenges they face in building and operating nuclear plants.

Because the AEC and NRC have been more interested in rapid licensing than in public concerns or safety issues, all construction license applications since 1967 have been approved. Only once has an operating license been denied to a completed plant, in 1984, and it was later granted. And it took a mishap like Diablo Canyon's reversed blueprints, discovered in 1981, for the regulators to suspend an existing license. Accusations of NRC coziness with industry (e.g. passing internal documents) have continued throughout its existence, erupting as recently as 1987 (see chapter 11).

Regulators wished to license plants, and they succeeded, owing to the tremendous centralization of decisionmaking. The AEC and later the NRC had complete control over licensing. Through the early 1970s the states had very little power, having been preempted in this domain by the AEC. In *Northern States Power Company v. Minnesota* in 1972, the Supreme Court held that states could regulate radiation hazards only when given that power specifically by the AEC, something even the AEC could not grant in the case of nuclear power plants (Cooper 1981). States were to gain more jurisdiction by the end of the decade, but through the early 1970s they had almost none. Even the independent ACRS, which was generally more cautious in approving plants than the AEC or NRC, only made recommendations to the AEC commissioners, and these were often ignored. The AEC represented a strong capacity on the part of the federal

---

[21] Herring (1936), Bernstein (1955), Lowi (1979), McConnell (1966), Davidson (1977), Ripley and Franklin (1980), Quirk (1981), and others have described the "iron triangle" and "capture" patterns in regulation. An iron triangle or "subgovernment" is a form of capture that involves congressional committees as well as the regulatory agency. The AEC, JCAE, and the nuclear industry are a classic triangle. Ford (1982, chap. 2) describes the close ties between the AEC and industry, quoting, for example, Senator Abraham Ribicoff as saying that it was "difficult to determine in the organization scheme of the AEC where the Commission ends and the industry begins" (p. 39).

[22] One consulting firm even advised utilities on ways to deceive NRC inspectors (*New York Times*, "Tips at Atom Plant: Keep Inspectors in Dark," 1 September 1985, p. K23).

government; limits on its actions were due to ideology more than lack of power.

American licensing and regulation are legalistic but sympathetic. There are many rules to be met, but the AEC and NRC have helped the utilities meet them. What is more, regulators can often use those formalities to discourage public intervenors. The regulatory system has provided no serious roadblocks to nuclear development in the United States. Rapid deployment makes sense under the assumptions of technological enthusiasm, namely that the technology is safe and improving. Safety requires competent utilities, and improvement requires competent manufacturers. In the early 1970s there was no reason to think the United States lacked these.

UTILITY FINANCING

Many utilities began to face a financial squeeze in the late 1960s, when inflation pushed interest rates upward. The problem was not the higher rates themselves, but the inevitable lag between the need for increased funds and the rate relief granted by state public utility commissions (PUCs). PUCs were accustomed to granting infrequent price changes, usually downward. When the costs of electricity production began to rise in the late 1960s, and to rise rapidly in the 1970s, PUCs could not respond promptly because of inadequate staff size and political pressures against rate increases. From 1965 to 1973 the resulting squeeze on utility profits, combined with generally increasing capital costs, raised the construction costs for utilities, which increased their reliance on external funding and resorted to various forms of "creative financing."[23]

Nevertheless, before the oil crisis no nuclear orders were cut because of financing difficulties since it was assumed that demand for electricity would continue to expand and would have to be met no matter what the cost.[24] What is more, state regulators were beginning to change their

[23] This discussion of utility financing and construction decisions relies on Campbell (1986, 1988), Gandara (1977) and Hyman (1985). Navarro (1986) and Ringleb (1986) discuss changing utility regulation at the state level. In 1965 investor-owned utilities generated 65 percent of their funds internally. By the 1970s this figure had dropped below 30 percent. See *The Morgan Guaranty Survey* (1969, 1971).

[24] In arguing that nuclear energy collapsed in the United States because financing for it dried up, John Campbell (1986, 1988) rightly says the key evidence is that reactor orders began to be canceled in 1972, one year before the oil crisis. Slack demand after the oil crisis would then only have exacerbated the financing crisis. But aggregate data on the financial problems of utilities do not explain the cancellations of 1972, none of which were made because of financial difficulties. Four utilities canceled orders for six reactors in that year, although none was canceled in 1973. In interviews, officials of these utilities (Baltimore Gas and Electric, Consolidated Edison, Florida Power, and New York State Electric and Gas)

practices to speed rate increases and allow construction charges into the rate bases, and in 1976 the Federal Energy Regulatory Commission began to allow construction charges in the rate base for the 12 percent of American utilities it regulated. Financing would hurt nuclear energy only if cheaper alternatives were found or if state regulators turned against it. Financing was a potential threat to nuclear energy, but before the oil crisis it had no effect.

Conclusions: Hidden Weaknesses

As late as 1973 nuclear power seemed destined for a great future in the United States. Enthusiasm remained high among utilities on the grounds that problems in manufacturing and regulation were being solved and costs would start falling. Reactor orders were at a record high. Oil prices had begun to move upward in 1971 for the first time in history. Critics of nuclear energy existed, but aside from changes in the regulatory structure, they were having little effect. Those changes, most notably the creation of the NRC, were expected to facilitate the spread of nuclear reactors. Commercialization of the LWR in the United States seemed to have succeeded in a short amount of time. This rapid and enthusiastic commercialization was so successful, in fact, that it created several pitfalls for itself, apparent only with hindsight. In the words of Luther Carter (1987), the LWR was a "technology ahead of itself." It got ahead of itself because technological enthusiasm allowed grave problems to be treated lightly or ignored completely. If problems were recognized at all, enthusiasts assumed that solutions would be found during the constant march of technological progress. That seemed a reasonable assumption, even though it was paraded as fact.

The first weakness of the American program was thus the rapidity with which it was put in place after 1963. There were significant development and dissemination costs that enthusiasts had pretended would not be there. American utilities paid the cost of being pioneers of commercial nuclear energy; Sweden, and especially France, came later and benefited from American trial and error. These costs, which on average caused total reactor costs to be twice as high as estimated, were typical for the com-

---

denied financing problems as a reason—naming instead local opposition; the use of alternative sites; and, to a lesser extent, uncertainties about reactor costs. The bond ratings for the utilities were excellent and stable during this period; Moody's gave them A (Con Edison), Aa (Florida and New York State), and Aaa (Baltimore), its three highest ratings. In succeeding sentences Campbell (1988, 101) says both that the utility industry was in a capital crisis in 1972 and that its capital investment peaked as a portion of total industrial investment. Many utilities had financial problems—although not the four utilities canceling reactors—but they continued to build generating capacity.

mercialization of a new technology (Bupp and Komanoff 1983), but enthusiasts had predicted that costs would fall, not rise. They had inflated most people's expectations. Disappointment would be inevitable when accurate cost analyses became available.

Technological enthusiasm also allowed utilities to overlook the real managerial challenges of nuclear energy. Just as they had assumed that technical problems were solved, they assumed they could operate nuclear reactors just like fossil fuel plants. Nuclear energy was seen as "little more than a novel way to boil water" (Bupp 1979, 139), and the euphemistic jargon for atomic reactors was "nuclear steam supply systems" or "NSSS." At this point there was little awareness that particular care and especially high standards should be maintained, that unusually rigorous management was necessary. Persistent attempts to site reactors in or near urban areas is just one indication that the risks were not well appreciated.

Third, utility financing was not assured. It depended on state PUCs, which were partly subject to political pressures. As long as the PUCs reorganized to avoid backlogs in granting rate increases, and as long as they remained favorable to nuclear energy, utilities could get the funds they needed. If investors began to mistrust nuclear energy, however, the costs of financing construction could rise. Then alternatives to nuclear fission might look more appealing. But these possibilities looked most unlikely as late as the oil crisis of 1973–1974.

Finally, the agency that should have brought the utilities out of their naïveté, the AEC, had abdicated this function. It had shifted its internal focus to breeder research on the enthusiastic assumption that the major LWR problems had either been solved or would be solved by industry. Its regulation of LWRs combined external formalism with indulgence, an unfortunate consequence of American regulatory traditions combined with technological enthusiasm. The AEC was thus unable to instill more serious management in the utilities, as long as they followed the rules. And the AEC itself was peopled by enthusiasts who didn't see the necessity for more care and caution. Licensing problems were already beginning to surface, leading to attempts to reorganize the system. This reorganization, however, while solving certain problems, ultimately opened the way to more serious threats to the American nuclear power program.

Premature deployment, poor utility management, financing difficulties, and hands-off regulation were only *potential* weaknesses in the early 1970s. Each of them should have become less important as time passed. The first country to develop nuclear fission and commercialize it seemed destined to continue leading the world in building reactors. The reasons that problems grew instead of disappearing lie in the period of the oil crisis, the subject of part three. More remarkable than the potential weaknesses, however, was the ability of a small group of technological enthusiasts to create a large and thriving industry from scratch. The AEC, the JCAE, and then the manufacturers were able, with little solid evidence, to create a bandwagon market for light water reactors: they reshaped technological reality. A thriving nuclear industry became a reality largely because these groups believed it could. It was an impressive accomplishment, but as Peter Bradford said, it was "the wrong kind of strength."

CHAPTER 4

Early Victory for Light Water in Sweden

WHEREAS the LWR developed from the American submarine program and was the natural choice of technology in the United States, in other countries it often had to displace indigenous reactor designs. Most other countries did not pursue light water technology before the mid-1960s, partly because it required high-grade uranium available only from enrichment plants in the United States and the Soviet Union. Sweden developed a reactor using natural uranium as fuel and heavy water as a moderator, one of the reactor types that seemed most promising in the early 1960s. Sweden had to abandon this heavy water design before it could fully adopt the American light water technology. Of all the countries that had developed their own reactor lines, Sweden made the transition to light water technology earliest and most smoothly.

EARLY HISTORY OF SWEDEN'S NUCLEAR PROGRAM

Sweden's nuclear energy program dates from the appointment of a royal commission in 1945, making the program roughly as old as those of the United States and France.[1] The commission evolved into a permanent Atom Committee, which advised the government on channeling research funds. The committee was unusual in not having direct representation from political parties; its thirteen members included six nuclear scientists, four industry representatives, and two defense representatives (Garris 1972, 37). Like France and the United States, Sweden had a strong interest in developing military nuclear technology, but it postponed this pursuit in 1958 and dropped it officially in 1968.[2]

Research funding was especially directed to two institutions, the Chalmers Technical School in Göteborg, and the Radiology Institute at Stockholm's Karolinska Hospital. Sweden's small size (8.3 million people)

---

[1] Other countries that set up nuclear research programs immediately after the war were Britain, Canada, Norway, and Belgium. The history of Swedish nuclear energy in the 1950s and 1960s is described in Wittrock and Lindström (1982 and 1984), Swedish Government (1970), Rhenman (1958), Holmberg et al. (1977), and Larsson (1985).

[2] It was not publicly known that Sweden had continued to study nuclear weapons until 1972, so that the revelation of this in 1985 caused a minor scandal. Garris (1972) details the official discussions of the 1950s and 1960s. The articles by Christer Larsson that chronicled the secret continuation are in *Ny Teknik*, 25 April, 2 May, 9 May, and 16 May 1985.

meant that Chalmers produced almost all the country's nuclear engineers, an "old boys" tie similar to the elite corps in France. In April 1947 the Atom Committee recommended establishing a joint government-business enterprise to coordinate all research activities, still largely confined to studying other countries' results. AB Atomenergi began operation in January 1948, with the state owning four sevenths of the shares, and 24 private companies owning the rest. This mixed ownership structure, unusual in Sweden, was deemed necessary for maintaining tight coordination and control while also attracting capable scientists and industrial capacity.

In 1949 AB Atomenergi began to develop a process to use Sweden's own uranium deposits, which were heavily mixed with shale. By 1954 it had refined enough uranium to power a tiny reactor, R1, that went critical in July. This was designed to be the first step in developing a "Swedish Line" of reactor technology based on natural uranium and heavy water. In contrast to the American light water design, this technology seemed to have several advantages. Instead of the heavily enriched uranium that was expensive to produce and available only from abroad in the 1950s, it used the natural uranium plentiful in Sweden. It also produced plutonium, which could be used not only in atomic bombs but in the breeder reactors already envisioned. Similar natural uranium designs were being developed by other nuclear pioneers, including France, Canada, and Britain. They seemed as economical as the LWR (Beaton and Maddox 1962, 99).

Enthusiasm for commercial nuclear energy heated up in 1955, in part because of the international conference on the peaceful atom at Geneva, which inspired many countries' atomic plans (Wittrock and Lindström 1984, 69ff.). AB Atomenergi grew to more than three hundred employees by the end of 1955, and its budget allocation was doubled for 1956. It continued to grow in the following years, reaching twelve hundred employees by the end of 1960 (Garris 1972, 287). It set up a committee with the State Power Board (*Statens Vattenfallsverk*, producer of half of Sweden's electricity) and ASEA, the country's largest producer of electrical equipment, to arrange the site of the first nuclear power plant for electricity, but organizational rivalries prevented a clear choice.[3]

The government established a state commission in December 1955 to outline the options for nuclear deployment. The three-man commission, working with a staff of three, completed its report in the almost record time of three months (many commissions take several years). There must

[3] The State Power Board wanted two pilot reactors, one to produce electricity and the other heat. Unfortunately the Swedish program eventually concentrated on electricity. By the 1980s many critics claimed that reactors producing heat as well would be both safer and more efficient, and Swedish research came to focus on this type of reactor.

have been a clear feeling that nuclear energy was an important option for Sweden. The report recommended 100 MWe of capacity by 1960, 3,000–6,000 MWe by 1970, and 6,000–12,000 MWe by 1975, split between heat and electricity production. (This was roughly the size of the program France was anticipating at the time, and it turned out to be close to what Sweden actually built.) The commission felt that nuclear authority was too decentralized, and it explicitly noted the centralization in the United States under the AEC. The ministries of Commerce, Communications, Education, Interior, and Foreign Affairs all had some jurisdiction. The commission recommended that a single state body be given authority over nuclear energy and that private companies be allowed to produce reactors once AB Atomenergi had developed them. In spite of a turf battle over precise responsibilities, the Riksdag (Sweden's parliament) passed most of the Commission's recommendations in the Atomic Energy Act in 1956, although an informal meeting later that year granted private industry further rights to build prototype reactors.

The Atomic Energy Act proposed another small research reactor by 1959 and a power producing reactor by 1963. The former was a light water reactor purchased from the United States (Allis-Chalmers) under the "Atoms for Peace" program, and it went critical in 1960. The third reactor, using Swedish technology, would be built at Ågesta and produce both electricity and heat for homes. This was to be a prototype for a larger reactor, also of heavy water, to be built at Marviken. Sweden's plans also called for a fuel reprocessing plant to complete the nuclear fuel cycle. France and the United States had similar ambitions for their nuclear industries, but few other industrial nations did. Sweden did not abandon its reprocessing plans until the 1970s; delay helped it avoid a costly mistake.

In Sweden military and civilian nuclear developments were linked in a unique way. Several of the political parties were riven in the 1950s over whether Sweden should develop nuclear weapons, with the governing Social Democrats especially split.[4] Opposition parties hoped to exacerbate this split by forcing the Social Democrats to take a clear position for or against weapons research; the Social Democrats resisted. Beginning in 1954 there were protracted public debates parallel to those in the 1970s over civilian nuclear energy. In 1958 politicians from the major parties, meeting privately, decided not to decide anything: neither to start an explicit weapons program nor to preclude having one in the future. A large civilian nuclear energy program was a convenient way to smooth over the issue, since much plutonium as well as technology usable in a

---

[4] Jerome Garris (1972) describes these conflicts and debates in great detail. He points out (p. 113) that "there was no unanimity either within the parties or among them."

weapons program could be developed under the heading of peaceful nuclear energy. One Defense Department scientist explicitly "argued that atomic bombs were quite simple to manufacture and could be produced as a bi-product of the civilian atomic energy program with little problem" (Garris 1972, 118).

Although weapons research continued secretly until 1972, public statements and policy renouncing nuclear weapons technology made some legislators happy, while it satisfied others by leaving open the possibility of picking up a weapons program again in the future. This policy of compromise, based on avoidance of conflict through delaying a firm, irrevocable decision, is a common Social Democratic governing tactic. In this case the maneuver allowed weapons development until 1972 to be mixed in with (and hidden by) civilian nuclear research. The irony is that Sweden's early nuclear program was thus partly a result of internal Social Democratic efforts to keep the party united, while in 1980 the hotly contested referendum on the nuclear program would be partly a result of exactly the same tactic.

Dropping the Heavy Water Line

In the early 1960s it seemed most reasonable for Sweden to continue using some LWRs at the same time it developed its own heavy water technology. The heart of the development of the Swedish reactor line was the reactor project at Marviken, one hundred kilometers south of Stockholm, a joint venture of AB Atomenergi and the State Power Board. Sweden's supply of uranium was not decisive in the choice of design: a natural uranium model seemed most promising to almost all European countries at the time. Britain, France, and other European countries were pursuing this line of research.

In spite of increasing enthusiasm and investment of resources, the reactor construction schedule faced delays and cost overruns. In fact it was the technological enthusiasm that led to greater disappointment when high expectations were not met. The completion date for Ågesta was delayed from 1961 to 1963 and then to 1964, and costs almost doubled. Unforeseen challenges in constructing the underground building and its safety features contributed. Cost overruns are the norm for new technologies, but technological enthusiasm fostered impatience.

Mild disenchantment with the Swedish heavy water line was reinforced by the rising tide of enthusiasm for LWRs emanating from the United States and advancing across Europe. In 1962 West Germany had chosen a boiling water version of the LWR for its first nuclear power plant. In the mid-1960s news of the American turnkey deals spread throughout the world's nuclear elites. As Bupp and Derian (1978, 64) say,

"The early turnkey deals and the economic promises they made in them were widely publicized in Europe and created the impression that a major commercial breakthrough had been made with nuclear power in the United States." Little was known about the LWR in Sweden, but hopes fastened on it.

Now Swedish industry began to suggest that AB Atomenergi turn its attention to LWRs, and in 1964 the State Power Board withdrew from the Marviken project (Holmberg et al. 1977, 25). At the same time a private company, ASEA, was developing its own version of the boiling water reactor. It sold one in 1965 to OKG (Oskarshamn Kraftgrupp), a company formed by nine utility companies expressly to produce electricity from nuclear fission. In 1966 an agreement for a thirty-year supply of enriched uranium was concluded with the United States. While it was a private utility that purchased the first LWR in Sweden, it was the large state utility that had doomed the Swedish line by withdrawing its support.

With no interest left on the part of electricity producers, the Swedish line was effectively dead. If the state could not maintain its own State Power Board's interest in the Swedish line, it certainly couldn't attract the smaller private utilities. Marviken's costs were higher than had been expected—roughly $100 million before it was officially abandoned in 1970 (Dröfer 1974). But high costs and technical difficulties would not have killed it; it could have been sustained by the same technological enthusiasm that carried the LWR program through difficulties in the United States. After all, LWR costs also turned out to be twice what was expected. Given sufficient time and funding, the Swedish line probably would have worked, just as the Canadians managed to develop a heavy water, natural uranium design (CANDU) that is now the darling of many nuclear observers.

To the extent there were technical difficulties with Marviken, their causes were similar to those in the United States nuclear program. Concern for profits led to commercialization before adequate testing could be conducted (Wittrock and Lindström 1982, 33). What killed the Swedish line was that light water technology seemed superior, so that research on the former was discontinued. Cost-minded utilities won against technological enthusiasts developing a new reactor line. Whereas in the United States technological enthusiasts swept cost-benefiters along with them toward commercialization of the LWR, Swedish cost-benefiters were inspired by the commercialization that had occurred in the United States and tripped up Sweden's own technological enthusiasts.

The strain between technological enthusiasm and the cost-benefit perspective is often expressed as conflict between industrial and trade policies on the one hand and policies of national self-sufficiency on the other. Wittrock and Lindström (1982, 31) view the abandonment of Marviken as a victory for the former over the latter. This analysis is plausible and will be used to compare Sweden with France, where the nationalist policies prevailed much longer. But to understand the logic of the two kinds of policies, as well as their conflict, it is necessary to keep in mind the policy rhetorics and worldviews behind them.

SWEDEN'S NUCLEAR PROGRAM

The Swedish government quickly took steps to ease the transition to light water technology. AB Atomenergi was being left without a clear purpose, so the government took it over completely in 1969 and restricted its role to research instead of research and dissemination. At the same time ASEA-Atom was formed, owned jointly by the government and ASEA, to build and sell reactor systems. During this reorganization orders began to flow from the utilities, most for Swedish boiling water reactors but two for Westinghouse pressurized water reactors (Table 4.1). Finally, in 1970, Marviken was transformed into a fossil fuel plant, and it was used for experiments on reactor accidents as well.

Sweden has a mix of public and private electricity producers. The government-owned State Power Board produces roughly half the country's electricity, while private companies and municipalities account for the rest. Thirteen of the largest producers belong to a trade association, Kraftsam (formerly CDL), which provides a forum for developing forecasts of demand growth and coordinating supply expansion. Ownership of

TABLE 4.1
Commercial Reactors Ordered in Sweden

Reactor	Year Ordered	Year On Line	MWe	Type	Owner
Oskarshamn 1	1966	1972	440	BWR	OKG
Ringhals 1	1968	1976	750	BWR	Vattenfall
Oskarshamn 2	1969	1974	580	BWR	OKG
Ringhals 2	1970	1975	800	PWR	Vattenfall
Barsebäck 1	1970	1975	570	BWR	Sydkraft
Forsmark 1	1971	1980	900	BWR	FKA
Barsebäck 2	1972	1977	570	BWR	Sydkraft
Ringhals 3	1972	1979	915	PWR	Vattenfall
Forsmark 2	1973	1981	900	BWR	FKA
Ringhals 4	1973	1980	915	PWR	Vattenfall
Oskarshamn 3	1974	1985	1,050	BWR	OKG
Forsmark 3	1976	1984	1,050	BWR	FKA

Source: Vedung (1979), Swedish Atomic Forum (no date).

Swedish reactors takes varied forms. The State Power Board owns the four Ringhals reactors. The largest private producer, Sydkraft, owns the two Barsebäck reactors. The other reactors are owned and operated by two companies established specifically for that purpose. OKG, which operates the three Oskarshamn reactors, is owned by eight private electric companies (including Sydkraft). The State Power Board owns three quarters (and another company one quarter) of FKA, which operates the three Forsmark reactors.[5]

Sweden, like most European countries, became accustomed to following the American lead in nuclear development, in some social arrangements as well as technological design. Sweden, France, and fourteen other European countries signed agreements in 1960 and 1963 to limit liability in case of nuclear accidents, just as the 1957 Price-Anderson Act had in the United States (OECD 1985). The Paris and Brussels Conventions limited liability to 120 million standard European currency units (roughly equal to U.S. dollars): 15 million of this would be covered by the nuclear operators, and the rest by governments (Jacobsson 1985). Countries were allowed to compensate victims beyond the minimum. Those nations that later elected to do so—Sweden, Germany, and the Netherlands—were precisely those where policies became less pronuclear. New policy choices changed the system of insurance; the availability of insurance did not influence those policies.

By the end of 1971 all four of Sweden's nuclear sites had received construction permits for at least the first reactor. By the time the oil crisis struck in October 1973, eight of Sweden's eventual twelve reactors had already been licensed for construction, and others were at advanced stages of planning. When these reactors began to come on line in the mid-1970s, they made Sweden the world's largest per capita user of nuclear energy. In addition, there were plans for far more reactors in the future. The trade association of Sweden's electricity producers recommended that eighteen reactors be brought on line by 1985 and twenty-four by 1990 to meet growing electricity demand (Centrala Driftledningen 1972, 17–18). As in the United States, the future looked very bright for nuclear energy in Sweden.

Licensing and Regulation in Sweden

The 1956 Atomic Energy Act that proposed a Swedish reactor line also set up the regulatory apparatus for nuclear reactors. This machinery is formally similar to American regulation of the same period. The Swedish

---

[5] Sahr (1985) provides details about the Swedish energy system in the context of an excellent discussion of political debates.

Atomic Energy Board, renamed the State Nuclear Inspectorate (*Statens Kärnkraftinspektion*, or SKI) in 1974, issues all licenses for "nuclear activities" and inspects all reactors. There are more stages to the licensing process than in the United States—from the first construction, to final design features, to fuel loading, to testing, and finally to commercial operation. The fact that SKI also "issues directives" through this process indicates a close working relationship between SKI and the utilities; it helps them learn to operate the plant rather than simply approving techniques after the fact. This emphasis is different from that of the American NRC (and AEC), where formal procedures involve less ongoing cooperation on operations oversight and research.[6]

From the beginning of the approval process, the license application is covered by the Building Act of 1947, of which section 136a requires the approval of the municipality for many large industrial plants, including nuclear reactors. This municipal veto power is a major legal difference between Sweden on the one hand and France and the United States on the other. In France there are local public inquiries, and in the United States public hearings, but nothing approaching a power of veto. Nevertheless no Swedish community used its veto power to block a nuclear plant, largely because site selection and construction began before organized opposition arose. As table 4.1 shows, all four of Sweden's reactor sites had been granted construction licenses by the end of 1971. In the 1960s there were few doubts on the part of the public and no antinuclear organizations to stir any latent ones. Once a site contained one reactor, others could be added without being subject to communal vetoes. As great as their veto power had been, the municipalities had none left when opposition to nuclear energy began to grow after 1973.[7]

While in France and the United States two permits are required—one for construction and the other for operation—Sweden has one primary license covering both, with separate additional examinations, consultations, and permissions at later stages. The spirit of this regulatory system has remained essentially the same since 1956, while the official legislation has continually changed. The 1977 Stipulation Act concerned the conditions under which fuel could be loaded, and the politics surrounding it will be discussed in chapter 8. In 1984 the Act on Nuclear Activities replaced both the 1956 and 1977 acts, but by this time all Swedish reactors were either operating or about to.

SKI was modeled after the American AEC and NRC in many ways, es-

---

[6] This discussion of the Swedish regulatory apparatus is based on the OECD NEA (1980, 1984) and on interviews with Olof Hörmander, head of SKI; Lars Gunnar Larsson, former head of reactor inspection at SKI; and Carl Erik Wikdahl of OKG.

[7] This veto power was, however, used to block the mining of uranium at Ranstadt in the mid-1960s.

pecially in the system of plant inspection and the standards applied. In addition there are many personal contacts between SKI and the NRC. But although in many cases the contents of the standards are similar, the role of standards in the Swedish regulatory system is different from that in the United States. While American regulation is external, hands-off, and formalistic, Swedish regulation is informal, internal, and cooperative. There are far more explicit rules and guidelines in the United States, while the Swedes rely on ongoing communication and persuasion. The Swedes can do that for two reasons. First, there are both trust and personal contacts between regulators and the electrical and nuclear industries. The personal contacts are largely due to the country's size and the fact that so many nuclear engineers come from Chalmers Technical School. The trust is due both to these contacts and to the fact that regulation is not seen in an adversarial or "external" fashion. The two groups see themselves as working toward the same end, rather than different ones, and there is little public suspicion about their working together to reach it. Second, regulators actively study safety and other problems at a very technical level, attaining as good a grasp of them as industry has. This continued research into the operation of reactors was precisely what the American AEC thought was unnecessary. SKI's competence further increases the trust between SKI and the nuclear operators.

Just as the American regulatory style has deep historical roots in the mistrust between business and government, so the Swedish style is based on close cooperation resulting partly from the public's traditional deference to state officials. Whereas the American system is grounded on the (often accurate) assumption that companies will cheat and resist, the Swedish system works by way of informal personal contacts, assuming (also accurately) that the actors agree on general goals. The Riksdag does very little to monitor regulatory agencies; it provides only the most general directives, and it exercises only sporadically its right to question agency heads. Agencies are trusted to be generally sensitive to government wishes.[8]

Steven Kelman (1981a) links regulatory styles to legal styles. The American adversarial system of cross-examination, in which two sides battle it out with a judge watching, contrasts sharply with the civil law system of Sweden, in which the judge takes an active role as representative of the state, personally questioning the witnesses and calling witnesses of her own (pp. 137–39). In Sweden the state is given more authority to ferret out wrongdoing; for example the prosecutor makes an opening statement, but the defense cannot. The regulatory parallel is that the state is trusted to oversee business cooperatively from within; it

---

[8] Steven Kelman (1981a) describes the role of personal deference in Swedish politics.

is trusted not to become implicated in the business perspective. The American state is never trusted like this. A cooperative regulatory system, with close personal ties between regulators and regulated, has the efficiency the Swedes claim but also the potential for abuse the Americans fear. Which of the two occurs depends on the discretion of the actors involved or on the outcomes of particular power struggles. In fact, there is no evidence of abuse in Swedish regulation.

CONCLUSIONS

In both the United States and Sweden there was conflict between technologist and cost-benefit worldviews that had to be overcome before light water commercialization could succeed. In the United States this occurred through the two main reactor manufacturers, who had a clear interest in spreading technological enthusiasm. But in Sweden (and, we shall see, France) the technologists developing the Swedish line encountered cost-benefiters who preferred the American line. The lack of industrial support for the former prevented the enthusiasts from moving beyond the early stages of development, so that their enthusiasm had to be transferred to the LWR. When it was clear that light water had won, then technological enthusiasts and cost-benefiters could align themselves behind it. By 1973 nuclear energy had a promising future in Sweden as well as the United States.

In comparison with France, Sweden made the transition from its indigenous natural uranium reactor technology to American-inspired light water technology early and easily.[9] France also had an important research organization that had to be reorganized substantially when its own reactor line lost out. France's electricity producer was also instrumental in redirecting attention and funding to LWRs. But these shifts occurred in France several years later than in Sweden, and they did not happen smoothly.

[9] Alfvén (1972) and the Swedish Government (1970) agree that the transition could have been more difficult, but the latter makes it sound particularly smooth (p. 6).

CHAPTER 5

The Difficult Transition to Light Water in France

IN SPITE of centralized political and economic control, France got off to a rocky start with its nuclear program. In the 1950s the French had concentrated on building a nuclear reactor based on natural uranium, like the Swedish model, but with a gas coolant and graphite moderator. For both countries there were two important factors: the unavailability of enriched uranium and the desire to produce plutonium for use in weapons (something this design does well).[1] France differed from Sweden in clinging to its own gas-graphite reactor line until 1969, delaying its adoption of a large reactor construction program. Renewed political will and a reorganization of the nuclear industry cleared the way for a successful program only in the early 1970s.[2]

THE GAS-GRAPHITE REACTOR

The Commissariat à l'Energie Atomique (CEA), established at the end of World War II, was parallel in function and design to the American AEC. After a shakedown in the early 1950s, in which physicists were widely replaced by engineer/bureaucrats, the CEA developed several reactors capable of producing electricity. In 1955 plans were laid for a series of reactors that the state-owned utility Electricité de France (EDF) would operate to generate electricity. The PEON (Production d'Electricité d'Origine Nucléaire) Commission was established to advise the CEA in this

---

[1] In France there was no public debate over weapons production, as there had been in Sweden. Indeed, most top politicians had little idea that the Commissariat à l'Energie Atomique (CEA) was producing a bomb (Scheinman 1965, 183). Throughout the 1950s, France, Sweden, and the United States each had a secretive system of nuclear research in which military and civilian applications were closely related.

[2] Several fine sources present the early history of nuclear energy in France. Weart (1979) describes the period around World War II. Scheinman (1965) covers the period from the war to 1958. Biquard (1966) is a biography of Joliot-Curie, the man who headed the CEA just after the war and was removed in 1950 because of his communist political beliefs. Goldschmidt (1982), Nau (1974), and deLeon (1979) cover the 1950s and 1960s in passing. Picard et al. (1985) is a good history of EDF from the inside, Gravelaine and O'Dy (1978) from the outside; Simonnot (1978) discusses the major decisions and conflicts from the perspective of interviews with many of the important figures; Saumon and Puiseux (1977) is a good summary in English. Louis Puiseux (1977, 1978), former top economist for EDF, has many insights into the politics of French electricity. The best work on the French energy system in any language is probably Lucas (1979). I rely on all these.

endeavor, with representatives from the CEA, EDF, the state, and industry. It soon recommended a series of prototypes that by 1965 would total 800 MWe of capacity. The PEON Commission and the CEA revealed the same kind of technological boosterism as the American AEC and JCAE.

In the late 1950s and early 1960s, EDF displayed the same blend of curiosity and skepticism about nuclear energy shown by electric companies in Sweden and the United States. It was involved in several joint ventures into experimental reactors, much like American utilities.[3] Its only policy position on reactor designs was that it was important to keep open several possibilities, but it "expressed no preference between gas graphite and American light water techniques" (Nau 1974, 82). The CEA concentrated its research on the gas-graphite reactor, partly out of national pride and partly from organizational interests, both reinforced by immense technological enthusiasm. Speaking of gas-graphite costs in 1964, the PEON Commission (quoted in Simonnot 1978, 240), wrote that "behind these figures there is not only a simple extrapolation, but also foreseeable technical improvement; one can hardly doubt the considerable progress still to come of nuclear compared to classical techniques." At this point the CEA could claim (1964) that the gas-graphite reactor would be cheaper than the LWR, because of a new form of fuel that would increase the energy obtained from each unit of natural uranium.

The case of the gas-graphite reactor shows how technologies are shaped, constrained, or encouraged by their economic and political contexts, since a lack of commercial support choked the French reactor line. Several commercial gas-graphite reactors were ordered in the mid-1960s (at Bugey and Saint-Laurent), and one was exported to Spain (Vandellos). But when a prototype for this series was brought on line in 1966, there were several embarrassing problems. There was much controversy over who was to blame. There were faults in the construction, but the builders blamed EDF, which preferred to retain architectural and engineering control itself, for not letting them see the plans in their entirety. The nuclear industry was emphatic on this point, since EDF's control meant that the industry would not gain the know-how to export reactors successfully. Ironically EDF, also blaming the construction process, tightened its grip on the contractors, centralizing the process even further.

In the mid-1960s EDF had the same casual attitudes to nuclear reactors that American utilities did; they were just another way to boil water. But

[3] In 1955 two experimental gas-graphite reactors were approved to be built at Chinon, constructed by the CEA but operated by EDF. In 1960 EDF collaborated with Belgium in launching a pressurized water reactor at Chooz. In 1962 the CEA decided to build a reactor at Brennilis, and EDF had the major research responsibility. Four other gas-graphite reactors were begun in the mid 1960s out of collaboration between the two organizations (deLeon 1979, 139).

unlike American utilities, EDF management took strong steps to control both construction and operations at its nuclear plants. Pierre Massé, president of EDF from 1966 to 1969, helped change those attitudes (quoted in Picard et al. 1985, 193): "The truth is we had neglected the difficulties of nuclear in every way. We had underestimated the need for materials of great purity and techniques of great safety. I remember a regional director who said 'It's very simple, a nuclear plant: it's like a thermal reactor in which there is a reactor instead of a boiler.'!" Taking full control over all aspects of construction was the first step in changing these attitudes; careful training of operators was another. Good management at EDF changed precisely those casual attitudes that were a hidden (and ultimately fatal) weakness of American utilities.

EDF and the private nuclear industry found a mutually agreeable scapegoat, the gas-graphite line. In attacking it, EDF could undermine the control of the CEA over nuclear technology, substituting its own. Private industry saw the American LWR as an opportunity to gain expertise independent from that of big state organizations like the CEA and EDF. EDF practiced delay tactics, not ordering gas-graphite reactors, or indeed any reactors, from 1966 to 1970. This lack of orders prevented further gas-graphite development at a time when LWRs were in the middle of their great bandwagon market in the United States. Industry attacked the gas-graphite reactor as not being exportable, but the reason was that no one had an interest in exporting them except for the CEA, which had no apparatus for doing so. When Argentina tried to buy one, it could find no one to talk to (Saumon and Puiseux 1977, 148).

The lack of industrial promoters dimmed the gas-graphite reactor's domestic prospects as well as its export potential. At the beginning of 1968, when EDF took bids for a gas-graphite reactor at Fessenheim, few manufacturers submitted any, and those who did submitted very high ones. Speaking of the gas-graphite reactor, Saumon and Puiseux (1977, 147) say, "This absence of an industrial partnership interested in promotion of a graphite-gas 'product' doubtless accounted for much of the difficulty in finding a market for it. . . . the builders did not believe in its export possibilities and did not want to take financial risks to develop it. This accounts for the high level of the bids for Fessenheim, but it became the pretext for EDF to stop the project." The opposite had happened in the United States. There heavy industry was enthusiastic about nuclear technology and had an interest in promoting it. As a result the perceived costs of LWRs—and certainly those presented in the turnkey bids—were much *lower* than the real costs. In France industry exaggerated the costs of the gas-graphite line.

French technological enthusiasts for the gas-graphite reactor were frustrated by their inability to gain industrial and political support. Had there been continued research monies and a flow of orders, the technology would have continued to improve. A similar lack of organizational momentum had damned the Swedish reactor design, but at a much earlier stage of development. By contrast, enthusiasm swept through American government, manufacturers, and utilities, and the resulting surge of orders encouraged the refinement of the LWR. One reason EDF gave for favoring the LWR was that there were 80 MWe of LWR capacity running or under construction compared to 8 MWe of gas-graphite technology—but almost none of the 80 MWe was yet in operation (Lucas 1979, 143–44).

In France enthusiasm for the gas-graphite line was torpedoed when EDF became more enamored of the LWR. This switch occurred because EDF recognized the technological enthusiasm of the CEA for precisely that, and EDF's economic calculations were too good to be taken in by the CEA's groundless optimism. But they weren't good enough for EDF to avoid being taken in by foreign claims just as optimistic and just as unsupported by data. When the PEON Commission gathered information on LWR costs in 1969, the figures ranged from $132 per kw capacity to $210. Every single figure was based on arbitrary estimates for plants then being ordered. Yet, like American utilities, the PEON Commission and EDF accepted these figures.[4]

EDF decision makers fell for one set of inflated expectations and not the other partly because they combined technological enthusiasm and cost-benefit thinking and partly because they wanted autonomy from the CEA. Like American utilities, EDF contains many engineers; it associates organizational expansion with technological progress, and it associates electricity production with economic growth. Hence it has a strong technological enthusiasm. But there was enough organizational rivalry that EDF didn't jump on the CEA's gas-graphite bandwagon. It instead used cost-benefit rhetoric against the gas-graphite reactor. A true cost-benefit approach would have recognized that the evidence for the LWR was almost as flimsy as that for the gas-graphite, but the LWR provided a release for the technological enthusiasts at EDF who liked the idea of nuclear energy but didn't like the CEA. EDF used cost-benefit rhetoric quite selectively.

Presidential Discretion

EDF's reluctance concerning the gas-graphite technology blossomed into a nasty conflict between EDF and the CEA, notoriously labeled "la guerre

---

[4] Bupp and Derian (1978, chap. 4) document the ungrounded cost estimates. One estimate given to PEON was for Diablo Canyon, ordered in 1966: $146 dollars per kwe. When it came on line twenty years later, its costs were almost $3,000 dollars per kwe, a wild increase, even accounting for inflation.

des filières." Proclaiming cost minimization as its goal, EDF claimed that LWRs produced electricity at lower costs than gas-graphite reactors. Its evidence for this was a comparison between the gas-graphite reactor at Saint-Laurent and—what else?—the LWR at Oyster Creek. Not only was the latter a loss-leading, turnkey project, but it had not yet even been completed. Surprisingly, the CEA didn't dispute EDF's cost estimates of 1,100 francs/kWh for the gas-graphite versus 880 for the LWR (Picard et al. 1985, 197). Instead the CEA argued that France should avoid dependence on American technology and American enriched uranium, with the assumption as well that gas-graphite technology would continue to improve. The language of the debate pitted a cost-benefit approach and technological enthusiasm, embodied in bureaucratic rivalry.

Additional lines of cleavage were laid bare in the conflict. A light water program meant a reduced role for the CEA in the construction and operation of nuclear plants, and more autonomy for EDF. In fact, there would be a serious lack of purpose at the CEA without its research on gas-graphite technology; the LWR represented a genuine organizational threat. There was also a conflict between the Corps des Mines, dominant at the CEA, and the Corps des Ponts et Chaussées, dominant at EDF. Simonnot (1978, 189) also mentions several others, among them conflict "between two relations to power: direct access, thanks to military work (CEA), and indirect access; between two types of power: weight (CEA) and extent (EDF); and between two attitudes to capitalists: collaboration (CEA) and competition (EDF)." Simonnot concludes simply that each side had "its own total and coherent vision of the world." The vision at the CEA was technological enthusiasm in a pure form: this organization had long been dominated by the Corps des Mines and more generally by Polytechnique graduates, exactly the groups most taken by technological enthusiasm. EDF was run by engineers and occasional economists, the latter especially important at the top.[5] EDF was capable of great enthusiasm for a technology, but especially that technology which seemed most economical. Once EDF's cost-benefiters accepted the assumptions (concerning the LWR) held by EDF's technological enthusiasts, this pronuclear bloc became relatively impermeable to information critical of nuclear energy.

Since worldviews and material interests line up in this conflict, it is difficult to discern the weight of each. But the CEA had doubts about the LWR independent of its own interest in the gas-graphite line. It examined two other reactor models in the late 1960s, Canada's CANDU and Euratom's ORGEL, rather than conceding the ground to light water technol-

---

[5] Marcel Boiteux, general director from 1967 to 1979 and then president until 1987, was an economist; Paul Delouvrier, president from 1969 to 1979, was an ENA graduate in the *Inspection des Finances*.

ogy. The argument for "national independence" had two sides at this point. The first was a certain degree of chauvinism, very strong in the 1960s (Gilpin 1968), that wanted to prove France could develop high technology as capably as the United States. This kind of nationalist technological enthusiasm would be epitomized, and receive its name, from a project then under way: the *Concorde* syndrome. The second, a commitment to the LWR, entailed dependence for ten or more years on American reactor technology and enriched uranium, at whatever price. The CEA's was a coherent position, based not only on the uncertainties of each technology, but also on certain assumptions about how the world works. It is not clear that pure organizational interests were primary.

One striking feature of the CEA-EDF conflict was each organization's confidence that it was pursuing the best interests of France. That both found national interests to line up with their own organizational interests does not mean they were insincere, but that each set of interests arose out of the policy style that dominated its organization. Autonomous technological development was good for France and the CEA; minimizing the costs paid for nuclear energy was good for France and EDF. Engineers and economists on both sides saw themselves as in the service of the country. Each policy style contains a clear image of national interest.

The conflict, after being played out over several years in committees, especially the PEON Commission, was finally decided at the highest political level, by the president of the republic. Through a combination of persuasion and political appointments, EDF had managed to capture a PEON Commission majority, so that the PEON report of 1968 suggested that the CEA continue its gas-graphite research but that EDF launch a LWR program and that the state encourage the industrial regroupment necessary for this. But EDF's economic argument, for which there was inadequate evidence one way or the other, was irrelevant to questions of national independence. So Charles de Gaulle continued to support the gas-graphite line in spite of pressure from EDF and the nuclear industry. This nationalist line was possible with a modicum of technological enthusiasm to reassure policymakers that the gas-graphite reactor would continue to develop and improve.

Soon after Georges Pompidou assumed the presidency in 1969, he reversed this decision, approving a policy in which EDF would build more LWRs without necessarily engaging more gas-graphite reactors.[6] Here the meaning of "national independence" shifted from technological autonomy to economic competitiveness. The rationale for the gas-graphite reactor

[6] This reversal was taken on 13 November 1969, at a meeting of major ministers as well as the heads of EDF and the CEA. For details on this specific decision see Picard et al. (1985, 199–202) and on the more general policy shift, Vilain (1970, 47–50).

had been that France would develop its own technology, and thus develop its own industry and technical know-how. In this way it could prove itself a top industrial and political power. With Presidents Pompidou and Giscard d'Estaing (former director of the Rothschild Bank and former finance minister, respectively) came the idea that France could prove its worth instead by competing on international markets (Camilleri 1984, 49). Heavy industry supported the LWR because it would be more competitive in export markets, a compelling argument if one accepted this new idea of independence. This new image of the place of the nation in world affairs, and of the national interest, was similar to that held by the Swedish elite. It is typical for Sweden but uncommon for France.

ELECTRICITÉ DE FRANCE

What enabled EDF to win the battle over reactor technologies, as it did so many other energy policy debates? Electricité de France *is* electricity production in France. It is charged with distributing all electricity, over 90 percent of which it produces itself. It is the largest of France's nationalized firms, with over one hundred thousand employees. It has often been called a state within a state, and there is evidence that it lays down the energy policy of the French state to a greater degree than it follows policy laid down by others. To follow the trajectory of nuclear energy in France is to examine the internal machinations and the external strategies of EDF.

EDF's origins have given it tremendous public prestige, especially with the Left. It was created by a 1946 law merging many electricity producers into one nationalized company. The bill was written by the Communist Marcel Paul, and the governing structure proposed for EDF was that favored by the Communist trade union, the CGT.[7] It consisted of a governing board with representatives from state ministries, from the trade unions, and from other interested groups, especially consumers. This structure was a political issue; de Gaulle himself opposed it, arguing that there should be less public and more state representation (Kuisel 1981, 211). From the beginning EDF was highly centralized, controlling both production and distribution, a trend that was strengthened through control of the company by members of the Corps des Mines and the Corps des Ponts et Chaussées. EDF even dominated its sister company, Gaz de France.[8]

---

[7] The Confédération Générale du Travail (CGT) has remained strong at EDF since 1946, never losing its first place among EDF workers. It lost much of its power during the red scare in the late 1940s and early 1950s, but the Communists were not hounded out of the company as thoroughly as at other nationalized firms (Frost 1983, 210).

[8] Bouthillier (1969) is the best source on these particular nationalizations, but Sturmthal

EDF has traditionally had a reputation for putting service above profits. It has traditionally been seen as technically extremely competent, yet as putting its skills to socially useful rather than merely profitable ends (presumably because of its nationalized status). It also enjoyed a reputation for treating its workers in exemplary fashion. All this began to change in the late 1950s, when the goal of nationalized firms evolved from maximizing production for rebuilding to paying their own way (rentabilité). As Robert Frost (1983, 312) points out, this shift foreshadowed a redistribution of power in the company from engineers concerned with technical efficiency to economists concerned with financial efficiency. Frost also describes two accompanying changes: internally, workers came to be seen as factors of production rather than artisanal human beings; and externally, profits replaced service as the goal of the company (especially after Roger Gaspard left the presidency in 1964). The Catholic trade union, the CFTC, bitterly criticized these shifts. Later to become the CFDT, the CFTC began to develop a critique of bureaucracy and profit-seeking from the point of view of democracy and humanity. Although this critique did not gel until 1968, changes at EDF help explain why the CFDT became an early critic of EDF's nuclear program.

Despite alienating a small part of the Left, EDF retained the support of the Communists—still strongly represented among employees and never overly concerned with democratic humanism—and of the general public. Like electricity producers everywhere, EDF was popular in the period of falling prices that continued until the 1970s. In addition, its combination of service and competence helped it attract able managers. Sheer competence was a major reason it was able to change the casual attitudes that saw nuclear fission as just another way to boil water.

A combination of skilled managers and structural position enabled EDF to prevail in the battle of reactor technology. Through several years of delays and hesitations, it undermined the gas-graphite program; and it would have been virtually impossible for the CEA to force EDF to order reactors against its will. One big bureaucracy facing another created a standoff in which no reactors were ordered. On top of that, EDF skillfully maneuvered to win the battle. In addition to maneuvering within the PEON Commission, the general director of EDF in fact anticipated the decision to abandon the gas-graphite line by several weeks, announcing it in public before the decision had actually been made (Picard et al. 1985, 200). This is but one example of a tactic often seen in French policymaking, in which a nonelected official makes a statement that could be em-

---

(1952) and Fletcher (1959) are also good. Briefer discussions can be found in Kuisel (1981, chap. 7) and Frost (1983), who has an interesting discussion of what nationalization meant for the different groups involved (pt. 2, sec. B, p. 2).

barrassing to the government but forces the latter to follow along. Structural explanations of French energy policy, in which the state takes decisions and can implement them because of the centralization of the electrical industry (Campbell 1988; Kitschelt 1986), may have the order of causation wrong. What is more, EDF had to campaign actively to win its battle, working behind the scenes to defeat a powerful and privileged state agency (and its own regulator); even so, victory came only with a change in the Presidency. Structural factors alone do not explain EDF's skillful manipulation, the preferences of Pompidou or de Gaulle, or the outcome of the CEA-EDF battle.

This pattern in which a French state enterprise is powerful enough to resist and thwart government policy is not unique to nuclear energy.[9] Harvey Feigenbaum (1985) has studied the state oil companies in France, concluding that they usually follow their own organizational goals regardless of whether these coincide with the intended policies of the state. The 1966 Nora Report strengthened this autonomy, recommending that state enterprises concern themselves with profits just as though they were private.[10] The two state-owned oil companies seem to have done just that, avoiding taxes, withholding information about oil prices from the state to keep those prices high, and refusing to give France special status during the oil embargo of 1973.[11] As Feigenbaum (26) puts it: "Ceteris paribus, profit maximization and managerial autonomy have tended to limit the ability of the state elite to control public firms so as to implement policies in the elites' own interest."

The autonomy of state enterprises adds a twist to widespread structural accounts of the state and nuclear energy, in which the French state chose a nuclear path and was able to stick to it in spite of public protest, while the American state chose the same path but had to back off. The original path may have been that of EDF, France's "state within a state," rather than that of the government. Descriptions that imply a monolithic French state may work for the period of conflict in the mid and late 1970s, but they ignore the long process through which the French state

---

[9] Several works describe the inevitable conflict between the interests of a French public enterprise and those of the larger society or state. Chevallier (1979) is a good general work, while Anastassopoulos (1980, 1981) focuses on conflicts in decisionmaking. Anastassopoulos (1986) names three categories of conflict: between the specific interests of the organization and the general interests of the state, between the concrete bureaucracies on each side, and between key individuals who interact with each other.

[10] The report, named after the commission's chair, Simon Nora, was requested by the Council of Ministers on 20 July 1965 and published 4 April 1966.

[11] See Feigenbaum (1985, chap. 3), and Stobaugh (1976). The 1973 actions may have been a spur for the highly critical Schwartz Report (1974), which originally described most of the abuses Feigenbaum discusses.

arrived at a consensus about nuclear energy, a consensus that was never complete.

THE TECHNICAL ELITE

Who were the men (this world contains almost no women) making decisions in French nuclear energy? Why were they so strongly pronuclear? Their attitudes are important for supplementing structural factors driving French energy policy. The word as well as the image of a "technocrat" arose in France, and that is still where it is most apt. Perhaps no other political system provides as large a role for people to exercise power on the basis of technical training and certification. The French educational system claims to find the most capable young citizens, and it ranks them in strict hierarchies according to what it claims is purely merit. There are two main families of technocrats: those who issue from the Ecole Polytechnique (commonly abbreviated simply as X) and whose top members belong to the Corps des Mines and the Corps des Ponts et Chaussées, and those who issue from the Ecole Nationale d'Administration (ENA) and whose "best and brightest" go into the Council of State, the Cour des Comptes, or the Inspection Générale des Finances. These corps, whether technical or administrative, are the mechanisms through which graduates from these schools maintain their contacts and privileges. Here I want to look at the technical elite, coming out of X.

The Ecole Polytechnique was founded in 1794 to provide the republic with engineers badly needed for civil defense; in 1804 it was given the mission of training military officers. These dual roles have persisted; on the one hand to produce people with technical expertise, whose bridges would not collapse, and on the other to produce civil servants, whose primary loyalties would be to the state. The best students in mathematics and the sciences generally go there, where they are officially civil servants under the direction of the Ministry of Defense. The school explicitly produces an elite whose power is based on technical expertise and a special relationship to the French state.[12]

The training of a *polytechnicien* is dominated by mathematics and certain applications of it. From the age of seventeen or eighteen, when they begin to prepare for the entrance examinations, these students study only two or three subjects. At X they also study some science, but rarely lab science. They are likely to study mathematical and theoretical physics, but not biology (Shinn 1984). They learn to grasp things quickly, to digest them, and to use them to solve problems. According to

[12] On the history of the Ecole Polytechnique, see Shinn (1980), Kosciusko-Morizet (1973), and, in English, Suleiman (1978).

Terry Shinn, X graduates always look at the big picture, both holistically and deductively, and opt when possible for grand solutions. They tend to make oracular pronouncements from on high, to present monologues. They are not engineers in the sense the word has in English; they are not interested in details, they lack a curiosity and love of tinkering, and they don't have engineering pastimes and hobbies.[13] Ezra Suleiman has pointed out how these top engineers leave details to their underlings, even while those details provide the rationale for their own position of power. He quotes one (1978, 168): "Our role as engineers of the Ponts et Chaussées does not consist in making calculations (this is the task of the forecasting engineers and of their collaborators), but to verify their legitimacy, to weigh the consequences of their eventual deviation from reality, to determine how much can be left to chance. . . . there too, there is a place for works of SYNTHESIS, which is perfectly worthy of the level of our corps." But their discourse is nevertheless technical; they legitimate their positions of power on the basis of technical and scientific facts.

X graduates are strong technological enthusiasts. They favor technological fixes to social problems, and have great confidence in the positive effects of technological development in general. The larger the task, the greater the coordination and planning problems, the more intellectually satisfying to X graduates the policy will be. Because they tend not to be familiar with the details, they lack a realist's doubts and caution. Finally, this kind of policy solution fits with both their image of how the world works and their own interests and position in that world. Nuclear energy in particular, being a very large-scale technological fix, has appealed strongly to X graduates.

The top students of each graduating class enter one of several professional bodies called corps. The most prestigious technical corps is the Corps des Mines, which accepts the top five or ten of each year's graduating class. They are a very tightly knit group, around three hundred at any given time, most of whom use the intimate second person, *tu*, with each other. They are fairly evenly split between the state, nationalized industry, and private business; their influence is particularly strong at the Ministry of Industry. After spending several years working for the corps itself, they can usually expect to head an important state agency or private company before they are forty. Having at some point in its history claimed expertise in pressure vessels, the corps sent many of its members into nuclear energy in the 1950s and 1960s. Many of the most im-

---

[13] Interview with Terry Shinn, March 1985. See Shinn (1984) for elaboration and Schwartz (1983) for a general critique of the French educational system, including a specific criticism of the grandes écoles.

portant names in the French energy system and nuclear industry of the last thirty years belong to members of the Corps des Mines.

The second most important technical corps is the Corps des Ponts et Chaussées (Bridges and Roads). Having traditionally been involved in economic development, as its name implies, this corps is still less Parisian than Mines, with around half of its fifteen hundred members outside the Paris area. Its prestige has waned somewhat from the days of great rebuilding after World War II, but it is still vital in urban development and the transport sector (Suleiman 1978, 214). Its members are more stable and traditional, but also more involved in politics. They became an important force at EDF during its large hydroelectric expansion in the 1950s and have maintained their power there in spite of the development of nuclear energy and the resulting inroads of the Corps des Mines.[14]

X graduates, especially members of these two corps, clearly control the "commanding heights" of industry, especially the nuclear industry. Philippe Simonnot (1978, 24-25) identified twenty-eight men as France's "nucleocrats" by virtue of their managing public or private organizations involved in nuclear energy; twenty are X graduates, nine of these Mines and seven Ponts. Of the five associated with EDF, three are Ponts, one is Mines, and one is an economist. The two who ran the CEA are both Mines; of two others highly placed at the CEA, one is Mines and the other merely X. A bureaucrat at the CEA (now headed by another Mines member) said that "X speak only to Mines, and Mines speak only to God."[15]

Competition between these two corps played a role in French nuclear policy conflicts, but this should not obscure the technological enthusiasm common to X graduates. One reason French policymakers strongly favored nuclear energy was that it fit so perfectly with their system for legitimating their own privilege. They held power because they had special technical knowledge that others did not, knowledge that promised economic plenty. Nuclear energy was precisely that kind of knowledge.

LICENSING AND REGULATION IN FRANCE

Pompidou's 1969 decision favoring the LWR decreased the CEA's control over France's nuclear program to the extent it increased EDF's. The CEA had almost no expertise with the new design, so that EDF and the reactor

[14] Each corps employs some of its members directly, while others are "detached" to other organizations. This detachment is a key mechanism for increasing the policymaking influence of the corps. As of 1981, the Corps des Mines had 299 active members, of whom 175 were in the service of the corps; Ponts et Chaussées had 1,234 members, 923 of whom worked for the corps (Bodiguel and Quermonne 1983, 102). The greater use of detachment in Mines helps account for its greater political influence.
[15] Interview with Michel Gras, February 1985.

manufacturer could claim—and develop—as much knowledge as the regulators. As a result the government made widespread changes in the administrative structures governing nuclear energy.[16] At the CEA a new administrator-general, André Giraud, made a series of internal changes in the early 1970s, including a 15 percent reduction of the workforce. The CEA's highly privileged access to the throne was removed, and it began to report to the Minister of Industry (as did EDF) rather than directly to the prime minister. As part of its new power, the Ministry established a Service Central de Sûreté des Installations Nucléaires (SCSIN) in 1973 to regulate nuclear activities, although SCSIN still relied on the CEA for technical support.

One of the CEA's expanding functions was to manage commercial enterprises covering various aspects of nuclear energy. The "-tomes," as these enterprises are known, were set up beginning at this time, often completely owned by the CEA but more typically as public-private ventures.[17] The CEA's research activities also shifted. Its weapons program (roughly half the organization) remained intact, but its energy program shifted from basic reactor design to safety research. The CEA was anxious to make its mark on the American technology, but the basic development work had already been done, so it launched into improving details, developing safer designs, and helping EDF improve reactor operating procedures. The CEA's new focus on LWR research came after the American AEC had begun to turn its attention elsewhere. Thus the CEA could take up research where the AEC had left off, perfecting those aspects of safety and operations that had been swept under the rug in the United States. France derived great benefits from borrowing the development work the United States had done on the LWR.

France's system for the regulation of nuclear construction and operations, centered on the CEA, took its present form in the early 1970s. The main change since then is a slight expansion of the number of groups and organizations involved in the regulatory process. As in the United States, there are two major licenses to be obtained, one for construction and one for operation. As in the American system, both authorizations most importantly involve interaction between the CEA and the utility, but various other approvals must be obtained along the way.

[16] See especially Glaize (1977) and Vallet (1986). This section also relies on the OECD NEA (1980, 1983b), Burtheret and de Cormis (1980), Cochaud (1981), and EDF (1982a), as well as personal interviews with Michel Gras of the CEA and Bernard Clavel, Patrick Girod, and Guy Peden of EDF.

[17] The CEA's major "filiales" are Cogema, which mines, enriches, and reprocesses nuclear fuel, and Framatome, which builds France's LWRs. In the CEA's annual report (French CEA 1983, 98–99), a chart of only the "principal members" of the CEA industrial group lists thirty-eight companies at various levels, from children down to great-great-grandchildren.

The main committees that handle applications at various stages are SCSIN at the Ministry of Industry and the Institut de Protection et de Sûreté Nucléaire (IPSN) at the CEA. Although SCSIN was formed in 1973 to reduce the CEA's power, it relies heavily on the advice given by the more technical IPSN; SCSIN either channels applications to IPSN or, when called upon for its own opinion, receives technical reports from IPSN. The SCSIN group that actually directs the analysis consists of roughly twelve people (typically young engineers, often from the Corps des Mines), drawn from not only the CEA but also EDF, Framatome, and the Ministry of Industry. They rely on IPSN committees that also have a mixed composition of this kind. Many stages are involved in these processes, including approval by several ministries (most notably Health), three drafts of the safety report (the middle one running to ten large volumes or more), an environmental impact statement, several reports to Euratom countries, a separate process for the approval of radioactive releases, and a public inquiry. Except for the lack of an antitrust review, these stages are similar to those found in American and Swedish regulatory processes. But behind the formal similarities lie several substantive differences.

The environmental impact statement, required for all nuclear sites since 1976, is similar to that demanded in the United States. The law, in fact, was closely modeled on the U.S. National Environmental Protection Act of 1969. In both countries there is considerable disagreement between antinuclear activists and utilities over what constitutes satisfactory environmental impact statements, and in both countries the utilities have been taken to court for inadequate statements. One EDF employee quipped: "We intended our law to be just like [the American] NEPA, and it is. But we never use our law to stop nuclear plants."[18] Neither does the United States; in none of our three countries have the courts been a successful route for stopping nuclear power plants. The antinuclear movements in France and the United States had occasional successes, but these were almost always overturned by the highest courts.[19]

French courts have less influence than American ones, more because of a consensus in the French elite than from different formal responsibilities. The highest court, the Council of State, does act as adviser to the government, giving opinions on proposed legislation, opinions nonbinding and usually not made public. But in addition to hearing appeals from lower courts, it acts as the protector of the citizen against abuses by the state and as a general upholder of legality. Despite its formal powers, it

[18] Personal interview, anonymous, winter 1985.
[19] This discussion is based on two interviews with Patrick Girod, Chef du Service Administration, EDF, Région d'Equipement Paris. Girod helped the government write the 1976 bill and has since helped EDF comply with it.

rarely strikes down government policies.[20] At least one observer (Gleizal 1980) claims that the Council's conservatism is due to the training, attitudes, and corps loyalties of its members, rather than to legal limits on its power.

The local inquiries required for nuclear installations are, roughly, a joke. Descriptions of the projects are made available to the public for a period of several weeks, and blank books are provided for people to record opinions and questions. Comments are collected and categorized, and a government engineer writes responses, usually terse and dismissive. This collection of documents is to be considered when the government must decide whether to grant a Declaration of Public Utility. It would be ludicrous to suppose that the public inquiries, in which the comments are usually strongly opposed to nuclear construction, have any effect on policy decisions. In the United States antinuclear intervenors in public hearings rarely had any effect on the decisions made by the NRC, but they at least managed to stage memorable and effective media events.

The internal operations of the CEA follow the Swedish pattern. There is close cooperation with EDF on safety and operations research, to the extent that many EDF employees have their offices at the CEA and vice versa. Most committees have members from both organizations—hardly imaginable in the United States with its norm of regulation from the outside. The CEA is able to conduct active research on safety and operations issues to a greater extent than Sweden's SKI, largely because of its enormous size. On the other hand the close EDF-CEA cooperation on licensing procedures, with little outside influence, follows the patterns of our other two countries.

EDF Turns Commercial

With the CEA now devoted to licensing plants, refining safety procedures, and tending its flock of industrial enterprises, EDF had free rein to build nuclear plants. At virtually the same time the CEA was reorganized, EDF signed a contract with the state (23 December 1970) that allowed it for the first time to market electricity actively, competing with other forms of energy. This was a significant change for the nationalized company—from a passive provider of services demanded by the public to a more active, even aggressive, force in shaping (manipulating, some

[20] Two of the most notable exceptions prove the rule by their extremity. The council ruled against the use of military courts to try quickly cases of treason after the Algerian War in 1962, and against de Gaulle's proposed referendum for direct presidential elections. On the Council of State, see Kessler (1968), Baecque and Quermonne (1981, chap. 3.1), and Pickles (1972, 293–99).

would say) those demands. These two turning points for EDF, as well as the logical connection between them, were summed up in its slogan "all electric—all nuclear."

All-Electric. EDF's commercial shift had its origins in the Nora Report on the profitability of nationalized firms. The goals of "rentabilité" and increased autonomy for management led to the idea of "contracts" between enterprises and the state to specify certain boundaries and goals but leave the enterprises free to pursue these as they wished. Simon Nora was heavily influenced by the accounting practices already in place at EDF, but his report gave extra impetus to EDF's concern for autonomy and commercial success. Its concern, in other words, was to be like a private industrial firm. This new ideology was widely accepted by state policymakers, making possible EDF's commercial emphasis in its 1970 contract with the state.

The commercial strategy was more successful in the residential sector than the industrial due to EDF's targeting of builders and appliance salespersons and a campaign extolling electric heating for being clean and precisely adjustable. Special rates were given to apartment buildings in which individuals had no control over their own thermostats, a strategy sure to increase waste. By 1975, 50 percent of new construction was all-electric, compared to 5 percent in 1970 (Saumon and Puiseux 1977, 153). Industry, on the other hand, in spite of a large push by EDF, was uninterested in electricity. EDF was only claiming that electricity was *as cheap* as fossil fuels, not cheaper, and energy costs all together were only 5 percent of industry's costs. There was little incentive to change. Nevertheless, its other successes assured EDF an expanding market.[21]

All-Nuclear. Orders for nuclear reactors began to flow regularly to Framatome, although at the moderate pace of one or two per year (see table 5.1). EDF's motive was now not simply cost minimization, since it was not clear that nuclear energy would achieve this. It was also the expansion of electricity (and hence EDF) at the expense of other energy sources, and nuclear energy was clearly the best way to build a great deal of electrical capacity. If EDF could sell the state on nuclear, its own future would be bright indeed. EDF's enthusiasm for nuclear was genuine: it felt France's best interest lay in economic growth, which depended on electricity, which in turn was best provided through nuclear energy. In a pattern typical of all organizations in the nuclear debates, national interests happened to line up with those of each debater's own organiza-

---

[21] See especially Gravelaine and O'Dy (1978, 26–42) and Picard et al. (1985, chap. 13) on EDF's expansion.

TABLE 5.1
EDF Nuclear Reactor Orders, to 1985

Year	Units Ordered	Year On Line	Reactor Type	Reactor Size
1956	1	1963	UNGG	70 MWe
1957	1	1965	UNGG	200
1959	1	1967	UNGG	480
1960	1	1967	PWR	280
1961	2	1967–73	mixed	70, 250
1963	1	1969	UNGG	480
1965	1	1972	UNGG	540
1966	2	1971–72	UNGG	480, 515
1970	2	1977	PWR	900
1971	4	1979	PWR	900
1974	16	1980–83	PWR	900
1976	8	1983–87	PWR	900
1976	4	1985–86	PWR	1,300
1980	4	1985–86	PWR	1,300
1981	2	1987	PWR	900
1983	4	1989–92	PWR	1,300
1984	2	1992+	PWR	1,300

Source: Interview with Henri Haond, March 1985.
Note: The dates at left represent contracts signed with manufacturers (Framatome, beginning in 1970), rather than dates construction began. In 1976 there were two separate contracts.

tion. The combined rhetorics of cost and technological enthusiasm were a powerful and convenient tool for EDF.

Now freed from the stalemate of the battle over reactor technologies, the PEON Commission could concentrate its efforts on promoting nuclear energy. The CEA and EDF, whose representatives were roughly half the commission (Lucas 1979, 141–42), then agreed on the need for a large nuclear program. As a result the Commission, preparing recommendations for the Sixth Plan (1971–1975), proposed at least 8,000 MWe, or roughly eight reactors, for the period. This goal was eventually incorporated into the plan. Paired with this in the plan was EDF's twin goal, the expansion of electricity at the expense of other energies: the plan's Commission on Energy recommended a "progressive acceleration in the growth of electricity consumption" (Puiseux 1977, 293). In 1972 PEON proposed an even larger nuclear program: 13,000 MWe of capacity to be installed from 1973 to 1977, the equivalent of fourteen or fifteen reactors, or roughly three per year.

In the early 1970s there was competition between two reactor producers similar to that in the United States between GE and Westinghouse. Framatome, largely a subsidiary of Creusot-Loire, itself part of the Empain-Schneider industrial group, produced pressurized water reactors under license from Westinghouse. SOGERCA, a subsidiary of the Compagnie Générale d'Electricité (CGE), produced boiling water reactors under license with CGE. Even though it negotiated with both companies, EDF eventually concentrated on pressurized water reactors and even canceled a contract with CGE. As compensation, another CGE subsidiary, Alsthom-Atlantique, was later given a monopoly on steam turbines to accompany Framatome's reactor vessels. This division of labor contrasts sharply with the American competition between reactor manufacturers that was allowed to continue unabated.[22]

These strongly nuclear policies originated with EDF, not the French state. According to Louis Puiseux (1977, 291), EDF's General Director Marcel Boiteux told close colleagues in 1970: "The French government does not have an energy policy. I am obliged to have one in its place." Almost all observers have noticed that "there is a remarkable similarity between the propositions of EDF in the late 1960s and early 1970s and the basis of the present national energy policy" (Lucas 1979, 139). In my own interviews, several EDF officials smiled when asked about the relationship between EDF goals and PEON Commission recommendations. They admitted that EDF had managed to stack the commission. One reason is certainly the immense number of EDF experts who fed information to the PEON Commission staff. Lucas (25) says, "EDF, with a large pool of able engineers and technologists, operated extremely skillfully within the PEON Commission—as it does within other important committees—to ensure that the decision of the group incorporated the views of the enterprise." Lucas (139) sums up the evidence, "There can be no reasonable doubt that the initiative for almost all aspects of the nuclear programme comes from EDF." Just as it had won its battle with the CEA, EDF gained the right to build nuclear reactors and actively market electricity. But before the oil crisis, it could not build as many reactors as it wished, due to the only government organization that resisted the large nuclear plans of the early 1970s, the Ministry of Finance.

THE MINISTRY OF FINANCE RESISTS

Because it monitors state spending both before and after budget approval, the Finance Ministry has immense power within the French administration. Alain Peyrefitte (1981, 182) has written about the French

[22] See CFDT (1980, 451–57) for the division of labor in the late 1970s, as well as Colson (1977, 33–34) and Camilleri (1984, 148).

state: "There are two stages to administrative omnipotence. The first is that exercised by the administration over everything outside itself. The second is the Finance Ministry's power over all the rest of the administration." The power of the ministry of finance rests less in the formulation of policy than in its implementation. Major policies are discussed and decided in the Council of Ministers, where the Minister of Finance has only one vote, and where many political considerations come into play. In the implementation, however, the Ministry of Finance exercises enough financial control to curtail many programs it opposes. Lucas (1979, 131) distinguishes the relationship between the two levels of policymaking, "*Les Finances* possesses in daily reality the means of technocratic control within constraints imposed by various decisions of the *Conseil des Ministres.*"

The general strengths and weaknesses of the Finance Ministry appear in the domain of energy policy. As nuclear energy became an important issue in the late 1960s, the Finance Ministry managed to increase its representation on the PEON Commission from two "appointed" to three "ex officio" members, although this was not enough to affect the recommendations of the commission. In its 1970 deliberations, the majority proposed that roughly eight reactors be installed under the Sixth Plan (1971–1975), while the representatives of the Finance Ministry pushed for only three. The larger proposal carried the day and was also incorporated into the Plan itself, but not without qualifications and dissenting opinions from the Ministry of Finance (Lucas 1979, 141–43).

No matter what the stated policy, the implementation of any large and expensive program could not avoid the Finance Ministry. There would have been several ways for EDF to finance its nuclear program: through money raised from its charges for electricity; through direct grants from the state; through money borrowed at favorable rates from the state; and through funds borrowed in the usual financial markets. Both the grants and the loans are administered by the Finance Ministry, whose agents monitor and often interfere in the operations of the nationalized enterprises. They can set up offices on the enterprise's premises, veto certain actions until approved by the ministry, and give advice on how various policies will be viewed by the ministry. Both the construction of the state budget and the channeling of the allocated funds are under the control of the Finance Ministry, in the latter case acting through the Treasury, which in turn can require additional justification of spending.

Figure 5-1 shows the sources of EDF's funds since the early 1950s. In only five out of thirty-two years has the majority come from electricity sales. The hydroelectric program of the 1950s and early 1960s was financed largely by direct public funds, while the nuclear program since the early 1970s has relied heavily on loans. Both sources required the

FIGURE 5.1 Sources of EDF Funds, 1952–1984

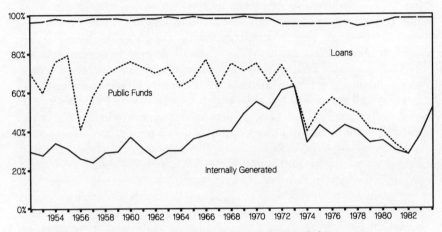

Source: Electricité de France, internal table provided by Jean-Michel Fauve

approval of the Ministry of Finance, which thereby managed to hold EDF's nuclear construction to modest levels through the end of 1973. Direct state funding dropped, and loans held roughly constant during the years prior to the oil crisis. Investment as a percent of sales dropped through 1973, after which it started to rise rapidly. Had the pace of the early years of the Sixth Plan continued, the 1970 PEON Commission recommendation of 8,000 MWe would not have been reached on time, much less the 1972 recommendation of 13,000 MWe. In spite of the stated policy goals of the government, as late as 1973 the Finance Ministry had successfully restricted the size of France's nuclear energy program.

The Finance Ministry wished to curtail EDF's nuclear ambitions because of both an organizational interest and its policy style. The ministry's central purpose is keeping the state budget in balance, so it opposes and trims projects requiring large outlays. But a good investment presumably saves money in the long run, so that economic rationality occasionally dictates large investments. Louis Puiseux (1977, 130) describes an EDF announcement that nuclear would cost 9.6 centimes per kWh as opposed to 13.2 for oil, and says this difference must be "diabolically seductive for a Finance Minister, whether of the right or the left." The fact is, however, that in the early 1970s nuclear energy didn't look like a good investment to a careful cost-benefiter. EDF had adopted the optimistic assumptions of the technological enthusiasts, but a cost-benefit analysis that refused to accept mere speculation would almost certainly not have embraced a large nuclear program. A modest one for research and development purposes—like the three reactors proposed by the

Finance Ministry—might have seemed appropriate. The ministry's cost-benefit perspective also appears in its 1969 opposition to EDF's commercial turning, presumably on the grounds that electricity could not cost-effectively replace other energy sources (Lucas 1979, 141).

Even though the Finance Ministry does not seem to have developed its own estimates for the costs of nuclear energy, its refusal to accept those provided by PEON Commission's and EDF's technological enthusiasts was well grounded. Not only were the commission's estimates for the early 1970s too low, but nuclear costs have risen continually ever since, instead of falling as the PEON Commission had predicted. The commission's report of April 1973 estimated that the cost of nuclear-generated electricity would be 10.0 centimes per kWh in 1977. When 1977 came around, the commission was making cost estimates 53 percent higher, even accounting for inflation. By 1982 cost predictions were being made that were 93 percent higher than the 1973 estimates. These discrepancies represent both overly enthusiastic original estimates and rising costs. In real terms costs rose 4.4 percent per year during the 1970s and early 1980s. Because of the lack of cost information in the early 1970s, one's assumptions about technological development—especially enthusiasm versus skepticism—were decisive.[23] But in the early 1970s it was impossible to guess whose assumptions would turn out to be correct.

Just as in the CEA-EDF controversy each organization felt its position was in the best interest of France, so too did the Ministry of Finance. It wished to avoid massive borrowing from abroad to finance a technological venture whose benefits did not clearly outweigh the costs. In the ministry's view of the world, it was in the interests of France to avoid this debt. Because EDF and CEA officials had more optimistic assumptions about technological development, they felt that France could hardly afford to delay its nuclear program. Each of the three organizations saw national interests as identical to its own organizational interests. Policy rhetorics and organizational missions lined up.

How could two institutions devoted to economic rationality, EDF and the Ministry of Finance, adopt such different opinions of nuclear energy? Many assumptions have to be fed into a cost-benefit analysis for it to arrive at concrete recommendations, especially concerning the likely development of technology. In its analysis of nuclear energy, EDF adopted the assumptions of technological enthusiasts, not surprisingly given its high contingent of engineers. The original enthusiasm for nuclear energy

---

[23] Georges Moynet (1984) of EDF outlines the cost underestimates for French nuclear power. The figures I use come from his table 1 (p. 149). EDF has not managed to stabilize the costs of nuclear energy, but it has become better at estimating the increases. Denis Fouquet and Eric Mignot of EDF provided useful information about nuclear costs and cost predictions.

can be traced to the education of the technical elites at the Ecole Polytechnique. The mathematical emphasis on grand plans, extensive coordination, and technological fixes to social problems all favored nuclear energy more than decentralized and mundane alternatives. This pronuclear fixation is common in the other countries too. What is different is that in France the technological enthusiasts from X have loftier positions than do enthusiasts in the other countries. Once that fixation was there, the French political structure could not have done much to snuff it out; at best it could restrain it.

On the other side was the Finance Ministry, dominated by Ecole Nationale d'Administration (ENA) graduates heavily trained in economics, with no *polytechniciens* to provide technological enthusiasm. Just as X is the breeding ground of the technical corps, the ENA trains the administrative corps. Founded in 1945 to facilitate the rebuilding of France's economy, ENA was explicitly modeled on X, and the use that its graduates (*énarques*) make of economic facts is similar to the use the X graduates make of technical ones. Enarques are well trained in economic analysis and statistics, but they use "hard economic facts" to bolster very broad policy positions. They no more do economic analysis than X graduates develop technologies, but their discourse refers to it and their legitimacy is based on it.[24]

Because hard economic facts are usually seen as following from hard technological ones, énarques often depend on the technical corps and their facts. Despite competition between the technical and administrative corps for influence over policymaking, they share an emphasis on "hard facts." They consider themselves realists, following the heartless dictates of economic and technological feasibility in the face of unrealistic ideological and political demands. Together they form the famous French technocracy, and when they agree they form an unstoppable force in French policy-making.[25]

If X graduates embodied technological enthusiasm, énarques are not precisely the parallel embodiment of the cost-benefit perspective. Although the ENA program consists largely of economics, including cost-benefit analysis, its emphasis is not on the careful, skeptical break down of all cost components into their smallest pieces. There is instead a con-

---

[24] Bodiguel (1978) is a thorough study of the social origins and careers of *énarques*, a good contribution to the large literature on social mobility and the reproduction of the elite in France. Kessler (1978) is a balanced history of ENA. Mandrin (1967) is a lively critique of "Enarchie" by a group of socialist party theorists writing under a pseudonym.

[25] Touchard and Solé (1965, 31) characterize—with some irony—the key belief of the technocrats: "One must be a realist." They caricature them as obsessed with productivity and cold-hearted; as admirers of America, drinkers of whisky and Coke, and patrons of supermarkets; and as Paris-centered.

cern with the broad picture, a perspective that necessarily does some harm to the minutiae of cost-benefit analysis. One could characterize the *énarque* education, then, as a cross between the cost-benefit and the technological enthusiast attitudes, with the former's concern for costs but the latter's sense of the laws of a broad system.

Those énarques who enter the Inspection des Finances develop a more pure cost-benefit policy style, since it is reinforced by their organizational tasks. The Inspection des Finances controls all state spending before the fact, with the purpose of keeping budgets in line, and this corps dominates both the Ministry of Finance and the banking sector. Its members remained skeptical of nuclear energy in the early 1970s.[26]

Conclusions: A Standoff

Technological enthusiasm dominated the CEA, convincing its members that the gas-graphite reactor could be developed into a cheap and reliable source of electricity. It gained support from top politicians in this belief by linking gas-graphite technology to nationalist technological ambitions. EDF, managed by a combination of high engineers and economists, was skeptical of this enthusiasm in its pure form, and it had an organizational interest in reducing the power of the CEA. In the 1960s EDF felt that light water was cheaper than gas-graphite technology, and these short term considerations outweighed gas-graphite's long run potential. Yet to adopt the LWR also required some technological enthusiasm, since the cost figures were unclear at best. In a situation of falling fossil fuel prices, it took great optimism to believe in the rapid drop in the costs of nuclear energy. Once a reactor line had been chosen, and EDF and the CEA were allies in promoting nuclear energy; their combination of technological enthusiasm and a cost-benefit perspective, as found in the PEON Commission, gave them unshakable confidence in nuclear energy.

The Ministry of Finance was subject to no such technological enthusiasm. Its vision was more purely that of costs and benefits, which in the 1960s and even the early 1970s did not add up in favor of nuclear energy. The ministry's institutional function of moderating state spending reinforced this skepticism. On the eve of the oil crisis, this standoff between

---

[26] Like X, ENA sends its top graduates into certain corps: the top 20 percent enter the Council of State, the Cours des Comptes, and the Inspection des Finances. The first of these provides the personnel for the court structure of the same name. The Cours des Comptes is the accounting agency for all nationalized enterprises, state organizations, and the social security system. For more detailed information on these three corps, see Rigaud and Delcros (1984, chap. 8) and Escoube (1971). Suleiman (1978) also discusses them, along with the technical corps, in the context of their maneuvering for power. Bodiguel and Quermonne (1983, chap. 2) provide interesting details about the operations of both administrative and technical corps.

technologists and cost-benefiters had kept the nuclear program far smaller than the PEON Commission recommended. But just as technological enthusiasts had managed to feed their optimistic assumptions into EDF's cost-benefit machine, so too would the Finance Ministry be forced to accept these assumptions when the oil crisis struck in 1973 and 1974.

CHAPTER 6

Commercial Success in Three Countries

WHAT DETERMINED nuclear policy before the 1973 oil crisis, when it was still developed through discussions between various state bureaucracies and private companies? There was controversy, but it was rarely public, rarely a topic worthy of media attention. The commercialization of light water reactors had succeeded in the United States, Sweden, and France before energy policy became a prominent public issue after 1973. How did commercialization happen? The debates in utilities and government agencies were similar in all three countries as the initial conflict between cost-benefiters and technological enthusiasts grew into a firm alliance. Differences in political structure played little role in this period, as all three countries established autonomous regulators and ambitious nuclear programs. The United States in particular seemed to provide an insatiable market, and reactor orders hit a record forty-one in 1973. Sweden was a smaller market, but it too seemed poised to continue its steady expansion of nuclear energy in spite of a 1973 decision to pause and rethink its plans. In France there had been a modest but steady flow of orders for one or two reactors a year. The CEA and EDF desired an acceleration of this expansion, but in 1973 the Ministry of Finance was still able to resist these ambitions.

Despite their contrasting political structures, the United States, Sweden, and France established similar systems to promote and regulate civilian nuclear power. The decentralized "weak state" was even more successful at commercialization than the two "strong states." In all three there were links to weapons programs, so that centralization and tight government control over research and development were universal. In France and the United States, the CEA and the AEC were laws unto themselves, untouchable by most normal political means. Regulation in all three countries was, until the early 1970s, a matter of close cooperation between the government bureaucracy promoting the technology and the electricity producers buying it. Public participation was limited and had no effect, even though there was greater formal scope for it in the United States. The AEC and JCAE were unusually autonomous, powerful agencies for the American state, and the AEC differed from its Swedish and French counterparts less in formal structure than in the peculiar combination of policy style and partisan ideological balance that led it to de-emphasize LWR research.

In all three countries private industry was interested in building and selling nuclear reactors, and there was rivalry between contending developers (GE versus Westinghouse, Framatome versus Alsthom-Atlantique, AB Atomenergi versus ASEA). With smaller electric markets, the French and Swedish states negotiated a division of labor. Multiple manufacturers were allowed to compete in the United States, and the rivalry between GE and Westinghouse encouraged a technological enthusiasm that led to the inexpensive turnkey deals beginning in 1963. Through dubious cost claims, the LWR was commercialized first—and perhaps prematurely—in the United States. A combination of economic structure—different market sizes and EDF's monopoly in French electricity—and conscious decisions by the state explain the divergent manufacturing industries. A combination of technological enthusiasm and marketing strategy (both were necessary) at GE and Westinghouse explain the inexpensive turnkey deals. Figure 6.1 shows both the much greater size of the American reactor market (capable of supporting at least two manufacturers) and the earlier deployment of reactors in the United States.

The cheap turnkey deals attracted utilities—first in the United States, then Sweden, then France—more effectively than state bullying had. Persuasion and some dubious data were necessary, since none of the three governments could force utilities to buy reactors they considered too expensive.[1] The doubts had not concerned risk of accidents, since

FIGURE 6.1 Reactor Orders in the United States, Sweden, and France

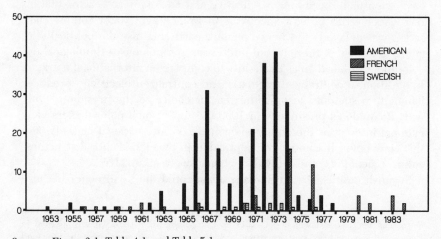

Sources: Figure 3.1, Table 4.1, and Table 5.1

[1] Charles Lindblom (1977) has described this as a general trait of mixed economies: the state cannot force businesses to invest, it must bribe them. The state has negative power to forbid, but not much positive power to build.

similar laws limited operator liability in each country; they had been about costs. But once utilities began to adopt the assumptions of the technological enthusiasts, forming an alliance of cost-benefiters and technological enthusiasts, this alliance's faith in nuclear energy was unshakable. As long as the cost-benefiters took their assumptions about the likely path of technological progress from the enthusiasts, both groups could see only the prospect of huge benefits from nuclear energy. There were no grounds or criteria for questioning nuclear energy that made sense to them, and no data existed to question the enthusiasts' assumptions of continually decreasing costs. In all three countries the energy systems—utilities, manufacturers, and policymakers—became nuclear enthusiasts.

The strength of the economic-technological alliance arose partly from the nature of the electricity industries in the three countries. Annual growth in electricity demand had been 6 or 7 percent since the 1950s in all three, and costs had fallen steadily, in part because of technological improvements. The "mission" of the electricity industries in all three countries was to provide the technological progress to produce the electricity that was at the heart of postwar economic development, itself often considered the greatest possible good for society. Decisionmakers in the three electric industries strongly held the values described in chapter 2 as the "dominant pro-growth paradigm."

Because of the enthusiastic response to the turnkey deals, all three countries eventually adopted the same basic reactor technology. France and Sweden faced a choice between their own reactor lines that might have eventually become cost-effective and an American reactor that already seemed to be cost-effective.[2] In France light water commercialization was delayed, and hence painful, partly because the political elite clung to a view of the national interest as autonomous technological development rather than as the ability to compete on international markets. In addition, a centralized regulator and centralized electricity producer fought to a standoff, leaving the reactor choice to the president. Only with Pompidou's presidency in 1969 did top French policymakers view national interest as the ability to compete on international markets, and this view never became as widespread as in Sweden. Political structure, cultural meanings, and individual choices were all involved.

Swedish politicians and bureaucrats adopted the LWR sooner because

---

[2] The LWR's export potential seemed greater because of the publicity it was receiving more than because of lower costs. Later analysts assume that the LWR was adopted in France and Sweden because it was patently cheaper, but this was not clear at the time (and might not have been true had research funds continued to flow to other designs). Polach (1969) reports a 1966 Euratom study that estimated the costs per kilowatt hour from gas-graphite and light water reactors to be almost identical, and those of several heavy water designs to be lower.

they have long had to respond quickly to international markets. In Europe, Peter Katzenstein (1985, 32) has argued, "small states in world markets" have developed political and economic structures that allow them to adjust flexibly to trends that larger countries are shielded from. These mechanisms include "an ideology of social partnership expressed at the national level; a relatively centralized and concentrated system of interest groups; and voluntary and informal coordination of conflicting objectives through continuous bargaining between interest groups, state bureaucracies, and political parties." The conflicting groups quickly agreed that the LWR looked more promising for export potential. Technological autonomy mattered less.

The light water reactor succeeded in France and Sweden because it seemed more economical than the indigenous designs, and cost-benefiters in the electric industries managed to stymie the native reactors promoted by technological enthusiasts. But these cost-benefiters also had enough technological enthusiasm to accept the LWR in a period when its cost advantages were doubtful. In all three countries the confident predictions of falling costs turned out to be wrong, and instead nuclear energy became steadily more costly. But at the time costs were anyone's guess.

Two factors that should have influenced nuclear policies but did not were the availability of alternative energy supplies and the degree of public support for nuclear energy. The United States has far greater fossil fuel resources than France or Sweden, although the latter both have extensive hydroelectric facilities.[3] Yet the United States pioneered nuclear power in spite of its alternative resources. The presence of alternatives became relevant only later, when nuclear power became controversial, and when the oil crisis encouraged careful cost comparisons between energy sources. Similarly, public opinion had little effect at this time, since it was probably pronuclear in all three countries.[4] Neither public opinion nor energy alternatives seemed to affect nuclear policies up to 1973.

The United States differs from the other countries in having pioneered the commercial use of light water technology, absorbing many of the inevitable costs of this process. To borrow two phrases from Thorstein Veblen (1915), the United States "paid the penalty of taking the lead," while

[3] Connolly et al. (1982, 238) ranked sixty-nine countries according to their fossil fuel resources. The United States was "high," while France and Sweden were both "low." But as late as 1982, Sweden was producing 55 percent of its electricity from hydro, France 26 percent, and the United States 13 percent (OECD IEA 1984, 384, 391).

[4] There are no good data for Sweden before 1976, but Holmberg and Petersson (1980, chap. 6) believe opinion was pronuclear up until 1974. French and American attitudes were strongly pronuclear. Jasper (1988) discusses the causes of public attitudes toward nuclear energy. For the trends over time, see figure 14.1 on p. 261.

France and Sweden exploited "the merits of borrowing."[5] Sweden and France could to some extent begin where the United States left off. Figure 6.1 shows that major commercialization (sufficient to gain manufacturing know-how) came in the mid-1960s for the United States, in the late 1960s and early 1970s for Sweden, and the mid-1970s for France. The nuclear programs were thus influenced by what Giddens (1984, 251ff.) calls "world-time": the extent to which an idea or process or technology has spread across countries through a kind of international learning process. None of our three countries is an independent case, since both light water technology and the antinuclear movements spread outward from the United States to the others. Thus the three nuclear programs differed according to when they were deployed.

The costs of going first should have been temporary, but the United States continued to suffer them because the AEC thought the development process was finished. The AEC's hands-off style, along with technological enthusiasm, encouraged the AEC to shift its research attention from the LWR to other technologies. In contrast to the AEC's abandonment of light water technology, France and Sweden developed regulatory systems characterized by continued research and cooperation between reactor operators and regulators. Regulators capitalized on being latecomers.

The AEC's change of focus had no discernable effect on reactor orders through 1973, but it may have affected utility operations. In the United States the AEC left operations to the utilities. In France EDF also escaped much regulatory oversight, but it maintained close ties to the CEA and cooperated in its research on safety and operations. American and French utilities both escaped some regulation, one by default and one by conscious effort; the success of nuclear energy was in their hands. EDF came to see that nuclear energy was more than another way to boil water. Many American utilities did not. Public ownership explains little of the nuclear decisions utilities made. EDF is a public enterprise, while electricity production in the United States is almost entirely private and in Sweden roughly half private. But EDF has the ability to set its own policy path as though it were private.

EDF's public status affects it more by providing a role for the Ministry of Finance to play in EDF's internal decisions. The ministry's cost perspective made it skeptical of EDF's exaggerated claims for nuclear energy, and its structural position in the state allowed it to restrain France's nuclear program through 1973. But the structural power of the ministry influences EDF decisions only when there is disagreement. In the case of

[5] Veblen means something slightly different, using these phrases to refer to the cultures and habits that develop around a technology, rather than the direct development costs.

nuclear energy, that power was used because two different policy styles happened to dominate the two organizations, and those two policy styles involved conflicting opinions about nuclear development. Formal political structure explains *how* the Ministry of Finance could interfere if it wished, but it is the conflict between policy styles that explains *why* it wished to.

Political and economic structures must be supplemented with cultural and individual factors to explain why the LWR was commercialized in the United States, Sweden, and France. The original turnkey breakthrough in the United States resulted from industrial rivalry, technological enthusiasm, and new company heads. Sweden's quick conversion arose out of corporatist cooperation driven by an image of national interest oriented toward exports. France's tardy conversion arose from the balance of power between the CEA and EDF, a conflict between elite images of national interest, and the discretion exercised by de Gaulle and Pompidou. A change of mindset was crucial in all three countries. For electricity producers to buy nuclear reactors, they had to incorporate in their cost-benefit analyses the optimistic assumptions of the technological enthusiasts. Utility decisionmakers had to believe that nuclear energy was or soon would be the cheapest way to produce electricity. Government agencies could not simply command them to believe this. Yet the change was not difficult because of the strong technological enthusiasm and technical backgrounds already widespread in the electricity industry. Utilities in all three countries were brought around by the apparent successes of Westinghouse and GE in producing cheap reactors. Not everyone was taken in by the turnkey deals—only those who had backgrounds conducive to technological enthusiasm, those ready to believe.

In all three countries, public officials created new organizations to promote nuclear energy, even without clear evidence that nuclear electricity cost less. When one regulatory or legislative framework failed to attract buyers for nuclear reactors, it was changed. The enthusiasm for nuclear energy came first; it created its own administrative and regulatory structures.

PART THREE

To Build or Conserve: Dilemmas Arising from Public Opposition and the Oil Crisis, 1973–1976

CHAPTER 7

The Reassertion of the Economic Perspective in the United States

THE OIL CRISIS of 1973–1974 changed the cozy nature of nuclear energy policymaking in the United States. Chapter 3 argued that the prospects for nuclear energy were good in 1973, since it looked as though minor financial and regulatory hindrances were only temporary. When the oil embargo hit in October 1973 and the price of imported oil quadrupled in the following months, these events seemed to cement nuclear power's bright future. President Nixon announced that the United States would pursue energy independence through increased reliance on nuclear energy, which would account for roughly 40 percent of electricity production by 1990. But the oil crisis set several other processes in motion that transformed energy policymaking and ultimately prevented the achievement of Nixon's goals.

First, energy policy, previously conducted quietly in iron triangle arrangements, came under continuous media scrutiny. It became a partisan issue on which political leaders had to take public stands and over which Democrats and Republicans could squabble. Instead of a question of how to help electric utilities build nuclear plants, nuclear energy policy became more a question of government intervention versus free markets in energy. This was the "dominant partisan cleavage" in American politics around which the two parties had fought for years, and it came to dominate the simple question of whether nuclear energy was a reasonable energy source.

Second, the new drama and importance of energy issues made the emerging antinuclear movement a visible force in policy discussions. Until 1969 opposition to nuclear energy consisted primarily of isolated citizens opposed to reactors near them but not to nuclear energy in general. Increasingly from 1969 to 1974, these locals combined forces with the large environmental movement and a growing number of scientists questioning particular aspects of commercial nuclear energy. When the oil crisis placed energy problems on the national agenda, nuclear opponents began to think of themselves as a national movement with a common purpose. As a result they attracted new members who were more concerned with democracy, technocracy, and revolution than simply with nuclear energy. The language of the new movement was often moralist and completely alien to the decision-making criteria held by the nuclear insiders, who dismissed activists as irrational and tried to exclude them from policymaking. The movement had less direct effect on policy outcomes than is often supposed, but it helped split the coalition of technological enthusiasts and cost-benefiters described in part two, and it indirectly increased the costs of nuclear energy.

Third, the oil crisis encouraged more rational energy planning than had previously occurred, involving rigorous comparisons between different energy sources. Cost-benefiters excelled at these calculations and so developed more influence on energy policy. Their skepticism toward nuclear reactors, which had been overwhelmed in the 1960s and had reemerged only slightly in the early 1970s, blossomed in the years after the oil crisis. In many cases cost-benefiters during the late 1970s shed the assumptions of the technological enthusiasts and adopted some of the assumptions of the antinuclear movement. The increased rationality injected into energy policy from an economic perspective had the effect of undermining the American commitment to nuclear energy.

The Rise of the Antinuclear Movement

The first nuclear reactor in the United States to face public opposition was Fermi 1 in 1957. Democratic members of the Joint Committee on Atomic Energy (JCAE), annoyed with the AEC for approving a reactor so close to a large city (thirty miles from Detroit), instigated opposition by the United Auto Workers. Like most partisan battles, the question soon came to turn on the government's role in reactor development. The president of Detroit Edison "announced that the safety issue was a subterfuge, that the key to the controversy was public versus private power" (Fuller 1975, 50). But Walter Reuther and the UAW, who weren't concerned with partisan conflicts, believed the debate *did* revolve around public safety versus cost-benefit calculations and technological development. This concern with the "Public" eventually fed into the moral ecologist policy style.[1]

Since Fermi, local "Not In My Backyard" (NIMBY) activists have been the backbone of opposition to nuclear energy in the United States. Indeed they were the only opponents throughout the 1960s. In 1964 local

[1] General works on the rise of the antinuclear movement include Mitchell (1981), Mitchell and Nelkin (1982), Lewis (1972), and Price (1982). On the scientific controversy, see Primack and von Hippel (1974, chap. 15), Ebbin and Kasper (1974), and Ford (1982). Case studies of local controversies include Jasper (1985), Stever (1980), Meehan (1984), Cohen (1981), Nelkin (1971, 1974, 1979), Evanoff (1982), Rose et al. (1979), Sharaf (1978), Johnsrud (1977), and Ames (1978). Also of interest are Mazur (1981), Del Sesto (1979, chaps. 5–7), Nader and Abbotts (1979), Gyorgy et al. (1979), and Gofman and Tamplin (1971).

opponents succeeded in dissuading Pacific Gas and Electric (PG&E) from building a reactor on Bodega Head in California, largely because PG&E had several other sites under consideration. When the Sierra Club, traditionally a conservationist organization, helped PG&E select another site, a major rift developed in the club's management and membership. The club endorsed the building of a nuclear reactor at Diablo Canyon to save another portion of the beautiful California coast, the Nipomo dunes. The board of directors was proud of its spirit of compromise, and one of them, Ansel Adams, wrote: "The conservation movement is becoming more adult, more aware of the universal problems of society. We are developing cooperative relationships with PG&E" (Evanoff 1982, VII-15). Many other club leaders and members rejected this policy of "reasonableness and compromise," and one of them wrote to the club's president: "In my book, the club has no obligations to be 'reasonable' about our natural beauties except to fight for their preservation for as long as possible. . . . experience seems to indicate that one can't be reasonable with the commercial interests. They have no conscience, so we should not have to compromise with them. The Sierra Club is not in the business of compromise, as I see it" (Quoted in Evanoff 1982, VII-24).

Executive Director David Brower and many of the club's members left to form the Friends of the Earth on the basis of this uncompromising philosophy. This moral ecological worldview held that there were certain environmental and human goals and values so important they could never be compromised. The traditional Sierra Club position, in contrast, was similar to a cost-benefit worldview in recognizing a large set of economic and environmental goals to be weighed against each other.[2]

Local opposition was successful not only at Bodega Head and the Nipomo dunes, but at Point Arena and Malibu, California; at Queens and Lake Cayuga, New York (Nelkin 1971, 1974, and 1979); and at several other sites. Local opposition was rare in the 1960s but increased steadily. From 1962 to 1966, 12 percent of the construction permit applications to the AEC were contested; from 1967 to 1971, 32 percent were (Rolph 1979, 102). But as the rate of intervention went up, the intervenors' rate of success went down. The early successes had been due to utility preferences for avoiding local conflicts rather than to any decisions in the regulatory, legal, and political systems. Utilities had felt that an individual site could be abandoned, since it would be easy to find others. As utilities realized they could not avoid conflict by simply choosing another site, they began to fight the opponents. When the cases went through the normal AEC hearings and political procedures, the utilities won.

[2] Mark Evanoff (1982, chap. 7) describes the Sierra Club controversy, and Devall (1970) analyzes the club's voting and politicking patterns. See also McPhee (1972).

A second source of support for antinuclear activities was the environmental movement that bloomed in the late 1960s. The memberships of existing conservation groups soared, typically by 900 percent during the 1960s, and hundreds of new groups were formed (Mitchell 1985, 25). Several major environmental laws were passed, including at least one, the National Environmental Policy Act (passed 1969, signed 1970), that affected nuclear reactor siting and construction. Because of tax-exempt status and funding from the Ford Foundation, new environmental law firms were able to take advantage of these laws and managed to win roughly half their cases during the 1970s (Wenner 1982). In addition to providing some grounds on which antinuclear activists could bring lawsuits to stop reactors, this growing environmental activity boosted the antinuclear movement by providing a network of activists and a sense of a national movement oriented toward vital issues. It also nurtured pro-environmental values and developed a language of moral outrage appropriate to them.

The environmental movement receded in public prominence around 1973, when media attention turned to energy as the central social problem of the day. Many activists as well as members of the public interpreted energy issues, especially that of nuclear energy, to be environmental in nature, so that many members of the environmental movement moved easily into antinuclear activities. They brought with them a sense of a national movement and a moralist policy style.

A third strand of the emerging antinuclear movement consisted of a handful of scientists who began to express doubts about the safety and health effects of nuclear reactors. Much of the controversy was generated within the AEC's large research empire itself, out of the tension between the AEC's concern for rapid commercial development on the one hand and professional pride by engineers and scientists on the other (see chapter 3). For most of the dissidents, the issue soon became the arrogance and insulation of the AEC. This scientific disagreement provided evidence for other opponents of nuclear energy to use, and undermined the confidence of politicians and the public in the AEC.

The "low-dose controversy" erupted in 1969 after the AEC assigned one of its researchers, Arthur Tamplin, to refute Ernest Sternglass's claim that 400,000 Americans had died because of low-level radiation. Although most of this radiation was due to weapons testing, the spread of commercial nuclear reactors would add enormously to the total. Tamplin did conclude that Sternglass was wrong, but he estimated that perhaps 4,000 people had nevertheless died from radiation. When the AEC pressured him to publish his critique of Sternglass but not his own estimates, Tamplin and his superior John Gofman resigned. They felt that scientific inquiry should be followed wherever it led and that it should not be directed by the dictates of industrial development. Both Gofman and Tamplin soon began to work full time against nuclear energy.[3]

Only two years later an even larger controversy rocked the nuclear establishment, the emergency core cooling system (ECCS) hearings. The ECCS is designed to inject large amounts of water into the reactor core (containing the fuel, typically uranium) when an accident has lowered the level of the coolant so that the core threatens to overheat and melt. Soon after the Union of Concerned Scientists became an intervenor in hearings over the Pilgrim nuclear plant in Massachusetts, it discovered that major tests of the ECCS currently in use had failed. In July 1971 the Union of Concerned Scientists published a report critical of the ECCS, saying there was little evidence other than abstract computer simulations that these vital systems would work when needed. The report attracted wide media attention, and local intervenors began to question ECCS adequacy in AEC hearings for individual plants.

To avoid delay in individual hearings, the AEC decided to hold roughly six weeks of "generic" hearings on the ECCS beginning in January 1972. Instead, the hearings lasted more than a year, becoming a forum for the Union of Concerned Scientists to bring to light intense disagreement in the nuclear establishment. Many engineers had privately expressed doubts about the ECCS technology and had been overruled by the AEC, and they expressed these doubts openly under questioning in the hearings. For the most part they continued to believe in nuclear energy, but they were good artisans and wished to do their jobs properly. They felt they weren't being allowed to build nuclear reactors that were as safe as possible. Unlike Gofman and Tamplin, these engineers were not critical thinkers or natural skeptics; they were merely thorough. They were technological realists who had been pushed too far by the pure enthusiasts. In this case adversarial hearings worked to bring out crucial information that had been buried for organizational reasons. The hearings created skeptics on both sides, each questioning the other's facts and claims.

The AEC did little to change its ECCS standards as a result of the hearings, and it seems to have made up its mind before the hearings were over (Primack and von Hippel 1974, 229). But the hearings revealed both confusion and the suppression of information by the AEC, and they encouraged politicians to reform the AEC and the JCAE. The hearings also helped bring antinuclear opponents together into a national movement,

[3] Lewis (1972, chaps. 3, 4), Metzger (1972), Nowotny and Hirsch (1980), and Pringle and Spigelman (1981, 311–18) analyze the low-level radiation controversy. Boffey (1969) describes Sternglass's work and the controversy that ensued. Tamplin's original paper (1969a) was published in revised form in the *Bulletin of the Atomic Scientists* (1969b). See Thompson and Bibb (1970) for the AEC's response. Gofman and Tamplin (1971) published additional epidemiological results that also showed some effects from low levels of radiation.

both through personal contacts (sixty opposition groups from twelve states participated) and by attracting a media wizard, Ralph Nader, to the cause. Participants and the media both began to speak and think about "The Movement" rather than about scattered opponents.

THE REACTION OF THE NUCLEAR ESTABLISHMENT

To the insiders who for decades had guided nuclear energy policies, the new outsiders making demands seemed to be nothing but irrational, emotional members of the public who hadn't an inkling of what a nuclear reactor was. The technological enthusiasts and the cost-benefiters had their disagreements, but they agreed in dismissing the antinuclear moralists. The few politicians involved in nuclear policymaking before the oil crisis were members of the JCAE and nuclear insiders themselves; they agreed with the engineers in labeling the outsiders as irrational. But as policymaking became more public after the oil crisis, new politicians who didn't share the technological enthusiasm of the JCAE became interested in nuclear energy. The vehement reactions of all involved are best understood by examining the policy styles they held.

The moral policy style behind much (not all) of the antinuclear movement was alien to the bureaucrats, economists, and engineers of the nuclear establishment in two basic ways. First, it held that certain policies and technologies were simply wrong, morally, and hence not to be pursued regardless of the cost advantages or the trajectory of technical development. The cost-benefit style favors those policies that maximize benefits and minimize costs, so that to exclude a possibility that may do this is irrational. The technological enthusiast sees great gains to be had from nuclear energy, so that to preclude this seems irrational and detrimental to society. The grounds for selecting policies are different for the three.

The second area of disagreement involves the kind of information and arguments that go into those decisionmaking processes. At their best, engineers have relatively clear criteria for efficiency in producing energy from raw materials, and economists have clear algorithms for comparing costs and benefits. Both break decisions into small parts that can in turn be analyzed accurately. Moralists have a more impressionistic, holistic way of evaluating nuclear energy, taking into account a broad range of connotations, implications, and symbolic meanings. A nuclear reactor implies a commitment to large scale systems and to economic growth, and these implications cannot be quantified and put into a cost-benefit analysis. Thus when an economist asks, "How cheap would nuclear energy have to be for you to accept it?" or an engineer asks, "How safe would the reactor have to be?" these questions make no sense to the moralist.

(Economists and engineers often asked these questions and derided antinuclear activists for not having answers.) The economists and engineers attempted to reduce nuclear energy to one dimension or another (cost, or risk of accidents), whereas moralists envisioned a whole range of aspects associated with nuclear reactors. Not only did their criteria for making policy differ, but the information to which those criteria should have been applied also differed. At both levels the different views talked past each other.

In debates between policy styles, each side, finding the other irrational, imputes hidden motives to them. The nuclear establishment first thought nuclear opponents were "well-meaning but poorly informed" (Weinberg 1971), but even when provided with better information, the activists didn't change their minds. The nuclear establishment could only conclude that the opponents were communists, agents of the coal industry, journalists trying to sensationalize stories, or professional activists who needed a cause to keep themselves employed (Metzger 1972, 38; Novick 1976, 228; Wolfe 1982). In other words, the nuclear establishment tried to give some account of the opponents that made sense in the establishment's own worldview; the opponents were seen as pursuing their own interests. Because this was true only rarely, the nuclear establishment began to dismiss the opponents as irrational, fearful, and ignorant. In the words of Dixy Lee Ray (1975), "Irrational fears of runaway nuclear energy don't stand up against scientific evidence."[4]

At the same time, opponents of nuclear energy believed that the nuclear establishment was motivated exclusively by self-interest. They didn't believe that the nuclear industry and regulators could be so blind to the faults and risks of nuclear energy, so they concluded that profits and bureaucratic pride motivated proponents. The nuclear establishment considered nuclear energy a great boon to humankind; this claim was so contrary to their opponents' worldview that the latter couldn't accept it as the true motivation behind nuclear energy. Many people on both sides had vested interests in what they were doing, but the majority also felt they were acting for the greater good of society. From their own perspectives, their actions were reasonable; but from the other side's, their actions seemed irrational.

Before the oil crisis, the few politicians concerned with nuclear energy policies universally accepted the optimistic assumptions of the alliance between technological enthusiasts and cost-benefiters. The only partisan issue on which politicians disagreed was that of government versus the

[4] At this point in the mid-1970s, the nuclear establishment turned to risk analysis as a way of discovering why nuclear opponents were irrational. In many countries surveys of risk perceptions were funded by the nuclear establishment.

free market. Even after the moralist outsiders had generated some public debate, most politicians were contemptuous of them. In a famous exchange between JCAE Chairman Holifield and AEC Commissioner Ramey (Metzger 1972, 38–39), both men dismissed nuclear opponents as extremists, professional "stirrer-uppers," phonies, and sensationalists, "with no scientific background or competence." "Kooks," said Chairman Holifield at another point. Neither of these men had any scientific background either, but they had large doses of technological enthusiasm.

As late as the oil crisis, the nuclear insiders who controlled nuclear policymaking were contemptuous of the growing moralist opposition. Within both the regulatory system and the JCAE that oversaw it, technological enthusiasm still ruled. However, the growing public debate over nuclear energy and the rise of energy as a partisan issue after the oil crisis brought the nuclear system to the attention of many politicians who were not insiders. Without adopting the moralist policy style of much of the antinuclear movement, some of these politicians questioned nuclear energy on economic grounds and attacked the system of insider politics that had governed nuclear energy policy. What attracted these outsiders to energy policy was the oil crisis.

THE AMERICAN RESPONSE TO THE OIL CRISIS

The oil crisis of 1973–1974 brought energy to the attention of policymakers and the public throughout the world. For three years international oil markets had grown tighter and tighter under the pressure of American imports, so that excess production capacity disappeared. Likewise oil-producing countries were gaining political savvy, economic power, and the inclination to use them. By 1973 they were preparing to renegotiate the price of oil, which had changed little since 1971 (Penrose 1976, 49). When war broke out between Egypt and Israel in October, the United States sent aid to Israel in spite of Saudi Arabian warnings against it. Arab oil-producing nations then launched an embargo on oil to the United States and other hostile countries (Sweden and France were not among them).[5] Although the embargo was ineffective beyond alarming policymakers and the public, it was soon followed by soaring oil prices that moved from $3.00 to $11.65 per barrel by December. France, Sweden, and the United States were all hurt by this increase, psychologically as

[5] Adelman (1972), Grenon (1973), Darmstadter and Landsberg (1976), Mendershausen (1976, chap. 1), Penrose (1976), and Veitor (1984, 193–202) discuss the background of the oil crisis; Lenczowski (1976) and Stobaugh (1979) are good on the crisis itself and its immediate effects. Ikenberry's (1988) excellent analysis of energy politics after the oil crisis, although labeled "institutional," relies as heavily on political struggles and the dominant ideological cleavage as on the constraints of political and economic structures.

well as economically, although Sweden's trade balance was least damaged.[6]

Presumably because it faced a possible embargo, the United States was the first country to articulate its plans for dealing with the increased price and potential shortages of oil. On 7 November 1973 President Nixon gave a speech (Nixon 1973) outlining Project Independence, a hortatory set of bold initiatives designed to provide energy independence by 1980. Resources would be mobilized in the same way they had for the Manhattan and the Apollo projects, Nixon said, relying on national pride to rise to the challenge. Canny politician that he was, Nixon saw the need for broad gestures and forceful ideas in what seemed a true crisis.[7] He used a pure rhetoric of technological enthusiasm.

Nuclear power was to play an important role in America's new energy independence. Nixon's project had three facets: to increase energy supplies, especially through nuclear energy; to conserve energy; and to develop new technologies through a large research program, especially for breeder reactors. Toward these ends a Federal Energy Administration was set up in December under William Simon, who because of his wide powers came to be known as the energy czar.[8] As befits the Manhattan Project comparison, the research and development effort was the centerpiece of the response to the oil crisis. The technological enthusiasm behind this focus fit well with a politician trying to make a grand gesture. In an effort to outshine Senator Henry "Scoop" Jackson, who had previously proposed a large energy research program, Nixon encouraged and took advantage of the technologism of the AEC, where the original proposal for a large research and development effort originated (de Marchi 1981a, 459). This competition between Nixon and a prominent senator—to push what is seen as a popular issue—is similar to the way that Senator Edmund Muskie and Nixon (each with an eye to the 1972 election) had "upped the ante" in proposing strong environmental measures in 1970 (Lundqvist 1980, 51–60).

Nixon's Project Independence speech, only three weeks after the start of the oil embargo, was not based on much analysis of the feasibility of various options, and it was short on details. Detailed proposals were not forthcoming, either, as the threat of actual shortages eased and as Nixon became preoccupied with the Watergate scandal. But the Project Independence staff, without developing cogent policy proposals, had an effect on how policymakers thought about energy problems by developing a

---

[6] The terms of trade figures for 1973–1974 were 0.98 for Sweden and 0.86 for the other two. Numbers below 1.00 represent a deterioration in the prices of exports compared to imports.

[7] Corrigan et al. (1973) and de Marchi (1981a, 458ff.) discuss the project.

[8] On the creation of the FEA, see Woolley (1973) and Fowlkes and Havemann (1973).

large econometric model of the American energy system, called the Project Independence Evaluation System (PIES).[9] As de Marchi (1981a, 463) describes it, "By September 15 [1974] the Project Independence Blueprint staff, mostly borrowed from other agencies, had grown to 500, and the team was hopeful that PIES might not only be useful in preparing the 1975 State of the Union Message but that it also might be accepted as the standard tool for subsequent analyses of presidential initiatives in the energy area."

Although PIES was only one of the large economic models being developed, it indicates one long-range effect of the oil crisis, the increased use of economic analysis in making energy decisions. By comparing the effects of different price changes, policymakers could presumably find an efficient balance between different energy sources. Recognizing the uncertainties in economic growth and energy demand would allow utilities to ask if they truly needed more nuclear plants. They could no longer simply assume a 7 percent growth in electricity demand each year. A cost-benefit perspective would be brought to bear on energy decisions more fully than ever before.

Bringing economic rationality to utility decisions has taken many years. Utility planning had often been done by engineers who simply extrapolated past trends. At the time of the oil crisis some utilities hired consulting firms with large staffs of economists, and eventually utilities developed planning departments that utilized economic data. But this process was slow and sporadic. The fault is not entirely the utilities'. Even the best economic forecasts vary widely, and PIES was widely disputed. But all utilities began sooner or later to analyze electricity demand. First, residential, industrial, and commercial demand differs; many uses of electricity could be curtailed in the face of higher costs. Second, a deep recession would slow the growth in electricity demand. As utilities studied the factors behind electricity demand, they would more accurately forecast future demand. In the two years following the oil crisis, most utilities clung to old expectations or simply admitted grave uncertainties in demand. But they paused to take stock.

In the two years following the oil crisis, American orders for nuclear reactors dropped precipitously, while delays and cancellations spread. This dramatic shift in utility attitudes toward nuclear energy is the result of the coincidence of increasing difficulties in financing construction work and the sudden lack of confidence in future electricity demand growth. Although full economic analyses were not yet in place, some utilities

[9] See the Federal Energy Administration, *Project Independence Report* (Government Printing Office, 1974). For discussions of this energy model and some of the alternatives, see Askin and Kraft (1976) and Tietenberg (1976).

could see their implications. Many state public utility commissions began to require utilities to prove the need for power before building new power plants, forcing greater rationality on them (Colton 1986; McKim 1977). The financial difficulties described in chapter 3 had squeezed utility profits without cutting orders for new reactors, because utilities had assumed that demand for electricity would continue to expand and would have to be met no matter what the cost. After the oil crisis, future demand was uncertain, and there were more cancellations than new orders. What had been a potential weakness before 1973 grew into a threat to nuclear deployment when utilities began to see alternatives to building new capacity.

The oil crisis cracked utility confidence in continued demand growth, which was nil in 1974 and only 2 percent in 1975. An enormous 130,000 MWe of nuclear generating capacity and 60,000 MWe of fossil fuel capacity were canceled or delayed in these two years.[10] According to Gandara (1977, 34) it was only after the oil crisis that "for the first time in twenty years the ability to finance was a critical factor in the construction program of electric utilities."[11] But in fact financial markets abandoned nuclear energy before electric utilities did; the latter would certainly have continued to build reactors had they been able to finance them. They couldn't because investors were doubtful. Utilities would have continued to expand capacity because many did not yet comprehend the new demand situation. This is clear from the tiny revisions they made in their demand forecasts: "Many utilities revised their expectation of load growth from the traditional 7 to 10 percent to 5.6 to 7.7 percent" (Gandara 1977, 34). Cost-benefiters, on the other hand, whether at the Council of Economic Advisors or in banks, had less enthusiastic growth predictions.[12]

Financial markets and PUCs forced electric utilities into more flexible investment strategies; canceling nuclear plants allowed greater room to maneuver in the coming years. Not only were nuclear plants more costly to build, but their lead times were several years longer than those of fossil fuel plants. The uncertainties of demand, seen clearly by those with a cost-benefit policy style, meant that flexibility was crucial, even if it involved building more costly fossil fuel plants several years later. Utili-

[10] Gandara (1977, 33–34). See Figure 3.1 for the pattern of cancellations.

[11] de Marchi (1981b, 521) reports the Treasury found that "by the end of 1974, 235 plant deferrals or cancellations had occurred. Of this total, some 68 percent of all nuclear plant deferrals or cancellations and some 48 percent of fossil fuel plant deferrals or cancellations could be related in part at least to financial distress." See also Campbell (1986).

[12] In fact, growth rates seemed likely to stay low for the remainder of the century. In 1980, for example, Exxon estimated that average annual growth in electricity demand would be 1.6 percent in the 1980s and 1.8 percent in the 1990s (Ahearne 1983).

ties were slower to come to this view, but one result of the oil crisis was that they eventually instituted more rational models for making just such decisions. Complex econometric analyses, well designed to capture changes in demand, began to replace the extrapolations of past trends that fit so well with the technological enthusiasm characteristic of electric utilities. The cost-benefit perspective, in the form of external pressure from financial markets and PUCs as well as internal improvements in demand forecasting, led to a wave of cancellations of reactor orders after the oil crisis.

PARTISAN POLITICS UNDER FORD

It fell to Gerald Ford to use the emerging PIES machinery to present a comprehensive energy bill to Congress, the Energy Independence Act of 1975. Its provisions included some relaxation of environmental regulations, the creation of petroleum reserves, price deregulation for newly discovered natural gas, and presidential power to increase tariffs on oil imports. The provisions that affected nuclear energy were those allowing rate relief to utilities for construction in progress—precisely what they needed in the face of financing problems for nuclear energy—and a plan to facilitate the siting of energy facilities. What held the provisions together was faith in the free market: "What the package did possess that was striking was consistency, born of a conviction shared by Ford and his advisors that moving to a free market in energy was the single best contribution they could make toward resolving the nation's energy problems" (de Marchi 1981b, 488).

The free market ideology of the bill is precisely what the Democratic Congress opposed most. The Democratic alternative that developed in the following months aimed at shielding the consumer from the sudden price increases that market forces would lead to, preferring to rely on oil allocation schemes and government programs to conserve fuel and on increased energy supply to do the same work as price increases.[13] One critic (U.S. Congress 1975, 629) testifying before Congress said, "The problem with the Ford plan is that it is the product of minds ruled by an ideological imperative to keep Government from meddling in business decisions rather than a rational response to real world conditions." Under the Democratic alternative, a new National Energy Production Board would have strong powers to allocate oil and generally direct the production of energy. The entire energy debate quickly shifted to the lines of the dominant ideological cleavage in American politics. The issue of what

[13] This Democratic alternative was released in February 1975 under the title *The Congressional Program of Economic Recovery and Energy Sufficiency* (Washington, D.C.: Government Printing Office).

technologies and resources would assure an adequate supply of energy was transformed into a question of the proper roles of the state and the market.[14]

The question of uranium enrichment was sidetracked by the same ideological debate. The AEC had projected that several enrichment plants had to be started in the mid-1970s to meet demand during the 1980s. The issues were whether there would be sufficient capacity and what technologies should be used. But the debate proceeded along the lines of whether the new facilities would be publicly or privately owned. Market extremists even argued that the three plants then owned by the government should be sold off. The Council of Economic Advisors pointed out that the wrong questions were being debated (de Marchi 1981b, 425), but the council itself contributed to the wrong debate by pushing for privatization.

More important to nuclear energy was Ford's proposal to aid electric utilities. By 1975 there were already massive cancellations of new power plants, most of them because of financial difficulties. Strong measures were proposed by the Federal Energy Administration, including national guidelines for electricity rate-making, government purchase of utility stock, a federal guarantee of utility borrowing, actual government construction of power plants, and the establishment of a large (up to $100 billion) Utility Finance Corporation (de Marchi 1981b). The sweeping nature of these measures developed within a Republican administration demonstrates that it was hardly inevitable that solutions to the energy crisis be phrased in terms of Republican free market versus Democratic government intervention. Only when proposals were made public and debated in Congress were they twisted to fit this ideological conflict.

For largely economic reasons the Ford administration did not pursue these measures for bailing out the utilities. The Council of Economic Advisors, which at the time had great influence under Alan Greenspan, opposed the proposals largely because it was not clear that past growth rates in electric demand would continue. The rapid expansion of capacity in the late 1960s and early 1970s was what had undermined the finances of the utilities, according to this analysis, so that the cancellations may have been a healthy corrective. It blamed the short-term liquidity problems on local regulators but argued that utilities were also local, so that the costs of poor service and capacity shortages would fall on those who had caused them. These cost-benefit arguments fit well with Ford's own free-market ideology. In the cost-benefit perspective, the contraction was a

[14] According to de Marchi (1981b, 499), the Ford program "was premised on full decontrol of oil, and the President intended to submit a decontrol plan to Congress by April 1. After discussion with the House leadership he agreed to a one-month delay; but then a cat and mouse game began in which Congress pounced each time decontrol showed its head."

healthy corrective to previous overbuilding based on technological enthusiasm rather than clear cost calculations, although in retrospect one can see that the Ford administration's decision against utility relief assured the stagnation of the nuclear industry in the United States.

The same choice of paths faced by American officials was open to French and Swedish policymakers. On the one hand there was the technological enthusiast's solution—to build more productive capacity, especially nuclear, that would assure energy independence. On the other hand there was the economic advice to expect slower growth in energy demand rather than simply extrapolating past trends, and to create the flexibility to adapt to demand uncertainty. Ford took the advice of the economists, because of his own ideology, because of the large influence of the Council of Economic Advisors and of economic analysis generally at this time in the United States, and because the technologists were absent from the cabinet.[15] The AEC had been disbanded in 1974, so that there was no longer a prominent spokesperson for nuclear development as there had been in Glenn Seaborg, James Schlesinger, and Dixy Lee Ray. The rhetoric of price had prevailed. It also prevailed in Sweden, but in France technological enthusiasts won the day.

A NATIONAL MOVEMENT

In 1974 and 1975 the emergence of energy policy as a key national issue brought the antinuclear opposition into prominence as well. In spite of the many plant cancellations, many were still under construction, providing targets. A national movement emerged, although it remained a coalition of diverse groups fighting in different ways to stop the spread of nuclear reactors in the United States. While a wide network of contacts had been developed in the scientific and environmental phases of opposition, it took the oil crisis to convince nuclear opponents, especially the local ones, that their activities concerned an issue of basic national importance. At roughly the same time, Ralph Nader hoisted the antinuclear banner, attracting wide media attention with a "Critical Mass" conference in 1974. The organization he founded and named after the conference was never the leader of the national movement, but it helped convince the media that such a movement existed.

The emergence of a national antinuclear movement attracted new groups for whom nuclear energy was a symbolic lightning rod for broader aspects of contemporary society. Ralph Nader saw nuclear energy as a failure of democracy (Novick 1976, 318); small revolutionary sects saw it

[15] Ford's discretion is clear if we compare him to Richard Nixon. The Council of Economic Advisors and the Office of Management and Budget also opposed Nixon's proposed support for nuclear energy, but he overruled them. See de Marchi (1981a, 410).

as capitalist profit-seeking at the expense of human lives; the counterculture saw it as technology out of control, growing beyond human scale. Previous opponents had attacked particular aspects of nuclear energy: plant locations, a lack of sufficient safety mechanisms, thermal pollution. The new activists had bigger targets like the state, technology, or capitalism.

The original antinuclear activists typically had middle-class expectations of the regulatory system. It would respond to their demands and transform or curtail nuclear energy. They soon became disabused of this idea, and they often became outraged at the system's unresponsiveness. Their critique of nuclear energy broadened and deepened so that nuclear energy became a symptom of technological policy beyond democratic control. They were thus often receptive to the ideas of the counterculture, even while they remained aloof from the counterculture activists themselves. The original reformist activists expanded the counterculture's idea that nuclear energy was immoral, adding their own sense of moral outrage at the political and regulatory system. Most segments of the antinuclear movement began to exhibit a strong moralist tone.

Although each segment of the antinuclear movement had its preferred tactics, the movement did whatever seemed to work. Intervention in hearings continued throughout the 1970s, but pressure on elected officials became more popular. Nineteen seventy-six was a big year for state referenda, with seven antinuclear propositions on ballots. California's was watched closely, and it inspired mild state legislation restricting nuclear development as a way of defusing the referendum itself.[16] All but the mildest of the referenda lost, typically by two-to-one margins, demonstrating what opinion polls were saying throughout the 1970s: antinuclear sentiment was in the minority in the United States. Even serious accidents like the 1966 meltdown at the Fermi reactor near Detroit and the 1975 Brown's Ferry fire had no apparent effect on public opinion.

The public hearings held by the AEC and the NRC were hopeless undertakings. Originally designed to legitimate the licensing procedures as democratic, the hearings were seen by each side as a chance to persuade the other. No one on either side ever changed a single opinion, with the result that there was immense frustration but no increase in mutual understanding. How could there be anything else? The very existence of the AEC or NRC implied that nuclear reactors should be licensed, perhaps subject to minor changes. Most intervenors didn't accept that premise, or at least felt that changes should be major rather than minor. The

---

[16] See Carter (1976a, 1976b). Pryor (1976) discusses the Sierra Club's difficulties in gathering signatures to put the initiative on the ballot until it was joined by a small, "religion-oriented" group, Creative Initiative. While the Sierra Club was effective at lobbying, Creative Initiative could inspire volunteers to spend long hours collecting signatures.

worldviews on the two sides were so foreign as to preclude serious accommodation. If the hearings were designed to take public concerns seriously, then they truly were a "charade" (Ebbin and Kasper 1974, 246) and a "sham" (Rogovin 1980, 139).

Disillusionment with the nuclear system also led three engineers to resign from GE and two from the NRC in 1976, giving a further boost to the antinuclear movement. The GE engineers listed many faults and potential weaknesses in boiling water reactors and claimed a major accident was inevitable; the NRC engineers felt they were not being allowed to study safety issues or take appropriate actions to correct safety risks. The resignations increased the credibility of the antinuclear movement, since now experts themselves were disagreeing in public. But these resignations, and others like them, might not have occurred had there not been a supportive movement to give them a platform. Direct action was only one wing of the movement, but for several years it was the most visible wing.

The antinuclear movement also encouraged whistleblowers—typically utility or construction employees who had seen innumerable violations, mistakes, and abuses in nuclear construction and operation—and who wanted to tell about them. In the case of Diablo Canyon, two things were necessary before whistleblowers appeared in force in the early 1980s. First, it was discovered that blueprints had been switched, leading to major construction flaws: this information gave credibility to whistleblowers. Second, the Government Accountability Project and local opponents spent a lot of money working with the whistleblowers: this provided them a place to go and something to do with their information. The results were disappointing, however. Most were fired and had difficulty finding new jobs, and the NRC paid little attention to their complaints. One local opponent complained: "The NRC didn't want to listen: they were too anxious to license Diablo. And most of the evidence mysteriously disappeared from PG&E when they went back to look for it. Plus these guys lost their jobs. They stopped coming around after a while."[17]

Disillusionment with legal interventions and referenda led to a new strategy in the summer of 1976, direct action. This form of civil disobedience involved taking an action, such as occupying a reactor site, without going through the normal, legal channels for stopping the plant. It reflected the importance of the counterculture in the movement, which regarded normal legal and political channels as hopelessly rigged. But a lot of middle-class intervenors participated as well. Inspired by the 1975 occupation of a German reactor site that succeeded in having the plant

---

[17] Personal interview, anonymous (1985), with a local activist who had raised money for the Government Accountability Project's efforts.

canceled, and pioneered in the American antinuclear movement at Seabrook in August 1976, over 120 direct actions occurred during the next twelve months (Mitchell 1981, 82).

Direct action relied on the intense moralism that came to characterize the antinuclear movement after 1974. It was "exemplary action" out of the Christian and Protestant molds, and these were Touraine's new saints showing the world a better way (see chapter 2). The activists' moral ecologism meant they were confident they were right, even though the majority of Americans disagreed. Not only were they right, but they felt obliged to follow a higher law than that of politicians and bureaucrats. If nature and human survival were at stake, trespassing and breaking laws of private property were permissible.[18]

The moralism of the antinuclear movement involves two ironies. First, it occasionally may have hurt the movement's ability to attract the American public. At least one activist, Richard Grossman (1976), attacked the movement for being so confident in the moral superiority of its cause that it didn't reach out to blue-collar and minority groups. His argument appeared under the title, "Being Right is not Enough." Second, activists' confidence that their special insight allowed them to break normal rules was similar to the arrogance of the technological enthusiasts who had pushed nuclear energy in the first place. Both the moralists and the technologists saw overwhelming imperatives that meant they could help society by circumventing the usual cumbersome political processes. The AEC and the JCAE placed themselves outside normal political accountability, as did the ecological moralists who occupied nuclear sites. Such is the tendency of totalizing worldviews.

In summary, the antinuclear movement burgeoned precisely as orders for new reactors disappeared.[19] It grew in membership and visibility in the three years following the oil crisis, and its tactics grew more diverse. The campaign had two broad fronts. At the more general level activists tried to change the legal and political framework in which reactors were discussed, through court cases, state regulations, and federal laws. But

---

[18] Civil disobedience to a policy or law one considers immoral has a long and honorable tradition in the United States, stretching through the civil rights movement back to the abolitionists (Cooney and Michalowski, 1977). Ronald Dworkin (1985, 107) argues that while civil disobedience is usually based on questions of moral principle, it can sometimes be used in practical disagreement over a policy. He misses the fact that the practitioners even in this case opposed a policy they believe to be immoral, rather than simply unwise. England's Greenham Common protesters, the example he uses, certainly felt that nuclear arms were not just unwise but also immoral (Blackwood, 1984).

[19] No student of social movements will be surprised to find that mobilization did not directly follow the siting of reactors. According to unpublished calculations by Robert Cameron Mitchell, roughly 15 million Americans had had a reactor sited near them by 1965, 35 million by 1967, and 55 million by 1973. The number was stable after that.

public opinion remained pronuclear until 1979, so referenda failed and most politicians saw little advantage in initiating laws to restrain nuclear expansion. It was easier for politicians to argue the merits of markets and government intervention than to evaluate nuclear reactors. Thus in the mid-1970s the antinuclear movement made little political progress. On the second front antinuclear activists tried to stop particular plants and to persuade the NRC to tighten its regulation. On this front too they had limited *direct* success in this period.

REGULATORY CHANGES

In the early and mid-1970s American regulation of nuclear power changed in several ways that were *indirectly* caused by the growing antinuclear movement. Chapter 3 described several ways the AEC's autonomy was curtailed: Congress blocked the AEC's attempts to reduce public participation in hearings; in the *Calvert Cliffs* decision, federal courts criticized the AEC for arrogance and demanded better environmental impact assessment; Nixon appointed James Schlesinger, an economist and bureaucrat from outside the nuclear establishment, as AEC chairman; and the AEC was broken into the NRC and ERDA. But the politicians who made these decisions thought the changes would facilitate nuclear reactor siting.

Another big change occurred in 1975 and 1976 as Congress reduced the power of the Joint Committee on Atomic Energy. The outsiders who were attracted to energy issues after the oil crisis were disturbed by how the JCAE-AEC insiders had made policy in the past. Even members of Congress accused the JCAE of promoting nuclear energy without regard for promising alternatives. The creation of the NRC and ERDA was one way to redress this imbalance and to reduce the JCAE's influence. What is more, many of the most powerful JCAE senators had retired or died since the committee's heyday in the 1950s and 1960s, and House reforms to reduce the power of committee chairpersons hit the JCAE hard. In the beginning of 1977 the committee was finally disbanded. Congress had created the framework for serious oversight of nuclear energy, but it did not yet use it.[20]

In addition, the AEC changed two standards in 1971, probably because of pressures from antinuclear protest. In March it began to require cooling towers to avoid the kind of thermal pollution that had bothered the Cayuga Lake protestors and others. In June it tightened its standards on radiation release along the lines suggested by Gofman and Tamplin. Throughout the 1970s and 1980s, the AEC and NRC have tightened their

[20] Katz (1984, 38–52, 215–17) discusses the decline of the JCAE.

standards for nuclear construction. The bulk of this shift has been due to increased knowledge about accidents and a concern that the total risk to the public not increase too much as more reactors are built (Komanoff 1981). One of the potential weaknesses described in chapter 3, that reactors were built before much was known about them, began to have an effect: standards had to be changed in the face of operating experience. But the increased knowledge has led to changed regulations partly because of political pressure; the antinuclear movement has had an indirect, modest, but undeniable effect.[21]

Proponents of nuclear energy proposed most of the changes in radiation standards and administrative procedures as a way of protecting nuclear deployment from protestors' attacks. As a result the changes did little to satisfy the antinuclear movement, but did they have other effects on the ordering and construction of nuclear reactors?

Regulatory changes could have slowed the spread of nuclear reactors in several ways. The AEC and NRC could simply have denied construction or operating licenses to utilities. Indirectly, they could have raised the costs of nuclear energy, especially through licensing delays or requiring costly safety features. At the same time, regulatory changes could have scared away potential financial support in the private sector, making plants more costly or impossible to build. Finally, the traditional regulators could have lost their power to other agencies, especially at the local or state level, who then could have blocked plants or raised costs. Because the main changes in the early and mid-1970s involved licensing, they are examined here. Because costs and changes in the regulatory system itself developed later, they will be examined in chapter 11.

In the last twenty years no nuclear reactor in the United States has been stopped because of its inability to obtain a license. A handful of construction permits have been temporarily denied or delayed, but these have typically resulted in design changes rather than project cancellations. One may have to look back to Bodega Head and similar cases in the 1960s to find proposed reactors abandoned because of expectations that construction permits would be denied. It was twenty years later (1984) that, for the first time, an operating permit was denied to a completed plant, for Byron 1 in Illinois, but it was granted after further hearings.

Throughout the history of the AEC and NRC, critics have complained that more attention was paid to promoting nuclear energy than to pro-

---

[21] Elizabeth Nichols (1987) has assessed the impact of the movement on reactor siting, finding that it had few direct effects. Unfortunately, she does not fully examine the indirect effects such as the increasing political pressure on regulators to tighten standards. Bupp's preface (1981) to Komanoff's extensive study claims that Komanoff does not prove that political controversy did *not* cause tougher standards.

tecting public safety. While the NRC has become more of a regulator than promoter since its creation, cases of promotion still persist. The two NRC engineers who resigned in 1976 both complained that safety concerns were being subordinated to promotional ones. One of them, Robert Pollard, said that "as a result of my work at the Commission, I believe that the separation of the Atomic Energy Commission into two agencies has not resolved the conflict between the promotion and regulation of commercial nuclear plants. Because I found that the pressures to maintain schedules and to defer resolution of known safety problems frequently prevailed over reactor safety, I decided I had to resign."[22] Several years later the Kemeny Commission report (set up after TMI) made similar criticisms, linking NRC behavior to patterns established by the AEC (Kemeny et al. 1979, 20). The similarity is not surprising, since most staff moved intact from the AEC to the NRC. As recently as 1987, members of Congress accused NRC employees of improper ties with industry.[23]

Except for opening many documents to public scrutiny, regulatory changes have not improved public participation in the licensing process. The NRC's own 1980 Rogovin Report used words almost identical to Ebbin and Kasper's (1974), in calling this aspect of the licensing process "a sham." The Union of Concerned Scientists (1985, chap. 3) has described ways in which the NRC expedites licensing to the detriment of safety and public participation, including the exclusion of certain kinds of evidence, the prevention of public challenges to certain kinds of NRC regulations, the postponement of safety issues until the operating license hearings, and the overriding of its own licensing boards when they delve into many safety issues.[24] The mildest conclusion possible is that the NRC has not thrown unnecessary roadblocks at nuclear reactor licensing.

In summary, the nuclear regulatory establishment has remained at least partly captured, even though one corner of the iron triangle (the

[22] Quoted in Fenn (1980, 58), whose report (prepared for investment institutions) was highly critical of nuclear energy from a cost-benefit perspective. To our long list of engineers who felt prevented from fully pursuing safety concerns, add Pollard, who left the NRC in February 1976, and Ronald Fluegge, who left in October. For Senate testimony by Fluegge and two other dissenting NRC engineers, see Baum (1980, 92–108).

[23] The several accusations of 1987 are discussed in chapter 11. A Common Cause study in 1976 found that "72 percent of the NRC's top 429 employees have been employed by private energy companies and that 90 percent of these employees came from companies with which the NRC had current contracts or licenses. Sixty-five percent of the NRC's consultants are working as well for companies that have received NRC licenses or contracts." (Fenn 1980, 58).

[24] Cases in which licensing boards were told to drop their pursuit of safety issues include San Onofre and Comanche Peak in 1981 and Zimmer and Indian Point in 1982. In the Indian Point case, the licensing board judge resigned in protest. See the Union of Concerned Scientists (1985, 70–77).

JCAE) was removed.²⁵ The NRC still gives more priority to promotion and licensing and less to safety than many of its own employees prefer. Several NRC commissioners have even complained of this.²⁶ Since 1971 commissioners have realized that to be effective the AEC or NRC must be above suspicion of collusion, but they have yet to achieve that. Despite occasional rearrangements, the complaints of the antinuclear movement remain valid.

CONCLUSIONS

The oil crisis began by boosting the prospects for nuclear energy, since it was a natural alternative to importing oil. But the crisis set in motion other forces that discouraged utilities from ordering new reactors or even finishing many they had already started. Utilities realized that demand would not continue to rise as before, and they reluctantly began to adopt economic analyses to predict how it would rise. In the short term they simply saw great uncertainties, and in the longer run some of them began to see advantages to managing demand. Financing for new construction, which had been growing tight even before the oil crisis, was expensive and risky with demand so uncertain. Thus two of the potential problems before the oil crisis, utility financing and the premature ordering of nuclear reactors, became real ones in the two years after it.

The oil crisis also spurred political debates and the antinuclear movement. In a political climate increasingly critical of nuclear energy, regulators continued tightening regulations and demanding retrofits for plants. This tightening was the result of two hidden weaknesses: the premature deployment of nuclear reactors before all the problems had been worked out and the hands-off regulatory style that refused to interfere in utility plans. Problems that should have been solved in the 1960s were corrected in the 1970s, at far greater expense and with more embarrassment for the nuclear industry. Increasing controversy over nuclear reactors was one reason they were corrected at all.

The oil crisis and the rising antinuclear movement (with the latter reinforced by the former) brought nuclear energy policy from the realm of iron triangles and closed doors into the glare of media coverage and par-

[25] Political scientists predict iron triangles will break up when the industries they cover become controversial (Ripley and Franklin 1980, 9). This is partly true of nuclear energy: when it became controversial, outsider politicians began to interfere and eventually disbanded the JCAE. Yet the NRC did not become one of the "new social regulatory agencies" pursuing consumer and public interests. It remained captured, partly because it dealt with a single industry, and partly because it retained, despite changes in structure, all the technological enthusiasm and most of the personnel of the AEC.

[26] Commissioners Victor Gilinsky, Peter Bradford, and James Asselstine, at least, have made this point, using language similar to James Schlesinger's 1971 speech.

tisan debates. Partisan conflict framed nuclear policy as a question of whether the government should intervene to stop the flow of plant cancellations or let the market correction take its course. The cost-benefit perspective strongly supported the market correction idea, arguing that demand uncertainties augured poorly for new construction. Not only was President Ford a Republican, and therefore reluctant to intervene in the economy, but he also was heavily influenced by the Council of Economic Advisors. When he did little to save nuclear reactor orders and when utilities began to do better demand calculations, nuclear energy in the United States may already have been doomed.

Observers at the time did not see nuclear energy's demise. They saw a pause while electricity demand caught up with generating capacity. This was possible; Ford could certainly have taken steps to revive nuclear energy. The weaknesses then appearing, of financing and premature deployment, need not have been fatal. In fact, many observers thought they were being corrected. But the biggest hidden weakness of all, poor utility management, had not yet been discerned or even had its full effect. Bad management is what kept the other weaknesses from being corrected. But it was not seen clearly until the late 1970s.

CHAPTER 8

Party Politics in Sweden.

NUCLEAR POWER became a major public issue in Sweden at the same time that it did in the United States, in 1973 and 1974, but more because of the antinuclear movement than the oil crisis. The main opposition party, the Center Party, became antinuclear in 1973 and caused a mild rethinking of Sweden's nuclear construction. The oil crisis at the end of that year intensified the reappraisal. It also brought the delay-oriented, cost-benefit perspective to the fore, since that fit well with the political strategies of the Social Democrats. Thus the Swedish reaction to the oil crisis was similar to that in the United States: to wait and see if demand would fall. But in 1976 the Center Party and its coalition partners won the Riksdag (parliament) elections, so that an antinuclear moralist became prime minister. From 1973 to 1976, energy politics were played out at the highest political levels.

Like the light water reactor itself, the Swedish antinuclear movement began partly as an import from the United States, but it quickly took a characteristically Swedish form. Like its American counterpart, it grew out of the environmental movement in the early 1970s and had a distinctive moral tone. But in Sweden there was a political party, the Center Party, that was open to both pro-environmental values and moral language, so that the antinuclear movement quickly found an important partner. That partner so influenced nuclear energy policy, in fact, that the antinuclear movement was largely absorbed and lost in it. Few American politicians were ever comfortable with the antinuclear movement's full range of tactics and arguments, but many members of Sweden's Center Party were. The Center Party provided a political locus for environmental and antinuclear sentiment that was absent in other industrial countries.

THE BIRTH OF AN ISSUE

The Swedish environmental movement reached its peak in 1972. American activities and scientific information were carefully monitored. A translation of Rachel Carson's *Silent Spring* had a wide impact in the 1960s, and an environmental group was founded in 1971 that became a chapter of Friends of the Earth (Jordens Vänner) the following year. Environmentalists were given a focus, encouragement, and a sense of urgency by the large international conference on the environment held in

Stockholm in the summer of 1972. During the conference Jordens Vänner published a daily newspaper on the environment that was distributed free to all hotels housing delegates. A Swedish physicist, Hannes Alfvén, who had lived and taught in the United States and developed concerns about the environment and nuclear energy while there, gained some notoriety in Stockholm when the Swedish government refused him permission to speak on nuclear waste at the conference.[1]

Just as the environmental movement came to prominence in 1972, Sweden's most respected newspaper, *Dagens Nyheter* began to question the strong progrowth direction of Swedish energy policy and eventually came to criticize the nuclear program. That a major paper was antinuclear came to be an important factor in the nuclear debate, and it was the only one that was. How did this happen? One editorial writer, Olle Alsén, was largely responsible. The paper's policy was to follow the position on a subject forged by the editorialist who covers it, and Alsén covered energy. Alsén came to question nuclear energy for two reasons. First, he had lived in New York in the 1950s and closely followed both the national debate on nuclear weapons testing and a local controversy over a nuclear waste dump. Second, he combined the natural skepticism of a journalist with a lack of training in either technology or economics. When presented with growth forecasts that went up and up, he asked, "How can that be?" Once he became skeptical, so did *Dagens Nyheter*, and its stance helped to legitimate opposition to nuclear energy.[2]

At the same time, a Center Party MP named Birgitta Hambraeus had been doing her own research into nuclear energy. Upon her election in 1971 the party had asked her to look into the question—partly because she was interested in environmental questions and partly because she could read the relevant documents, most of which were in English (she had graduated from Vassar). As in the United States in the early 1970s, nuclear energy was seen in Sweden as an environmental challenge. When Hambraeus became concerned by the idea that nuclear wastes would have to be watched and guarded "in perpetuity," she called on Alfvén for more information. Hambraeus first acted, in 1972 and 1973, as an individual member of parliament. She was a member of the committee governing nuclear energy and, aided by a committee chairman with a grudge against the minister of industry, encouraged it to hold hearings at which both pro- and antinuclear views were aired. As a result the committee unanimously asked the full Riksdag to reinvestigate its nuclear energy policy.

With little controversy the Swedish parliament passed in 1973 a two-

[1] My account of the early antinuclear activities relies on interviews with Lennart Daléus, Birgitta Hambraeus, Per Kågeson, Per Gahrton, Olle Alsén, and Sven Anér.
[2] Personal interview with Olle Alsén. Holmberg and Asp (1984, 35ff.) also describe the energy positions of *Dagens Nyheter*.

year moratorium on nuclear power plants beyond those already planned and approved. Much like the reorganization of the AEC in the United States, this policy was a way for the Social Democrats to defuse the antinuclear movement while appearing to look reasonable, and yet not stop the nuclear program they already had. Confident that antinuclear sentiment was limited, nuclear proponents were willing to make concessions, just as American utilities in the 1960s had been willing to abandon certain controversial reactor sites. The concession of the moratorium was minor, since it did nothing to slow the pace of Sweden's nuclear development.

Public controversy about nuclear energy was not yet widespread, so it had not become a partisan issue. It played no role in the September 1973 election campaign, for example. Had the moratorium been seen as a Center Party initiative, it would not have been so readily passed. It would take the oil crisis to shift nuclear energy into the dominant partisan cleavage.

The Center Party Becomes Antinuclear

Hambraeus introduced Alfvén and other scientists to the leader of the Center Party, Thorbjörn Fälldin, impressed by what he heard, asked Alfvén to address the Center Party congress of June 1973. Even with a party majority still uninformed about nuclear energy, the party decided to take a position against it. The Center Party, especially in its youth and women's sections, soon developed close ties to environmental groups. These groups retained their autonomy, and the prominence they had gained in 1972 allowed them to raise issues that would be taken seriously in the media. But public criticism of nuclear energy was increasingly dominated by the Center Party, and the Center Party brought several foreign antinuclear activists to speak in Sweden. Its close relationship to a major party distinguishes the Swedish antinuclear movement from the American and French movements.

The Swedish Center Party was the only major political party in the industrialized West to take a clear antinuclear stand in the 1970s. In part this position followed long-standing party concerns, and in part it was seen as a means of attracting young, urban voters. Equally important was the similarity between the moral style of the party and Fälldin and that of the antinuclear movement.

The Center Party had once been called the Agrarian Party and remained a farmers' party until the early 1960s. The decade of the 1960s was one of rapid economic growth and increasing urbanization and centralization of the population. The government built almost 1 million apartments in and around the cities—a huge number in a country of 8 million people. This urban growth was closely—and proudly, for their part—identified with the policies of the Social Democrats, who had governed since 1932. This was a grand success from the perspective of the "dominant pro-growth paradigm." The Social Democrats strongly favored the economic growth and technological change that form this paradigm. The Center Party, in contrast, had an affinity with the new environmental paradigm. The Swedish Institute (1987) puts it well: "The green wave sometimes smacked of neo-romantic, revivalist advocacy of a return to the simple rural life. . . . When the green wave came, the party was able to don an expression of innocence. All the other parties could justifiably be held responsible for the welfare society's obsession with material plenty. The Centrists for their part had long warned of the consequences of depopulating the countryside." Ecology provided a critique of Social Democratic policies, and nuclear energy came to represent a strong ideological cleavage between the major party of the right and the major party of the left. As in the American debate over free markets and government intervention, this cleavage reshaped the question of nuclear energy policy.

Another aspect of this contrast between Social Democrats and the Center Party may involve the level of sophistication of each party's voters. The Social Democrats were more urbane as well as more urban. According to Steven Kelman (1976), the Center Party attracted many voters who were relatively undereducated, generally discontented, and deeply suspicious of the intellectuals and technicians they saw as running Sweden. They disliked nuclear energy because they were afraid of it and because it represented another imposition from Stockholm. They generally felt disenfranchised, seeing all politicians as sophisticated Stockholmers, and they felt that opposing nuclear energy was a way of opposing the whole Swedish system.[3]

Concern for decentralized, regional development had held the environmentalist and agrarian tendencies in the Center Party together in the 1960s, as did their common distaste for the Social Democrats. In the late 1960s the Center Party had briefly favored nuclear energy as a means of encouraging regional development, but this fancy was easily abandoned when environmental questions about nuclear appeared. As the party embraced environmental issues, it lost the support of some hi-tech farmers from large farms in the south of Sweden, but it gained support among the young, urban middle class. By 1973 opinion polls showed that 29 percent of the voters supported the Center Party, and it received 25 percent of the vote in the elections of September 1973 (Holmberg and Petersson 1980, 247). This support was greater than the combined support for the other two bourgeois parties. Its voters came from a range of oc-

[3] In the United States, where nuclear reactors have rarely been associated in the public mind with central state policies, rural populations tend to favor nuclear energy strongly. France is more similar to Sweden, since reactors are associated with decisions from Paris. Different connotations accompany nuclear energy.

cupations, with less than one quarter of them from agriculture, compared to more than three quarters in the late 1950s (Särlvik 1977, 91). The shifting party composition was not due to a loss of support among farmers, who largely remained faithful, but to rapid increases in support among the urban middle class. The main attraction for them was the party's environmental concerns.

Environmental concern by itself would not have led to the moral, uncompromising antinuclear stand of the Center Party. In its cautious, bureaucratic way, the Social Democratic government created the machinery to clean up Sweden's air and water without making pollution a moral issue (Lundqvist 1980). The Center Party's moral tone arose out of its traditional references to the Swedish land and "people" and to the personal style of Fälldin, party leader since 1971. Fälldin's popularity came from his image as a simple sheep farmer from the North who spoke no foreign languages and whose political action grew out of intense moral conviction. Like many Center Party politicians, he spoke slowly and simply and was perhaps suspicious of science and technology. He was the perfect opposite of Prime Minister Olof Palme, who was aristocratic, polished, quick, urban, urbane, a brilliant orator—and not completely trusted by many Swedes.[4]

Fälldin's own initiative moved the Center Party from environmentalist to antinuclear. He had personally asked Birgitta Hambraeus to look into the nuclear question, and he brought the party around at its 1973 congress. Few doubted Fälldin's sincerity in his antinuclear stand. Even though others in his party may have been delighted to find an issue on which the Social Democrats might be vulnerable, the essence of Fälldin's political style was not to be so calculating and utilitarian in partisan politics. Moral outrage over the Social Democrats' blind faith in technology and economic growth fit his worldview and inclinations very well.

Antinuclear Activities

The growing Swedish antinuclear movement in the mid-1970s was similar to the American movement in several ways. It used the same arguments against nuclear energy and could draw on the same data that questioned nuclear safety, since these were largely borrowed from American sources. The American Union of Concerned Scientists was important in expanding publicity around the emergency core cooling system hearings of 1972 and in gathering, writing, and circulating technical critiques. The Union of Concerned Scientists could undertake activities impossible in Sweden. Its direct mail techniques could find fifty thousand Americans willing to contribute money, so enough resources could be pooled to

[4] For more on Fälldin, see Kelman (1976), Link (1976), and Lewin (1984, 314–39).

maintain a serious effort at information gathering. Sweden, with one thirtieth the population of the United States, could not accumulate such a critical mass of funding. There were also fewer physicists and nuclear engineers to draw on. In the United States there were enough dissidents to form some technical opposition, but not in Sweden.

Although antinuclear protest began in the United States earlier than in Sweden, both movements achieved national prominence in the period from 1971 to 1974. Environmental movements in both countries and growing technical controversy in the United States played a role in both cases. But in the United States new sites were still being adopted for nuclear plants in this period, whereas Sweden's four sites had already been selected. Two other proposed sites were put on hold and eventually dropped. The result was that the American antinuclear movement had easier local targets.

Like its American counterpart, the Swedish antinuclear movement used whatever tactics it thought might work. These differed from American tactics somewhat because the political and regulatory structures provided different points of access for public participation. American antinuclear activists devoted more time to regulatory hearings, although these became less important to them in the late 1970s. Swedish activists spent more time in electoral and parliamentary politics, since the Riksdag was the center of nuclear policymaking after 1973. Both movements mounted large demonstrations. Ten thousand Swedes protested uranium mining near Sköde in June of 1976; twenty thousand Swedish and Danish protesters marched to the Barsebäck reactors in September 1977.

The same elements made up the Swedish and American antinuclear movements except that the former had fewer local or technical groups. The environmentally concerned were joined by many members of the counterculture and a few left-wing revolutionaries. There was less local protest in Sweden for two reasons. The first was that siting decisions had been made before doubts about nuclear energy appeared in the early 1970s, so the choice of where to put the plants could not be a terrain for conflict. Local communities in Sweden have the right to veto construction projects like nuclear plants, and if siting had taken place in the mid-1970s rather than the late 1960s, these vetoes would probably have been used. The more important reason for the lack of local protest is the Swedes' willingness to rely on national representatives to fight their battles. This traditional tendency results from Sweden's small size and high degree of organization. From trade unions to the temperance union, every interest and perspective has a national organization to advance its cause. The country is small enough that the national government does not seem as distant as that in the United States.

The Swedish antinuclear coalition contained fewer technologists than the American or French movements. One reason was the country's small size, so that it was difficult to gather enough dissenters (not to mention the funds) to form a stable group. Another was that dissenting engineers were not fired as in the United States, where they could often find jobs only in the antinuclear movement. Unlike the CFDT in France, no Swedish trade unions were antinuclear or willing to fund and protect technical protestors. Finally, the Center Party had never attracted many intellectuals, so that it could not easily turn to antinuclear scientists like those so important in the French and American antinuclear movements.

Sweden's antinuclear coalition not only had few local and technical groups, but it faced competition with the Center Party for activists. Few individuals moved from one to the other, but many activists joined the Center Party, whereas counterparts in the United States probably worked for environmental and antinuclear organizations. The Center Party provided something for an antinuclear activist to do, whereas neither the Democratic nor the Republican Party provided this in the early 1970s. The Center Party was especially attractive to someone who desired political power, while antinuclear organizations kept the activists more concerned with ideological purity and internal democratic "process." The Center Party youth group in particular was active. Because the Center Party and antinuclear groups worked closely together, an aspiring activist could easily join the former.

Many members of the Center Party had backgrounds that could have led them into activist groups—and in the United States probably would have. Fälldin himself had a background similar to that of rancher Ian McMillan, one of the first to criticize Diablo Canyon and now a folk hero of California's antinuclear movement. Birgitta Hambraeus was precisely the kind of woman who founded and ran American intervenor groups like Mothers for Peace. Well-educated but not at first in a demanding career; often with humanist, service-sector husbands (Hambraeus is married to a priest; many of the Mothers for Peace to professors), and often new to less urban areas; these women become interested and active in local environmental issues. Their values were strongly pro-environmental, and their policy styles typically moralist. These "housewife warriors" have been the backbone of the environmental and antinuclear movements.[5]

Lennart Daléus differs from many Swedish environmentalists in that he switched from environmental organizations to the Center Party, although not until 1981. Having served as president of Jordens Vänner and later the Environmental Federation (a coalition of groups), he had jobs first in semigovernmental organizations like the Swedish Academy of Sci-

[5] Schnaiberg (1985) gives a general description of these activist women, while Levine (1982) and Gibbs (1982) describe the central woman behind the Love Canal protests.

ence and later in governmental ministries, at which point he joined the Center Party and began standing for various offices. He was clearly attracted to the party by his desire to get things done, the same reason Hambraeus gave for not leaving the Center Party to join an environmental party formed in the 1980s. The infrastructure of a political party and a Parliamentary seat were quite appealing.[6]

Even though a young activist might join either the Center Party or the environmental movement, the two retained some mutual suspicion. Party activists tended to see the ecologists as impractical, destined to worry about internal process too much to accomplish their practical ends. Likewise ecologists suspected that partisan politics often distracted from the goals of saving the environment. Every antinuclear movement was a coalition of different elements, but this cleavage was larger than others.

The Center Party's ability to attract activists is partly the result of general characteristics of Swedish politics. All the political parties send many grassroots activists to the Riksdag, in contrast to the lawyers and businessmen who dominate, for example, the U.S. Congress. One reason is the intimate parliamentary representation possible in a small country: a Swedish MP represents roughly twenty-five thousand people, compared to over five hundred thousand for an American congressman. A third reason is the confidence of Swedish citizens that the Riksdag can redress grievances, as mentioned above. But attracting political activists is one thing, attracting moral ecologists another. Only the Center Party, with its own moralist style, attracted many activists sympathetic to the moral ecological worldview.

The Swedish antinuclear movement was born from the combination of information and skepticism imported from the United States with the moral political style of the Center Party. All the early opponents to nuclear energy had links to the United States; antinuclear literature was imported to Sweden without even being translated. In the mid-1970s antinuclear views, even American ones, had a wider audience and more influence in Sweden than in the United States, since they were taken up by a political party and covered by the news media. In the United States they were still on the fringe.

REACTIONS TO THE ANTINUCLEAR MOVEMENT

Nuclear insiders in Sweden had roughly the same reaction to the rising opposition as their American counterparts. The opponents were seen as ignorant, irrational, or both. The Center Party's antinuclear stand did nothing to change this view, since most big Stockholm newspapers and

---

[6] Personal interviews with Daléus and Hambraeus.

many urban intellectuals dismissed the party as a collection of backward farmers. According to one of Fälldin's energy advisers (Abrahamson 1979, 34), "In the view of these technocrats, among them some ministry staff members, how could one take seriously a dumb farmer's views on nuclear reactors—the highest technological achievement of Western industrialized society!" It is true that the Center Party had an uncertain relationship with academics and scientists. But the Swedish political elite's solution to ignorance was different from that of the American nuclear elite (AEC and JCAE). Instead of trying to exclude them from the decisionmaking process, Swedish political leaders wished to educate the public in order to increase support for nuclear energy.

Swedish elite reactions differed from those in the United States for two reasons. First, Swedish elites expected to shape mass attitudes and educate the public. Steven Kelman (1981a, 151–53) has described the willingness of organizational leaders, especially in trade unions, to try to shape member attitudes on various issues. He reports (1976, 117) a survey of political leaders, which found that only 8 percent thought it proper to change their positions when their constituents disagreed. The language for influencing public opinion is accordingly stronger in Swedish than English: *bilda opinion* and *skapa opinion* mean to form and to create public opinion. Hans Dahl (1984, 101–3) has shown that many state functions are based on the Romantic image of the *Folk* as something to be educated, cultivated, and cherished.[7] Thus the Social Democrats were not worried about general public opinion, but they were not willing to ignore it. Besides, antinuclear attitudes were still thought to be in a small minority, even though no one thought to conduct a survey of them until 1976.

Social Democrats could not have ignored antinuclear opinion, but they might have tried to reshape it into pronuclear attitudes. They were prevented by the prominence of the antinuclear Center Party, an electoral force that could not simply be excluded from policymaking. From 1973 to 1976 there was an even balance in the Riksdag between Left and Right—175 votes each—so that the Social Democratic government could hardly be cavalier in developing policies. Nor could Social Democrats avoid having that issue symbolize basic differences between Center and Social Democratic ideologies, much as they tried to make nuclear energy seem part of a broader question of rational energy choices. The symbolic weight of nuclear energy overpowered them. In the United States in 1973, opponents of nuclear energy had little representation in political

[7] Anyone who has watched Swedish television, with its heavy dose of educational programing, has seen this tradition at work.

parties and could be dismissed. Not so in Sweden. Mass attitudes mattered little; organized attitudes mattered a lot.

The Social Democrats' reaction to controversy over energy policy in the spring of 1973 took the traditional Swedish form—the creation of several commissions to investigate the hazards of nuclear energy and a program of study groups for private citizens to learn about and form opinions about various energy forms (Liljegren n.d., 3). Many members of the government simply assumed that when citizens learned more about nuclear energy they would come to support it.

On the eve of the oil crisis, Sweden had already embarked on a reconsideration of its energy policy. The state commissions had begun work, but the study groups were not scheduled to begin until 1974. The moratorium on nuclear energy was more a tactic for quieting opposition than anything else. It had no effect on the reactors then planned, and since it only extended to 1975, it would have no effect on future construction. It was an important sign, though, that the Swedish government intended to resolve the nuclear controversy with its usual slow search for consensus. The government hoped that delaying the hard decisions would, as usual, allow them to be made more easily later. It expected the study groups and commission reports to reassure nuclear opponents so that Sweden's nuclear development could continue as planned.

THE SWEDISH RESPONSE TO THE OIL CRISIS

When the oil crisis occurred, Sweden was already conducting a reconsideration of its energy policy, but it still took Sweden longer to formulate its long-term response to the new energy situation than either France or the United States. The same partisan political considerations that had led to a rethinking of nuclear energy in early 1973 also complicated and delayed the development of a coherent energy policy over the following two years.[8]

The Social Democrats' energy policy was developed in response to the Center Party's call for a complete ban on nuclear energy and a reduction in the annual growth of electricity demand to 1 percent. In a Riksdag evenly split, the government could not ignore the main opposition party. The Social Democrats had to look forceful, but they were afraid of losing

[8] Several good sources cover Sweden's response to the oil crisis: Vedung (1982, chap. 3; 1984) sets it in the broader context of Swedish energy policy; Pearson and Nyden (1980) compare the Swedish and Dutch responses; Lönnroth (1977) sets the 1975 Energy Bill in the context of total electricity supply and describes possible future directions for energy policy; the Swedish Ministry of Trade (1975) provides straightforward details of the short-term measures taken; and Sahr (1985) gives a compelling interpretation of the political and bureaucratic intrigues during the entire period.

votes if they pushed for more nuclear energy. They tried but failed to prevent nuclear power from becoming a partisan issue. They failed because they had to develop an energy plan, and the other parties, especially the Center Party, then attacked it. The dominant cleavage in Swedish politics is that the Social Democrats, with almost half the electorate, propose, and the other parties respond. The Center Party would reject whatever the Social Democrats did. And nuclear energy had come to symbolize the two parties' attitudes toward the environment.

While over a year was necessary to develop a long-term plan, short-term policies to ease the prospect of oil shortages during the embargo came immediately. They included a campaign to change public attitudes and habits, price controls, fuel rationing, a ban on the heating of garages and storage areas, and a reduction in public lighting and temperatures in public buildings. This program was much more comprehensive than those instituted in France or the United States. Strong and quick measures were possible in part because a Committee on Economic Defense had existed since World War II for precisely such emergencies, even though its normal activities consisted of stockpiling key goods and resources.

As in France and the United States, top political leaders saw the importance of a strong public response, which in Sweden came in the form of a speech by Prime Minister Palme on 19 December 1973, Sweden's first such nationally televised address (Pearson and Nyden 1980, 417). The pressure for such a statement, much less Palme's style than it was Nixon's, came from the even balance in the Riksdag and from polls showing that Social Democratic popularity had slipped since the September elections. But the speech contained relatively few concrete proposals for the long term. This lack can be seen in Nixon's Project Independence speech as well, although not in Messmer's nuclear plan announced a few months later.

In spite of strong language and short-term measures, it took Swedish policymakers over a year to produce a plan to deal with higher energy prices in the long run. Or perhaps it was because of these strong, short-term measures that Sweden felt it could take time to plan its strategy slowly and carefully. First, the government established a mechanism for gathering technical information about Sweden's energy options—an Energy Council that met several times during the winter of 1973–1974 and three times during the following winter. Both political parties and interest groups were represented on the council, which heard testimony from energy experts with a variety of perspectives. It was also one component of a "strong" response by the Social Democrats, since it was chaired by Prime Minister Palme and was partly designed to show that the government was doing all it could to uncover and consider all energy options (Sahr 1985, 44).

In the fall of 1974 the political parties also attempted to assess and develop public opinion by establishing roughly eight thousand study groups of around ten members each. Thus approximately 1 percent of the population participated, studying the basics of electricity production, reading about alternative sources, and discussing policies with fellow citizens. Each party collected questionnaires from its study groups concerning the general path energy policy should take.

This attempt to plumb party opinion was hazardous for the Social Democrats, both because their party was somewhat divided over nuclear energy and because it would fall to them as the governing party to develop a concrete energy bill. Many party leaders, most of whom were pronuclear, felt that with increased information party members would become less antinuclear, but this did not happen. A poll in January 1975, just after the study groups had ended, showed that 19 percent of party members wanted to give up on nuclear power, 45 percent wanted to use it in the last resort, 24 percent thought it was needed for the welfare of the public, and 12 percent didn't know.[9] Perhaps sensing this balance of opinion, the party did not include a direct question about nuclear energy in its survey of study groups. On the other hand, there was strong support for energy conservation—specifically for reduced growth rates in energy use, although not for actual reductions. The study groups reflected the government's desire to shape opinion as much as to learn from it.

The 1975 Energy Bill, outlined in a speech by Palme in February and passed into law in May, had two central components: a small expansion of nuclear energy (from eleven to thirteen reactors), and a large reduction in the growth rate of energy demand (Swedish Government 1975). Given the results of the study-group questionnaire, it was the conservation proposal that the Social Democrats emphasized in all public discussions. The goal of conservation would be to restrict the growth in energy demand to 2 percent per year: 1 percent for households, and 3 percent for industry. Representatives of industry protested the growth restrictions through the *remiss* procedure, but they were largely mollified by a vague averaging procedure and the adoption of 1973 (a year of high energy demand) rather than 1974 (a year of depressed demand) as the base year for calculating the goals. Close continuing contacts between government and industry representatives also helped overcome the latter's resistance to reduced energy growth. The number of thirteen for the nuclear reactors seems to have been a numeric compromise on the part of Social Demo-

---

[9] Cited in Sahr (1985, 51). The poll, conducted by the Political Science Institute of Göteborg University, showed Social Democratic *voters* to be as antinuclear as party *members*.

cratic ministers, some of whom argued for fifteen and some for the eleven already planned (Sahr 1985, 61).

Cost-benefit rhetoric played a key role in the internal mechanisms that developed the energy bill. How did the government produce the bill? Not only were many cabinet meetings devoted to energy issues, but an Energy Policy Council set up in early 1974 was composed of undersecretaries from relevant ministries. Its function was to coordinate the actions of the various ministries (Industry, Housing, Trade, and Education), especially in formulating the energy bill. The Finance Ministry became involved only later, in sharp contrast to the situation in France, where the Ministry of Finance was an important actor from an early stage. One reason is that Swedish nuclear energy was financed with somewhat less help by the state. Five of the twelve reactors were owned by private utilities and financed on the open market. But this still left four reactors completely owned and three reactors largely owned (75 percent) by the State Power Board, whose budgets had to be approved by the government. A more important reason is simply that the Ministry of Finance has less power and autonomy in Sweden, where ministers are a more tightly knit group subject to the control of their party. Swedish party discipline would not allow a finance minister as much veto power as in France.

Discussions by government policymakers differed from parallel debates in France and the United States in two ways: the rapidity with which conservation was seen as a serious option, and the realization that it was hazardous to try to predict energy demand very far in advance. The two are connected, since if one believes that conservation can work, one becomes wary of predicting future demand. The government adopted a proconservation position quickly for several reasons. Sweden had a tradition of policies designed to increase energy efficiency, through special housing loans, building codes, and transportation policy.[10] Second, Swedish industry was willing to accept conservation policies, as described above. Third, the Social Democrats were partly responding to a demand from the Center Party for even greater conservation (an annual energy growth rate of only 1 percent). The Center Party's concern for conservation stemmed from its search for alternatives to Sweden's nuclear program.

In addition to a fondness for conservation, the government had other reasons to adopt the cost-benefit advice to delay any construction of new power plants. The prime minister and the Energy Policy Council had as

[10] Schipper and Lichtenberg (1976) have argued that Sweden uses around 60 percent as much energy per capita to produce roughly the same per capita wealth as the United States, partly because of higher energy prices, but also partly because of policies targeted specifically at energy efficiency.

advisers a group of energy experts with their own organization, the Secretariat for Future Studies (*Sekretariatet för framtidsstudier*). Members of the secretariat argued strongly that energy demand could not be predicted accurately far in advance, so further discussion and decisions should take place in 1978. In spite of initial objections from the Ministry of Industry, this proposal was incorporated in the final bill. Without any formal power the secretariat had great influence by using a cost-benefit rhetoric politicians were ready to listen to. The government was open to conservation and delay because these fit well with its own political tactics. Cost-benefit recommendations supported the common Social Democratic tactic of avoiding irrevocable decisions, a tactic necessary for maintaining consensus both within a diverse party and between different parties and interest groups.

Cost-benefiters were pointing out the difficulties of predicting energy demand in both France and the United States as well. The American Council of Economic Advisors successfully presented these hazards as a key argument against a bailout for the utilities. The French Ministry of Finance used the same argument with less success against EDF's mammoth program. In all three countries the electricity producers, who were producing the forecasts and were accustomed to extremely predictable demand growth, never fully accepted the possibility of new uncertainties in the future. The worldview of a technologist encourages just this confidence in lawlike regularities and infinite progress; that of a cost-benefiter at least considers the effect of prices on demand and at best embodies a general skepticism about the utility of proposed projects. Only in Sweden did energy policies formulated after the oil crisis actually demand a reconsideration several years later (in 1978).

Policy discussions changed sharply when they moved from internal government bureaucracy to the sphere of partisan competition and public debate. Once the energy bill was outlined to the public in February 1975, it was almost exclusively the nuclear proposal that was discussed. Robert Sahr (1985) has pointed out that "energy policy" was taken to mean "nuclear policy," despite strenuous efforts by the Social Democrats to concentrate the debate on total energy policy. Once the proposals left the bureaucratic realm and entered the political, they were subject to the public's need for distinguishing between the political parties. They were seen as part of the dominant partisan cleavage. The difference between eleven and twelve and thirteen reactors, not terribly important for energy supply, became critically important as a symbolic difference between the parties. Ideological differences between political parties diverted attention in public discussion to that part of the bill where such differences, albeit slight ones, could be found. In the United States the question of government intervention versus the free market dominated other energy issues; in Sweden it was the number of nuclear reactors a party wished to build.

The energy bill's conservation and nuclear proposals provided a clear contrast between politicized and nonpoliticized policymaking in Sweden. Developing a conservation plan involved sustained discussions between government and industry, attempts to balance competing goals, and several of the compromises for which Sweden is famous (Rustow 1955). The Social Democrats were concerned to find a least common denominator of public opinion, partly to hold their own party together. In essence, the conservation proposal was developed in a way that fit the traditional image of Swedish policymaking by consensus. But for nuclear energy there was less internal debate, and party members' attitudes were not fully solicited. Even so, policymakers must have been somewhat dismayed to be attacked for proposing a mere two additional reactors, hardly a dramatic response to the oil crisis. After all, France's Messmer Plan had called for thirteen reactors in the first installment alone. Nevertheless, the Social Democrats were attacked by almost all the other parties for these two additional reactors. All the caution and compromise in the world could not have avoided dispute and polarization over such a politically sensitive topic as nuclear energy.[11]

The Swedish reaction to the oil crisis started out looking characteristic of the Swedish political system as described by innumerable commentators. The 1975 Energy Bill was a middle way that emerged from months of slow deliberation, the gathering of information from contrasting perspectives, personal contacts between industry and government, the simultaneous assessment and development of public attitudes, and even a balance of cost-benefit and technological enthusiast worldviews. The bill emphasized uncontroversial conservation, deemphasized controversial nuclear energy, and called for a reconsideration in only three years. The essence of the Swedish reaction to the oil crisis, then, was caution about overbuilding, similar to the position Gerald Ford adopted in the United States. But there was also an element foreign to stereotypes of Swedish policymaking—proposals for a modest expansion of nuclear energy, hardly different from already existing plans, became a lightning rod for partisan ideological debates.

A MORALIST WINS POWER

After the Swedish Parliament passed the 1975 Energy Bill, the nuclear debate waned, but it reappeared as a strongly partisan issue in the cam-

[11] American commentators (e.g., Kelman 1981a) often romanticize Sweden's machinery for developing policy consensus. This consensus is often achieved by excluding positions that are truly dissenting, as the case of the Energy Commission in chapter 12 shows.

paign leading up to the September 1976 general election. Because the largest parties on each side, the Social Democrats and the Center Party, had different views about nuclear energy, it continued to be a natural focus of contention. It also contributed to the defeat of the Social Democratic government for the first time since 1932. The new government was a coalition of the Center Party and the other two bourgeois parties, the Liberals and the Conservatives. Fatefully for the new government, the latter two parties favored nuclear energy, the Conservatives quite strongly.

Although nuclear energy was only one of several issues that removed the Social Democrats from office, it was the most visible one and the one many politicians credited or blamed for the defeat. For most of the campaign the nuclear question had seemed unimportant. Social Democratic support had already withered from a series of scandals early in 1976 and from the party's Meidner Plan for union ownership of company stock (Kelman 1976). Thorbjörn Fälldin, the Center Party leader, made it clear he would not participate in a government that allowed nuclear energy to continue (Liljegren n.d., 19), but the issue was not discussed much in the spring and summer before the election. The three bourgeois parties did not wish to bring attention to their sharp disagreements over nuclear energy, and the Social Democrats were also happy to concentrate on other issues. Not only was their party membership split on nuclear energy, but party leaders believed the issue had helped and would continue to help the Center Party. They closely watched the California referendum of June 1976 (see chapter 7), and concluded from its pronuclear result that nuclear power would not grow into an important issue in the campaign (Kelman 1976, 116).

But for several weeks before the election on 19 September, Fälldin spoke almost exclusively about nuclear energy, and his moral, uncompromising position showed clearly. Kelman (1976, 122) quotes Fälldin and points out how unusual such moralism is in Swedish politics: " 'No minister's post in the world is so attractive that I am prepared to compromise with my conscience,' Fälldin told the parliament, with a dramatic flourish. Swedish politicians just don't make statements like that—such things are left to hot-blooded French or Italians—and Fälldin's position attracted attention for just that reason." Because Fälldin seemed to be gaining support for his position and because critics claimed the bourgeois parties would never be able to agree on energy policy once in office, the Liberals and Conservatives said little or nothing about nuclear energy during the campaign. Meanwhile Center Party advertisements hammered away at it.

The moralism of the antinuclear position made the Social Democrats especially vulnerable on the nuclear energy question. Sahr (1985, 83–84) argues that this "moral dignity" distinguished nuclear from other political issues, so that Social Democrats could vote against their party on this issue without feeling like traitors. For them, the uncompromising moralism of their nuclear opposition had to take precedence over the usual give and take of politics. In most elections, the Social Democrats have a surge of support in the final weeks of the campaign because they appeal to party discipline, but in 1976 they were unable to do this because party discipline is a weak motivation compared to moral conscience. Nuclear power became an issue only weeks before the election, but this may have been just the right timing for it to unsettle the Social Democrats (Link 1976, 44).

A net shift to the right of only 1.4 percent of Swedish voters since the 1973 elections changed the government. It is difficult to assess the importance of antinuclear attitudes in this result. The 1976 election was not driven by the same economic issues that influence voting in most Swedish elections. Olof Petersson points to three unusual factors: a general feeling that it was time for a change of government after 44 years, the Left-Right cleavage over the Meidner Plan, and nuclear energy. Party switching was higher in 1976 than before, but this was partly a long-term trend upward. In trying to separate the short-term effect of the nuclear issue from the other long-term trends, Hans Zetterberg called the former the trigger and the latter the motor for the electoral defeat of the Social Democrats. The Center Party actually had less electoral support in 1976 than in 1973, but one can argue that its decline would have been even greater without the nuclear issue. Nine percent of the Center Party's voters in 1976 had voted for the Social Democrats in 1973 (Swedish Central Bureau of Statistics 1978, 136), accounting for roughly 2 percent of the total vote, enough to carry the election. But the reasons they switched cannot be determined. It is reasonable to conclude that nuclear energy did play some role in the defeat of the Social Democrats.[12]

Although analysts debate the role of nuclear energy in the change of government, politicians on both sides felt it had been crucial. Lars Liljegren, one of the Social Democrats' most prominent energy policymakers, wrote (n.d., 20), "It was because Fälldin succeeded in persuading voters that the Center Party's energy policy was feasible that Sweden acquired a Liberal-Conservative government." This perception was enough to keep nuclear energy a closely watched and hotly debated issue

---

[12] See Holmberg et al. (1977) and Petersson (1978a, 1978b). I have also relied on an informative conversation with Olof Petersson, who mentioned Zetterberg's distinction.

after the elections, and its prominence was reinforced by the strong disagreements between the three new coalition partners.

The election left the Center Party with 86 seats in the Riksdag, the Liberals with 39, and the Conservatives with 55. The bourgeois bloc thus controlled 180 seats, compared to 169 for the Social Democrats (152) and Communists (17). The Conservatives and Liberals gained seats over 1973, while the Center Party lost four. The Center Party still dominated the coalition, however, so Fälldin was asked to form a government.

It was most unusual for any country to have an antinuclear prime minister, and this happened because of a series of chance events. First, the close balance between the Left and the Right shifted slightly to the right; second, the Center Party was the largest of the bourgeois parties; and third, the Center Party leader was antinuclear. Without one of these conditions, the nuclear issue would not have become so prominent. The bourgeois bloc won by fewer than one hundred thousand votes in 1976, and that was the last year the Center Party was the largest of the bourgeois parties. None of these outcomes could have been predicted by political structure, but none is truly accidental. Fälldin and his party were antinuclear because of their agrarian history and moral political styles; they grew because of these and helped the Right win the 1976 elections. But Swedish history could easily have unfolded quite differently.

Conclusions

In the two years following the oil crisis, Swedish and American nuclear policies were similar. The antinuclear movement was similar in the two countries: small but growing, using a variety of tactics, based on a coalition of groups, and showing a moral political style. There were even close ties between the two movements and a constant flow of information from the United States to Sweden. In both countries the oil crisis shocked politicians, who responded with strong words in the fall of 1973. But in the next two years top politicians in both countries decided to rely on slower demand growth rather than new construction as a response. This cost-benefit policy fit with broader political goals of both governments. In Sweden the Social Democrats wanted to delay a hard decision and let the threat from the Center Party fade; in the United States Republicans were reluctant to interfere in market processes.

Although in both countries nuclear policy was now being made through partisan politics rather than bureaucratic negotiation, nuclear policy became perhaps *the* central issue for partisan conflict in Sweden. Antinuclear sentiment had quickly captured the main bourgeois party, in large part because of the moral political style of its leader. Nuclear promoters tolerated a two-year pause in new orders because they saw no broader threat to nuclear deployment, much as American utilities were willing to abandon specific sites in the confidence of finding new ones. But as nuclear energy came to symbolize differences between the Social Democrats and the Center Party, controversy grew. That controversy helped to make an antinuclear moralist the prime minister of Sweden.

CHAPTER 9

Technological Enthusiasm at the Top in France

> The adversary is the great technico-economic apparatus, so powerful that it can impose even an energy policy which no scientific, technological, or economic argument could force one to recognize as being superior to any other policy.
> —Alain Touraine (1981, 21)

As IN SWEDEN and the United States, a growing antinuclear movement in France combined with the oil crisis to put nuclear energy at the top of the political agenda. In spite of the enthusiasm of Electricité de France (EDF) and the PEON Commission, France had a modest nuclear program before 1973. The antinuclear movement was also limited. France's reaction to the oil crisis was placid in the short run, but the long-term solution was to unleash EDF's nuclear ambitions. The standoff between the technological enthusiasts and the cost-benefiters was resolved in early 1974 when the prime minister and president took the side of the technologists. The resulting program of extensive nuclear construction, the Messmer Plan, triggered extensive public debate and organizing. Political parties had to take positions on nuclear energy, and the antinuclear movement attracted many scientists, technologists, and economists who had not previously been antinuclear. From 1974 to 1976 there grew a coalition of groups with widely different tactics and reasons for their protest. The segments of the coalition seemed tolerant of each other, and there seemed a good chance that the major opposition party, the Socialists, would adopt a more antinuclear stance.

The French government tolerated the antinuclear movement during this period because it felt the movement would dissolve as the public learned more about nuclear energy. But the government did not provide the needed information, indicating it felt the antinuclear movement was irrational as well as ignorant. The government did reduce its nuclear plans slightly in 1975, due more to continuing pressure from the Ministry of Finance than to the antinuclear movement, but it was soon too late to reconsider the massive Messmer Plan. Even the proliferation of energy models and comparisons and their increased sophistication were not allowed to challenge EDF's nuclear program. Politicians came to see the Messmer Plan as a symbol of the French state's ability to get its way, so that nuclear energy policy became a closed system precluding any public participation.

RISING OPPOSITION IN FRANCE

The French groups and individuals who opposed nuclear energy in the 1960s and early 1970s were similar to those active in the United States in the same period, and they were influenced by the latter. There was slightly less antinuclear activity in France than the United States before the oil crisis of 1973–1974, solely because the pace of reactor siting and construction was slower. Like Sweden, France intentionally imported American light water reactors and unintentionally got the visible antinuclear movement that accompanied it.[1]

In 1957 Jean Pignero, a high school teacher in the provinces, began extensive efforts to reduce the medical use of X rays. From the start Pignero and the radiologist he recruited to his cause had to rely on technical information from the United States, but they continued their efforts through the 1960s. Much of their activity, in fact, consisted of translating technical documents from the United States. Just as the lone voices of Sternglass, Gofman, and Tamplin later influenced AEC radiation standards, so Pignero affected French radiation standards—which were gradually tightened—although in each country the government agencies denied any direct effects of the critics.[2]

In spite of Pignero's efforts, general public concern with nuclear energy and radiation developed only with the environmental movement at the end of the 1960s. From 1969 to 1971 environmentalism burgeoned in France as in Sweden and the United States. In a battle over the Vanoise national park that began in 1969, traditional conservation organizations like the French Federation of Societies for the Protection of Nature (FFSPN) became more political and were joined by radical ecology groups

[1] Of the many works on the French antinuclear movement, some of the best include Fagnani and Moatti (1982, 1984a, 1984b), Fagnani and Nicolon (1979), Lucas (1979, chap. 7), Moatti et al. (1979), and Nelkin and Pollak (1981). Touraine et al. (1982) incisively expose the divisions within the movement through two study groups they organized. Representative works from the ecology movement that combine analysis and polemic are André Gorz [Michel Bosquet] (1977, 1980), Faivret et al. (1980), Ribes et al. (1978), Samuel (1978), and Vadrot (1978). Allan Michaud (1979) is an exhaustive study of the ecological "discourse," and Chaudron and Le Pape (1979) examine the emergence of the antinuclear movement out of the ecology movement in the early 1970s. Critiques of nuclear energy that emerged from the antinuclear movement or helped form it include Amis de la Terre (1978), GSIEN (1977, 1981), CFDT (1979, 1980), Colson (1977), Giry (1978), Laurent (1978), and Puiseux (1977). Anger (1977), Jund (1977), Moatti et al. (1981), and Nicolon (1979) describe local siting conflicts.

[2] See Pignero (1974) and especially Allan Michaud (1979, 104–52).

founded by young activists. Journalist Pierre Fournier began writing a regular column on ecology in 1969; a French branch of the Friends of the Earth (Les Amis de la Terre) was founded in 1970; and a group of prominent mathematicians started a magazine, *Survivre*, critical of government uses of science in 1970. Other organizations and journals soon followed, notably *La Gueule Ouverte* in 1972, *Le Sauvage* in 1973, and an ecology press agency (APRE) in 1973. By 1976 there were ten thousand environmental and nature groups in France with a hundred new ones forming each month (French Secrétariat Général du Haut Comité de l'Environnement 1978, 163).

French environmental activities were parallel to and influenced by the American situation. The FFSPN enlarged its activities but was also joined by more radical organizations, much as David Brower expanded the political activities of the Sierra Club but eventually left to form the Friends of the Earth. Many French ecology groups were founded largely by people who had spent time in the United States or Canada in the late 1960s. Les Amis de la Terre was founded by Alain Hervé, who had just returned from the United States, and E. Matthews, an American lawyer living in Paris (Chaudron and Le Pape 1979, 34). *Survivre*, soon renamed *Survivre et Vivre*, was started by A. Grothendieck, recently returned from Canada, and to a lesser extent by Pierre Samuel, who had spent 1969–1970 teaching at Harvard. All these men, as well as many others, carried the environmental debate back to France.[3]

Several segments of the ecology movement launched antinuclear activities in 1971. Fournier's column, then appearing in *Charlie-Hebdo*, began to cover the risks of nuclear energy, drawing mostly on American materials. In June 1971 *Survivre et Vivre* began to run articles on nuclear energy. Most importantly, the Parisian ecology movement began to develop links with local opponents of specific nuclear reactors. The result was demonstrations against two reactors, at Fessenheim and Bugey, both in the east of France. A local group had been formed to oppose Fessenheim in 1970, and fifteen hundred people marched to the reactor site in April 1971 (Amis de la Terre 1978, 301). The same month Fournier and others organized a similar group to oppose Bugey, calling themselves the "Bugey Guinea Pigs." Many members of the counterculture were attracted to the group's events, so that in July 1971 fifteen thousand people from around France attended an antinuclear festival near the construction site (Amis de la Terre 1978, 302). The ecology press as well as "the spirit of 68" aided the mobilization.

[3] My discussion of early environmental activities is based especially on Chaudron and Le Pape (1979), Allan Michaud (1979), and personal interviews with Pierre Samuel and Dominique Allan Michaud.

Inspired by the Bugey Guinea Pigs, citizens formed dozens of local antinuclear groups around France. Most of them were not even near proposed reactor sites, of which there were still few (Chaudron and Le Pape 1979, 49–50). These local activities did not arise out of fear of a nearby reactor, as similar activities did in the United States and as local opposition did in France after 1974. Before 1974 local antinuclear organizations in France resulted partly from close ties to Parisian ecology organizations and partly from countercultural values. Some ties to Paris were personal; several environmentalists (including Fournier and André Gorz) had country homes, sometimes near potential reactor sites. More importantly, many citizens in the provinces read Parisian magazines and newspapers, so that they were well aware of the growing environmental debates. Finally, a philosophy emphasizing decentralization, protection of the environment, and suspicion of the state and its technological projects held together the two components of the antinuclear movement.

The French ecology movement became radicalized—tending to reject all nuclear reactors on moral grounds—by 1971, earlier than movements in other countries. In December of that year the Fessenheim group organized a meeting of antinuclear activists from several European countries, at which the emerging ideological difference was obvious. *Ecologie* (No. 1, June–July) spoke of two types of antinuclear activism: the pre- and the post-Bugey. The former proposed legislation and worked through normal political channels; the latter felt its strong moral position allowed and demanded immediate action.

In 1972 and 1973 antinuclear organizations and activities expanded steadily, although not so dramatically as in 1971. Several traditional nature societies held debates on nuclear energy, and the number of ecology publications and special issues increased. At the end of 1972, twenty organizations began a campaign for a temporary moratorium on nuclear construction. In spite of growing media coverage and debate, there were few large demonstrations in 1972 and 1973, and a day of rallies planned for May 1973 drew a disappointing turnout. Nevertheless, as the Amis de la Terre (1978, 305) said about the 1973 events, "The ties, the networks of addresses and information that had been created in 1971 at Bugey and Strasbourg [Fessenheim], were reinforced and extended through the common action; no formal structure emerged, but the bonds created were solid enough to permit, the following year, the rapid, effective launch of the Dumont campaign."

THE CFDT

At the same time that environmental groups were beginning to criticize nuclear energy, activists in the Confédération Française Démocratique du Travail (CFDT) were also expressing some concern. Because these doubts developed into an official position of opposition to France's nuclear plans in 1974, and because the CFDT was possibly the only major trade union organization in the world to take such a stand, it is important to trace the roots of this concern. Three factors stand out: an interest in working conditions in the nuclear industry, the highly ideological and moralistic political style of CFDT activists, and the union's structure.[4]

The CFDT's decentralized structure and ideology meant that the union and its leaders followed the positions on particular issues developed by union members who were experts on those issues. Thus the CFDT members who were active in the nuclear industry—at EDF, the Commissariat à l'Energie Atomique (CEA), and La Hague—were the ones to articulate a stance toward nuclear energy for the whole union. What is more, the activists around Paris, especially those working for the CEA and certain parts of EDF (its research division) could coordinate their activities most easily and would later write the union's books on nuclear energy. A rather small number of activists and union officials could prescribe an antinuclear position for the CFDT, even though the membership probably had mixed attitudes toward nuclear energy. There is a parallel with the structure of *Dagens Nyheter* in Sweden, which became antinuclear largely because its editorial writer on energy did. The structural account needs to be filled in with an explanation of why key activists in the CFDT became antinuclear.

Throughout the 1970s the CFDT was the main trade union of workers at La Hague, France's central factory complex for reprocessing nuclear fuel. As a result of its work there, the union became accustomed to criticizing the nuclear industry and to attacking the industry's denials of all risks. The CFDT claimed that when the CEA shifted focus from research to commercial operations around 1970, it began to cut corners and let quality slide at La Hague at the expense of the health and safety of workers and the public. It is true that radioactive discharges increased in the early 1970s, and most visitors to La Hague left with the impression of a badly managed enterprise.[5] Its experience there taught the CFDT a skeptical attitude toward the nuclear industry.

Yet it was not La Hague activists, who had a clear organizational interest in attracting members by attacking nuclear industry management,

[4] My understanding of the CFDT comes from personal interviews with Michel Labrousse, Roland Lagarde, Philippe Roqueplo, Jean Tassart, and above all, Jean-Claude Zerbib, as well as Nicolon and Carrieu (1979), the CFDT (1982), Hamon and Rotman (1982), Schain (1985), and several of the chapters in Kesselman (1984). See also the occasional remarks in Mouriaux (1983) and Reynaud (1975).

[5] Gene I. Rochlin, personal conversation; Carter (1987, 316–22); and Lucas (1979, 80–85), who describes particular charges and provides the figures on radioactive discharges.

who formed the core of the union's opposition to nuclear energy.[6] This came from Parisian activists, employed by EDF and the CEA, who had organizational interests in *promoting* France's nuclear program. These CFDT members combined the skepticism developed through the La Hague experience with the moralistic political style typical of most CFDT activists. Most of the union's energy activists had been attracted to the CFDT either through Catholic activities or through political activism in the late 1960s. On the one hand these people had been trained as engineers and often favored technological solutions to social problems; on the other hand their political beliefs were formed in contexts that led them to view policies morally in terms of right and wrong. Even though nuclear development was in their professional interest, the form it was taking seemed immoral to their political selves. For these men ideology and political style won out over occupational training and self-interest.

The moralist policy style of the CFDT (in sharp contrast to the economism of Force Ouvrière and the communist ideology of the CGT) comes from its roots in the Catholic workers' movement and the student movement of 1968. Both tendencies criticized consumer society and pro-growth values, and both asked about the ultimate ends of human society. As a result the union has attracted many intellectuals as well as technicians and engineers. According to Roland Lagarde, "The CFDT asks the question, production for what? Not only about working conditions, but what are the effects on society as a whole? There is a long union tradition of saying, 'Wait a minute; this product is no good.' Part of it is pride in our work."[7] The CFDT combines moral questions with the artisanal impulse that also made some American engineers antinuclear. But in France these dissenters had an organizational apparatus for expressing their criticisms, so they could move beyond brief accusations to elaborate written critiques.

The CFDT's decentralized structure, combined with activists' moral political styles and skepticism toward the nuclear industry, encouraged the union to question the nuclear program beginning with the adoption of the LWR in 1969. An additional layer of skepticism existed for the CFDT members who were employed by the CEA, since their gas-graphite design was being abandoned. France's lack of knowledge about the new technology seemed to them dangerous, and CFDT members from EDF often agreed. But given time for development and familiarization, these problems could be solved. When the 1974 Messmer Plan ignored the need for this development period, however, the CFDT increased its opposition.

[6] Remember that membership in French unions is always voluntary, and many workplaces have lively competition between several unions.
[7] Personal interview, May 1985.

In summary, when the oil crisis struck in late 1973, French opposition to nuclear energy consisted of a small but growing network of environmental activists and the CFDT. Activities were centered in Paris and were countercultural and intellectual at the same time. As a result, the press debates were lively. Nuclear energy was one issue that concerned the environmentalists, but it was by no means the main one. In fact, there may have been some uncertainty about what issues and what tactics would best further the movement's goals. The level of environmentalist activity was equivalent to that in Sweden and the United States, and the nuclear component of that organizing was probably only slightly less. Like the ecologists in the other countries, those in France were motivated more by morality than by material interests. They believed nature was at risk and required urgent action to save it. But the state did little to heed their warnings.

The French Response to the Oil Crisis

In the first months of the oil crisis, the French government expressed little outward concern, but in March 1974 it announced an enormous acceleration in the pace of nuclear construction. Nuclear energy was to be the only element in France's response to the oil crisis, and the Messmer Plan outlined the largest construction program the French economy and heavy industry could handle. To increase industry's productive capacity, only one reactor design, the Westinghouse pressurized water reactor, was to be used, even though France had no operating experience with this design. The technological enthusiasts had finally persuaded top political leaders to ignore the cost-benefit concerns of the Ministry of Finance. France's "nuclear wager" was on.

French politicians seemed casual in their immediate reactions to the oil crisis. In the United States, President Nixon felt it was important to respond quickly and forcefully—Project Independence was announced within weeks of the embargo—but in France political leaders were hardly willing to condemn OPEC's price increases. One reason for the difference was the contrasting relationships the two countries maintained with the Arab world. France was in the most-favored category of oil importers, while the United States was the main target of the Arab actions, the friend of Israel that had inspired the embargo. But even more the difference lies in the *perceptions* in the two countries of their relations to the Arab world. The economic impact of the oil embargo was comparable in the two countries. The percentage cuts in oil supply and the price hikes were identical; the terms of trade were hurt by an equal percentage. But the United States defined itself as the enemy of the Arabs and of OPEC, while France continued to view itself as their friend.

France has continued to play an important part in the economic and political life of its former colonies long after their independence. This continued interest is partly due to French economic interests, like the uranium of Niger and Gabon, and partly to the opportunity to oppose American foreign policy under the banner of anti-imperialism. French attitudes toward the Third World have been colored especially by France's relationship to Algeria. Because of the paternalism at the root of this relationship, France has never fully forgiven Algeria's rejection of its governance, but it has maintained special ties to Algeria despite the revolution. The special relationship between France and Algeria often serves as a model for France's relations with all of North Africa, not only because increasing numbers of North Africans have immigrated to France, but also because the French lump these immigrants together under the rubric *Arabs*.

France's image of itself, as the protector of Arab interests against American imperialism and support for Israel, helped shape the French reaction to the oil crisis. France could attack American policy, emphasize its own friendship with the Third World, and cultivate its own oil supplies at the same time. Prime Minister Pierre Messmer stated: "Economic decolonization is heading for its inevitable success—the French government has said so for more than ten years. The oil companies can no longer solve the problem of supply and price. We must find an alternative organization."[8] The French pushed for a European response independent of American policies, going as far as arranging contacts between Arab spokespersons and European policymakers (Mendershausen 1976, 72). France favored direct negotiation with the oil producers, resisting the combative American strategy of setting up a large organization of oil importing nations and refusing to join the International Energy Agency (IEA) created by the other OECD countries.[9]

France's "friendship" with the Arab world affected the impact of the oil crisis on France in two ways. First, France was in a favored position in the short run. It had fewer restrictions on oil consumption than most other European countries. These included a speed limit, mild gasoline rationing, and restrictions on heating and lighting (Prodi and Clô 1976, 101). But French officials could not force oil companies, even the nationalized French ones, to deliver full oil supplies during the crisis, with the effect that France's oil consumption was 5 percent lower than the year

---

[8] Messmer is quoted in Mendershausen (1976, 71–72). This is an excellent source on French reactions to the oil crisis, and my discussion relies on it throughout.

[9] Mendershausen (1976, 59ff.) describes the French and American responses to OPEC as polar opposites, with the former accepting the new power of OPEC and wishing to deal with it, and the latter trying to break it. American desire to take the lead in cracking OPEC also played a part in France's opposition to this strategy.

before the crisis.[10] Its supply situation was only slightly better than those of Sweden and the United States. More importantly, the rise in oil prices was identical for the three countries. Given these small economic differences, the second factor appears even more important: French pride in the fact that it was officially favored. Taken as proof of the wisdom of French policy toward Arab countries, this favored status encouraged the French "to take a nonchalant attitude toward the immediate energy supply problems of their own economy" (Mendershausen 1976, 71). In the fall of 1973 the French government simply directed the oil companies to continue full supplies, not expecting their request to be ignored. When in November the cuts came anyway, the government still denied any problem: "Until the late winter of 1973–74, the French political leadership maintained that the country's great stocks and undiminished supplies from its Arab friends made driving restrictions and similar drastic measures unnecessary. Only in midwinter was the population admonished to save energy and to take some marginal economy measures" (p. 79).

The surface nonchalance of the government during the oil crisis contrasts sharply with the enormous size of the nuclear energy program approved by the Council of Ministers on 5 March 1974. The Messmer Plan called for EDF to begin construction of thirteen new nuclear reactors in 1974 and 1975, representing a huge shift in energy policy and a full commitment to nuclear energy. This would represent almost 13,000 MWe, twice the capacity that EDF then had in operation or under construction. By the year 2000, nuclear energy would account for 50 percent of France's total energy needs—even more than the 50 percent of electrical needs envisioned for that year by Nixon's Project Independence. The government presented the Messmer Plan as France's answer to the oil crisis, arguing that nuclear energy represented the only path to energy independence for a country that lacked fossil fuel resources. How could this massive program have followed months of minimizing or even denying the extent of the energy problem?

Such a large nuclear program could be launched so rapidly in France because the planning, enthusiasm, and technical infrastructure were already in place, just waiting to be put in motion. The PEON Commission, whose intense nuclear enthusiasm was described earlier, had recommended an equally large nuclear effort (13,000 MWe) even before the oil crisis, in its report of April 1973. The difference is that the earlier version spread the construction over five years rather than two (Simonnot 1978,

---

[10] Stobaugh (1976, 190) describes the impotence of the French government when it ordered oil companies to favor French deliveries. Feigenbaum (1985) details many futile attempts by the French state to coerce the nationalized oil companies into doing what it wanted.

269). In the winter of 1973–1974 the PEON Commission was again asked to make recommendations concerning the growth of nuclear energy. It quickly developed the 13,000 MWe figure, as opposed to 8,000 MWe for the same two years in the earlier plan. The 13,000 MWe plan was chosen because it seemed to be the maximum effort French industry could make toward nuclear energy. The government adopted the new recommendations in their entirety, after only three days of consideration—at least according to a popular rumor.[11]

The French state did not have to "develop" an energy program at all; EDF, Framatome, and the CEA already had one they were anxious to accelerate, and they had been constrained only by the resistance of the Ministry of Finance. In early 1974 this ministry's doubts were overridden at the highest levels—by Prime Minister Messmer more than by President Pompidou, who was already quite ill. In addition to an intuitive political sense, much like President Nixon's, Pierre Messmer had a personal incentive for allowing such a bold but risky move. It was already clear from Pompidou's illness that there might be a presidential election soon, and Messmer was a potential candidate for the conservatives.

In the United States, Project Independence could call for increased reliance on nuclear energy, but there were few organizations ready to push a large program. Reactors were already being canceled because of the financial difficulties of utilities. In France the utility, the manufacturer, and the regulator were straining to launch a sizable program. Potential financial difficulties for EDF were not clear, although the Ministry of Finance was pointing them out. Messmer was key, and he leaned toward the side pushing a dramatic move. This is not a case of a strong French state taking a decision and then setting up the apparatus to accomplish it. The French president and prime minister have the power to close off certain paths, to choose which will be pursued, but they have less power to create new ones. They often act merely as the switchmen for the locomotives of others. As Lucas (1979, 163) summarizes, "So, great as the power of the President is, it is the power to select which of the available competing institutional forces will be released, and is therefore the power to choose among the possible states attainable by the release of those *institutional* forces; it is not the power exactly to fashion the forces necessary to achieve the preferred state from among all those which might be *technically* attainable." The Finance Ministry's continued resistance to the nuclear program in future PEON Commissions demonstrates that it had been overridden rather than persuaded.

In France, Sweden, and the United States, the cost-benefit worldview

[11] Louis Puiseux (1981, 313) says this, adding that this is just a rumor. The degree of irony in this qualifier is not clear!

cautioned restraint in the nuclear programs, while technical enthusiasts saw nuclear power as the key to energy independence. Nuclear power had the allure of a dramatic and modern technology, much like the Apollo missions or the Concorde. Messmer jumped at it, while Ford retreated. First, Messmer faced a possible election in two months, and Ford in two years. Second, the moves that would be needed to support nuclear energy in the two countries were quite different. In France it was an elegant announcement releasing certain institutional energies. In the United States it would have been a complex scheme involving utility financing and rate relief for something called CWIP (construction work in progress), hardly the stuff of myth. But in the end, we find choices made by top political leaders on the basis of their own preferences and policy styles. They had to choose which argument to believe.

The triumph of technological enthusiasm at the top of the French state also reflects the resurgence of the idea that national interests lay in independence based on political suasion, maneuvering, and power instead of in competitiveness based on efficient reactions to international markets. The latter image had seemed to triumph with Pompidou's decisions against the gas-graphite reactor and for the more marketable LWR. As Pompidou's health failed and Messmer's discretion expanded, the more traditional image of France as a nation reemerged. This shift should not surprise us when we recall that Pompidou was a banker and Messmer a career military man and former colonial officer.

Personal decisions by top political leaders must still be placed in the context of the entire political system. In France a technological solution and rallying cry of energy independence were popular with many policymakers, partly because of the high placement of engineers in the French policymaking elite and partly because of the technological enthusiasm that characterizes them, an enthusiasm that spills over to other members of the elite. The rhetoric of technological enthusiasm falls on fewer receptive ears among American and Swedish policymakers. Costs and prices are a more familiar language to the Americans; political processes, negotiations, and consensus to the Swedes.

As in the United States, French electricity producers, indeed the entire nuclear industry, were reluctant to accept the new style of economic analysis that questioned old assumptions about demand growth and posed conservation as an alternative to nuclear energy. The PEON Commission had a unique solution to the conflict between these worldviews, arguing in its February 1974 report that if electricity were allowed to replace other energy sources, its growth could continue unabated. In criticizing this "backward logic," Simonnot (1978, 272) says: "In other words to respect the 'law' that electricity demand doubles every ten years, it would be necessary for electricity use to penetrate the energy system enough to compensate for increased energy efficiencies. What counts is that the law be met. That is fetishism."

Just as the Messmer Plan appealed to technological enthusiasm throughout the French political system, Messmer's emphasis on energy independence for France fits the goals of many French officials. Once a large nuclear construction program was begun, the best way to defend it was to make it seem inevitable, and the best way to do that was to portray it as the only way to achieve energy independence. Before the oil crisis the main argument EDF used for nuclear energy had been its low cost. After the oil crisis, many independent observers began to scrutinize energy costs and to question EDF's evidence, so that energy independence became a safer rationale.[12]

Was the Messmer plan simply the most rational response to the oil crisis, given France's lack of fossil fuels? Only in part. The oil crisis threatened energy supplies only in the first months, when the government's response was casual. More importantly, the subsequent rise in oil prices threatened trade balances, and several responses were possible. France could have (1) paid the higher prices and hoped they would eventually fall again (as they did) and increased its exports to compensate; (2) developed renewable energy resources; (3) discouraged consumption, especially through conservation; or (4) deployed nuclear reactors. An option open to the United States but not France or Sweden was to exploit domestic fuel supplies. Option 1 appeared too weak a response for a government about to face elections, and option 2 seemed to have limited potential in the short run. Options 3 and 4 pitted technologist and cost-benefit policy styles against each other. Technological enthusiasts continued to see the challenge as one of supply, while the cost-benefiters saw tradeoffs behind balance of trade problems. The French government's obsessive concentration on a single option was hardly inevitable and certainly risky. It was a victory for the technological enthusiasts.

In France, Sweden, and the United States the same arguments were

---

[12] In this period the French public did not share the political elite's concern for an independent French policy or for independent energy resources. In November 1973 a poll found that 65 percent of the respondents favored European solidarity and only 25 percent favored good relations with the Arab countries. The French government chose the latter course. In May 1975 a survey of French people who had viewed a television program about nuclear energy found that only 3 percent mentioned energy independence as a striking argument for nuclear energy—seventh on a list of nine. It took French policymakers years to sell the public on the independence rationale, and they succeeded only when the political conflict over nuclear energy died down after 1981. Fourgous et al. (1980, 77, 95–96) present the results of the two surveys. Réal (1975) was an early argument for nuclear energy on the grounds of national independence, while Durrieu (1977) claimed it would actually hurt energy independence (arguing instead for diversification). For a full elaboration of the view that won in the 1980s, linking nuclear energy and independence, see Dorget (1984).

used for and against a large nuclear commitment in the months following the oil crisis. On the one hand technological enthusiasts argued that nuclear reactors were needed to assure an energy supply independent of the Middle East; on the other cost-benefiters argued that higher prices would lower consumption and that the economics of nuclear reactors were still uncertain. In France the technological enthusiasts triumphed completely. The Messmer Plan, calling for dozens of nuclear reactors and sprung on the public with no previous discussion, could not help but cause controversy. It was a fat target for both opposition political parties and the growing antinuclear movement, which was well placed to profit from the public debate.

REACTIONS TO THE MESSMER PLAN

The oil crisis and the Messmer Plan made nuclear energy a visible public issue in France through three channels: opposition political parties; the existing ecology movement; and scientific, technical, and economic experts who questioned aspects of the Messmer program. Partisan debates twisted the energy decision into a question of how much big industry would benefit. The ecology movement added political and economic issues to the safety and environmental lines along which it had defined nuclear energy. Technical experts were drawn into the debate because widespread reactor construction would begin before there was any operating experience with the new reactor design, and economists were drawn into it because they felt the French economy could ill support such a large program.

All the nongoverning parties as well as a few from the governing coalition found reasons for criticizing the Messmer Plan. The nuclear program was a perfect channel for attacking the conservatives who had governed throughout the Fifth Republic. The Messmer Plan was developed with no participation by the public or opposition parties, typifying the high-handed methods of the ruling coalition. It promised to benefit heavy industrial groups that Pompidou had already been accused of being too close to. It also involved an American reactor design, reinforcing a feeling that Pompidou was not pursuing France's independent interests as doggedly as de Gaulle had. What is more, de Gaulle himself had vetoed this light water design in 1968 as undermining French technological independence. As a result, the Left often interpreted the Messmer Plan as a symbol of Pompidou's subservience to international capital.

Because of the symbolism of the Messmer Plan, partisan debate did not form over the benefits of nuclear power as an energy source, but over the lack of public and parliamentary discussion of the new program and over the immense benefits that would accrue to private manufacturers of reactors and turbines.[13] Compare partisan debates in the United States, which tended to revolve around free market forces versus government intervention, and in Sweden, which implicitly involved decentralization and explicitly the simple number of reactors.

Neither of the two main opposition parties, the Socialists (PS) and the Communists (PCF), took a generally antinuclear position, but both were hostile to the government's specific program. Two smaller parties of the left, the Unified Socialist Party (PSU) and the Revolutionary Communist League, took a more active role in opposing nuclear energy, but even they were slow to adopt the call for a moratorium. Most of the political parties were cautious in 1974, but they discussed nuclear energy heavily in 1975 in preparation for the parliamentary debate scheduled for May. That debate provided the main evidence for partisan positions in the two years following the oil crisis.

The Socialists were divided on the general question of nuclear energy, so they concentrated happily on specific aspects of the Messmer program.[14] During two days of meetings in late 1974, party leaders and activists could reach no compromise between those who favored nuclear energy as a means to economic growth and those who opposed it. Nicolon (1979, 111) believed that the officials higher in the party hierarchy were more likely to favor nuclear energy, while grassroots activists tended to oppose it; the former were generally "old left" socialists, and the latter the "new left" from 1968.[15] The two wings of the party could agree that big industry would gain too much profit from the Messmer Plan, so this became a major theme in their criticism. In the early months of 1975 the party hierarchy resisted the idea of a complete moratorium, but it attacked (1) the size and suddenness of the Messmer Plan, as well as its concentration on one reactor design; (2) the municipal contracts with which EDF could buy off municipal governments through tax aid and public construction projects, and (3) the lack of public discussion, including the assembly debate in May, which was clearly not designed to change government policy. The Socialists proposed the nationalization of the large industrial groups involved in nuclear energy as well as a system of safety regulation that would be completely independent from the nuclear

[13] Nicolon and Carrieu (1979) present the best discussion of partisan positions on nuclear energy, but also see the *Revue Générale Nucléaire*, April–May–June 1975, pp. 131–34 for excerpts from the 15 May assembly debate.
[14] Highlights of the controversy within the Socialist Party can be found in Touraine et al. (1982), Nicolon (1979), Hernu (1976), Sorin (1979, 1981a), Bell and Criddle (1984, 38–39), and Fagnani and Moatti (1984b).
[15] This split in attitudes represents what Webster (1986, 386) calls "a recurrent dilemma for the Left: how can it reconcile an attitude towards technology which regards it as inherently progressive with the fact that it is used and is being used as a weapon against labor?"

industry and its profit motives. In summarizing the May debate, deputy Jean Poperen made several points (*Revue Générale Nucléaire* 1975, 132): "We are going very fast and very far along the road to integrating large sectors of French industry with multinational companies under American direction. . . . One of the conditions for a true nuclear policy is the nationalization of the great corporations like Pechiney-Ugine-Kuhlmann, Creusot-Loire, CGE, and their subsidiaries."

While the PS threw words at the Messmer Plan, the small PSU took several actions, helping to organize various antinuclear activities during 1975. By October it was prepared to call for a moratorium on nuclear construction. The PSU saw nuclear energy as a symbol of the wrong kind of growth, growth in profits and technocracy and a police state and not growth in democracy and the standard of living of workers. The PSU was just large enough to be crucial for a leftist electoral victory: in 1969, its presidential candidate Michel Rocard had received 3.6 percent of the first round vote. As the PS gradually absorbed the PSU in the late 1970s, the latter's positions on nuclear energy should have influenced the former's.

Ecologists trying to draw the socialists into a coalition with antinuclear groups pointed to the lack of democratic process in energy policy and the industrial domination of the nuclear program, two criticisms widely voiced by the left. In *Le Nucléaire sans les Français* (1977) Jean-Philippe Colson elaborates this argument, portraying the Messmer Plan as a capitalist boondoggle imposed on a recalcitrant people. His subtitle raises both lines of criticism: "Who Decides? Who Profits?"

The PCF never expressed any hesitation in its support for nuclear energy in general, but it played the anticapitalist themes as strongly as the PS in attacking the government program. The ideology of the party was strongly oriented toward economic growth, and nuclear energy was seen as an important part of the infrastructure of any future socialist society. The PCF's opposition to the government program was vocal, however, and rested entirely on the claim that national interests were being subordinated to the profits of large, private—usually multinational—firms. Both the choice of the Westinghouse LWR and the rapid pace of reactor deployment served no purpose other than to increase industrial profits. The result was not only an increased risk of accidents but also an international nuclear cartel that would have France at its mercy. Nationalization of the nuclear industry and a slower pace of reactor deployment were key elements of the PCF's solution. To the PCF, nuclear energy was a rich symbol of immense technological progress (means of production) hindered and misled by private ownership and profit-seeking (relations of production).

So partisan debates unfolded largely as a question of who would own the nuclear industry and who would profit from it. Whether nuclear energy was a useful energy source and in what proportion it should be mixed with other sources were not as important. As in Sweden and the United States, the pre-existing ideological cleavages between French political parties changed the nature of the nuclear question. The strong Left-Right axis along which French parties position themselves assured that nuclear energy would be seen at least in part as a capital-labor conflict. The governing parties reinforced this cleavage by continuing the program of privatizing the nuclear industry that they had begun with the reorganization of the CEA in 1970. In addition, we shall see below that nuclear energy was important to President Giscard d'Estaing in part because it furthered and symbolized the integration of French industry into the world economy—precisely what many leftist critics felt would cause the *subordination* of French industry to multinationals. For both sides, nuclear was more than simply a good or bad energy source.

Like the political parties, antinuclear groups responded immediately to the Messmer Plan. Before the oil crisis, nuclear energy had been one of several key issues for the French ecology movement, but the Messmer Plan soon became the central focus. Groups like the Amis de la Terre and journals like *La Gueule Ouverte* were naturally poised to organize and publicize antinuclear activities in 1974 and 1975, but nuclear policy became such a rallying point in French politics that they were joined by a variety of other groups. Economic, technological, and scientific opponents coordinated their actions with the Amis, forming a moderate wing of the antinuclear movement. Beside this wing was a more radical segment that had little use for experts or electoral tactics. Local opponents maintained close links with both wings in Paris, so the French antinuclear movement was much more centralized than the American movement. Finally, the antinuclear movement began to attract "revolutionaries" of the left, who were concentrated in Paris but had few ties to other segments of the movement. During the two years following the announcement of the Messmer Plan, the movement did not show much conflict between these varied elements.

The ecologists continued their demonstrations and their call for a moratorium. In March 1974 the Amis held a bicycle rally in Paris attended by ten thousand; in the summer they organized an "information caravan" along the northern coast. Several site demonstrations were also held in July, August, and September (Amis de la Terre 1978, 308), but attendance was unexceptional. When, in May, the state television authority canceled a scheduled film about nuclear energy, activists began a letter-writing campaign that culminated in a November sit-in and march. This incident—the canceling of a rather neutral film that gave time to both anti- and pronuclear speakers—attracted the attention of activists and members of the public interested in civil liberties and democratic processes.

The ecology movement also entered electoral politics. René Dumont ran for president in the May elections, although he received only 1.3 percent of the vote. Yet in the four-week campaign between Pompidou's death on 2 April and the first-round election on 5 May, the Parisian ecology organizations strengthened their personal and organizational ties. In addition, many members of the public heard environmental ideas for the first time through media coverage of Dumont. In June a group of activists tried to institutionalize the organization Mouvement Ecologique, which had formed behind Dumont, but they faced strong opposition from others who believed formal structures damaged the spirit of the ecology movement. The only agreement was on the demonstrations to take place that summer.[16]

Now that nuclear energy was a public issue throughout France, local opposition groups began to form without the encouragement of national ecology organizations. Almost every site that was chosen in 1974 or later met with local opposition. There were several strands to this opposition: farmers protecting their land and way of life; urban dwellers with second homes; and the semi-urban professionals whose counterparts had organized much of the local opposition in the United States. In Alsace and Brittany there were also strong regionalist sentiments that saw nuclear energy as one more imposition by the Paris elite on the provinces (reminiscent of some of Fälldin's supporters in Sweden). Depending on their distance from cities like Paris and Grenoble, specific demonstrations were often swelled by urban sympathizers who came for the day.

Although public participation in antinuclear demonstrations remained limited during 1974, public debate grew rapidly. In addition to the ecology press, mainstream magazines and newspapers, *L'Express* and *Le Monde* among them, began to cover nuclear issues in detail. The CFDT publicly questioned the size of the Messmer Plan and held meetings in factories across the country. In November a committee of the National Assembly itself criticized the government for its lack of public information, although—or perhaps because—its report was confident that the public would accept nuclear energy if it understood it.[17] During the following week (15–17 November) an international meeting of the environmental group Nature and Progress was held in Paris, part of which was devoted to nuclear energy. Among others, Arthur Tamplin (from the low-dose controversy in the United States) spoke, and he remained in France

[16] On the 1974 presidential elections, see *Le Monde* (1974) and Penniman (1975). Amis de la Terre (1978) describe the events of 1974.

[17] "Rapport sur la Situation de l'Energie en France," Annex to the verbal proceedings of the National Assembly meeting of 5 November 1974, no. 1275.

after the meetings to address rallies at several local sites. By December the government felt it must respond to the growing public debate in order to allay fears.

In response to the rising antinuclear movement, Minister of Industry Michel Ornano announced on 2 December 1974 that he would make public details of the French nuclear energy plans, and that the regional assemblies would be able to debate specific locations for reactors. The documents that the government released contained a list of thirty-five sites under consideration for reactors. The intent of the Ornano Plan was to curtail criticism of the government's secrecy and to use the regional assemblies to identify sites where there would be unusual opposition (Bupp and Derian 1978, 110). The heavy nuclear commitment was not to be reconsidered. Like Nixon's shakeup of the AEC in 1971 and its replacement by the NRC in 1974, the Ornano Plan was meant to expedite reactor construction by making it more acceptable to the public. As in the American case, it had the opposite effect.

The local and the national segments of the French antinuclear movement, never so distinct as in the United States, were drawn together tightly by the Ornano announcement. Several important reports and compilations of information, spurred by the discussions throughout 1974, were published in early 1975. At the same time, the specification of potential reactor sites triggered extensive local activity, which naturally drew on the national organizations as sources of information. The timing of the development of both parts of the movement could not have been better. What is more, outrage over the lack of public participation and information added the question of political process to the nuclear debate in the press, previously focused on environmental issues.

Throughout 1974 scientific researchers in and around Paris, especially at the College de France and the University of Paris at Orsay, had discussed the Messmer Plan, and a CNRS commission had publicly criticized the state's secrecy in the matter.[18] The same informal network gathered signatures on a petition pointing out the number of unsolved challenges of nuclear energy and the enormous size of the proposed commitment. Originally known as the "Appel des 400" when it was published in February 1975, it was later signed by roughly four thousand scientists. In March some of the same scientists—a group from Orsay—published a more detailed report that sold forty thousand copies.[19]

---

[18] The Centre National de la Recherche Scientifique (CNRS) is the national organization that funds virtually all research in the social and natural sciences. My discussion of the scientific opposition to the Messmer Plan is based on conversations with Jean-Paul Schapira and Monique Sené. The discussion of the CFDT's role is based on interviews with Jean Tassart, Michel Labrousse, and especially Roland Lagarde and Jean-Claude Zerbib. See also GSIEN (1977) and the CFDT (1980), both of which were originally published in 1975.

[19] Later published by GSIEN (1977).

The "Appel des 400" was widely discussed in scientific circles as well as the press, and it spurred scientists to organize further. In December 1975 the Groupement de Scientifiques pour l'Information sur l'Energie Nuclaire (GSIEN) was formed, with about three hundred active members. Other groups were established, especially in the Rhone-Alps and in Geneva, and thousands of scientists signed various petitions opposing aspects of the French nuclear program. American models were important. The Amis de la Terre helped to organize GSIEN and continued to work closely with it, and GSIEN was partly inspired by the Union of Concerned Scientists. In 1975 GSIEN's promise was as great as that of the Union of Concerned Scientists: the importance of Paris in French politics, science, and media gave GSIEN easy access to ecology organizations, to the press, and—if asked—to policymakers; likewise the initial membership of three hundred scientists compared favorably to American antinuclear participation, especially as a percentage of all researchers.

The scientists of GSIEN cooperated closely with the activists of the CFDT described above. Although the CFDT had already questioned certain aspects of France's nuclear program, it increased its activities after the Messmer Plan was announced. Whereas the CFDT's previous opposition to the LWR had been inspired especially by CEA employees, who had an organizational interest in opposing the LWR, opposition to the Messmer Plan was supported by CFDT members throughout the nuclear industry. Technical details were now less important than economic and political ones. The size of the nuclear commitment, at a time when France's first reactor of the new model (Fessenheim) was still three years from operation, seemed too risky to the CFDT activists. The costs were uncertain, and any safety flaws that operating experience might reveal could vastly increase them. During 1974 the informal Paris network of CFDT activists had produced a series of reports on nuclear energy that stimulated public meetings and press conferences, and these became available in 1975 by publication in book form. In May the CFDT proposed a three-year moratorium to allow further research and reflection.

The CFDT passed much technical information to GSIEN, and many individuals belonged to both organizations. Along with the Amis de la Terre, the groups formed an important moderate wing of the antinuclear movement. Technically informed, willing to compromise, and not supporting the moralist line that nuclear energy is inherently wrong, these groups could have played an important mediating role in the nuclear debate.

Other moderates who could have mediated the conflict were the economists who questioned the cost assumptions behind the Messmer Plan. The same month the "Appel des 400" appeared, the Institut Economique et Juridique de l'Energie (IEJE) of Grenoble published a report strongly questioning the plan's economic basis.[20] The institute, widely respected by energy experts, had no particular bias for or against nuclear energy, but its careful appraisal of the economic justifications of the nuclear program found them groundless.[21] Some of its criticisms applied to nuclear reactors in general as an energy source, but most dealt with the Messmer Plan in particular. The IEJE first pointed out that official demand forecasts were inflated, since demand would not continue to grow as rapidly as in the past. Second, there was inadequate evidence to show that nuclear energy was cheaper than alternatives, especially for home heating. Many costs had not been included in the analyses, including those of insurance and of dismantling used plants. Likewise, there was no evidence for the assumption that the more reactors France built, the lower would be the construction costs per unit. Finally, there were social costs that cost-benefit analyses had difficulty in incorporating—for example, types of pollution. These were the same questions that economic skeptics in Sweden and the United States were raising at this time.

The IEJE was not saying there were clearly superior alternatives to nuclear energy, but rather that the evidence was inconclusive. The nuclear establishment had been wrong in saying that nuclear energy was justified by economic calculations. The IEJE analysis implied that nuclear power should be one energy source among many, not that it should be abolished. The IEJE also argued that, given the cost uncertainties and the grave risks, a large commitment to nuclear was imprudent. Hence the institute's especially strong criticism of the Messmer program. If demand growth and economies of scale were doubtful, then the cost arguments for an enormous investment in nuclear were gone. And if there was no rigorous way to decide if the risks of nuclear energy were worth taking, this question should be left to the public to decide, so the IEJE sharply criticized the government for its lack of public consultation. The institute finally criticized the Messmer Plan for pinning France's energy future on only one energy source and only one technological form of that source. Any problem with that technology—an American import little studied in France—could destroy France's main energy source. The IEJE summarizes the lack of economic justification for the Messmer Plan (IEJE 1975, 34): "Thus the classic instruments of cost-benefit analysis appear completely unsuitable to making energy choices explicit. At the present time it is impossible for cost accounting to integrate in a serious way all the

---

[20] IEJE, *Alternatives au Nucléaire: Reflexions sur les Choix Energétiques de la France.* Grenoble: Presses Universitaires de Grenoble, February 1975.

[21] EDF immediately labeled the IEJE "antinuclear," in a process of marginalization described further in chapter 13. But the fact that the IEJE refrained from further participation in the politics of nuclear energy implies a neutral stance, as does the cogency of its economic arguments.

social costs, certain or not, associated with nuclear development. It is impossible, except *a priori*, to justify to public opinion the decision for an intensive program of nuclear reactor construction. . . ."

Just as the safety critiques of GSIEN and the CFDT contributed to an ongoing debate in the press, so the IEJE report encouraged public debate over the economics of nuclear energy. Many economists recoiled from the Messmer Plan, believing it was "too much, too fast." Not only were the cost advantages unproven, but economists feared the effects on France's economy of drawing huge investment resources into one sector and of accumulating heavy foreign debts. Economists saw the oil crisis as a problem of trade balances, in contrast to the technological enthusiasts who saw it as one of domestic energy supplies. One example from the public discussion is an article in *Le Monde* (27–28 May 1975) by Claude Henry, director of the econometrics lab at the prestigious Ecole Polytechnique. He discussed the debate between the IEJE and EDF, and concluded that EDF had still not provided enough evidence to justify its large nuclear program.

Louis Puiseux provides another example of an antinuclear shift in thinking by economists. A close associate of EDF General Director Marcel Boiteux, Puiseux was in charge of economic forecasting for EDF as it developed its nuclear plans in the early 1970s. But Puiseux listened to the debates that unfolded after the oil crisis, and concluded that the critics were right on many counts. In particular, the usual cost benefit analyses overlooked many social costs of nuclear energy that made it look like a dubious endeavor. Important to Puiseux, as it should be to any economist, was the lack of an insurance market that would produce an estimate of the risks involved. In 1977, while still at EDF, Puiseux published *La Babel nucléaire*, a many-sided critique of nuclear energy that pointed out many of its hidden costs and risks, and he soon left EDF to teach at the Ecole des Hautes Etudes.[22]

As in Sweden and the United States, observers with a cost-benefit policy style were skeptical about responding to the oil crisis with a large nuclear program. In France, economists had a larger program to be skeptical about, so that they were even more outspoken, and many abandoned the technological enthusiasts. There was still insufficient evidence to estimate the real costs of the LWR, and this gap became more obvious the more closely they examined the evidence. The gap had always existed, but the oil crisis subjected nuclear energy policy to closer scrutiny. Second, the cost-benefiters began to accept some of the assumptions of the antinuclear movement—especially concerning hidden costs, the

[22] I rely here on a personal interview with Puiseux and an article in *Le Nouvel Observateur*, 21 November 1977. See also Puiseux (1977, 1984).

probability of accidents, and the need for new reactors. Given the lack of hard evidence, these assumptions generally determined one's attitude toward the Messmer Plan. Most economists had willingly accepted the assumptions of the technological enthusiasts, but when they examined the evidence closely, they saw little reason to continue doing so.

In early 1975 the moderate segments of the antinuclear movement seemed to be coalescing in the context of a lively public debate in the press. Experts were raising doubts about the safety and economics of the Messmer Plan, and along with large organizations like the Amis de la Terre and the French Federation of Nature Protection Societies (FFSPN), they formed a reasonable, moderate opposition. In January even the FFSPN (whom the Amis [1978, 312] call "calm and solid") began to support a moratorium on nuclear energy. The Amis, the FFSPN, the CFDT, and GSIEN worked closely together. At the same time dozens of additional local groups had formed after Michel Ornano had specified the sites being considered.[23] These groups could use the arguments and evidence of the experts in their attempts to pressure local officials. Most officials near proposed reactor sites became antinuclear or were soon replaced by ones who were (Lucas 1979, 180). The new local groups, unlike those before 1974, were formed independently of the national ecology organizations, and they typically contained many people with no activist experience (Moatti et al. 1981). The membership of the new groups was similar to that of local antinuclear groups in the United States. But the ecologists, locals, and expert critics had enough in common to cooperate successfully. It also seemed they were getting a sympathetic hearing from the PS and the PSU.

But as nuclear energy became the main political issue in France, the antinuclear movement also attracted new groups of "revolutionaries" whose main target was the state and for whom nuclear energy symbolized the state's activity as an agent of international capital. In the United States the same groups joined the antinuclear movement in this period for the same tactical and opportunistic reasons, but there were fewer of them than in France. The French antinuclear movement generally welcomed or at least tolerated the new groups and their more violent tactics, and it grew steadily in size and diversity.

Violence, rare in the United States and Sweden, increased in the French antinuclear movement from 1975 to 1977. Sabotage had been present almost from the start, ranging from the destruction of official documents to the bombing of EDF buildings. Farmers were often behind the

[23] The public inquiry for the Blayais reactor at Braud attracted 26,000 written comments (the only form available) in December 1974 (Nicolon 1979, 247). In April 1975, Flamanville and Port-la-Nouvelle held local referenda—non-binding, of course.

violence, relying on their traditional means of resisting agents of the Parisian government.[24] But in spring 1975 the reactor (not yet loaded with fuel) being installed at Fessenheim was bombed, and it was widely thought a leftist group was responsible. Surprisingly, moderate ecology groups like the Amis de la Terre did not condemn such actions and even expressed some sympathy with the sentiments behind them. The Amis (Vadrot 1978, 83) explicitly recognized that a successful social movement requires a variety of groups and tactics: "Sabotage against the Fessenheim reactors is not an isolated act: it is embedded in an antinuclear movement that grows every day under a variety of forms. . . . it is normal to oppose reactor construction by all means that do not endanger human lives."

The increasing use of violence against symbols of the nuclear establishment indicates the growing presence in the antinuclear movement of the far left (Lucas 1979, 174–78). This included small sect-like parties, of a Trotskyist or Maoist persuasion, some members of which espoused violence. The fringe left also included *autonomes*: anarchist leftists who rejected almost all social organization, including that of political parties. The autonomes, who often squatted in urban buildings, and who spent some of their time pirating radio waves, were thought to be behind most of the antinuclear violence that was not clearly rural sabotage. The far left was never integrated with other segments of the antinuclear coalition, but in 1975 and 1976 they were at least tolerated. The leftist presence in the antinuclear movement both encouraged and was encouraged by an increasing focus on the Super-Phoenix breeder reactor at Creys-Malville. Because the breeder involves plutonium fuel and higher concentrations of radiation, it spurred concern not only with health and safety but also with nuclear weapons proliferation. This latter connotation was especially important to the far left.

Through 1976 the differences in the antinuclear coalition did not become open conflict. The moderate wing led by the Amis, the CFDT, and GSIEN coexisted peacefully with the radical ecologists and the revolutionaries. The movement was also on the verge of linking its efforts with those of the Socialists, and it already had the support of the PSU. In spite of this solidarity, the segments of the movement had varied images of what nuclear policy meant. For economists the nuclear program was largely a dubious investment; for technicians and scientists an unnecessary health and safety risk. For ecologists it was a risk to the environment and a choice for economic growth at the expense of human values. For

[24] Attacks on nuclear and EDF installations continued long after the revolutionary fringe had disappeared. From 1980 to 1984, the Golfech reactor site alone was attacked thirty times (Marsh 1984b).

the revolutionaries, the PCF, the PSU, and the PS, the Messmer Plan was the subordination of the state to the interests of capital, a search for profits at the expense of human needs. For the PCF and PS it also represented the conservative government's industrial policy. For Bretons and Alsatians, nuclear industry was simply one more intrusion and imposition by Paris on the provinces, and for farmers it was the disruption of their way of life. For all these groups, nuclear energy also came to represent the arrogance of the government in refusing all serious public debate and participation in technology policy.

GOVERNMENT ACTIONS, 1974–1976

For three years following the announcement of the Messmer Plan, the French government continued its firm commitment to nuclear energy, but it responded to public doubts in several ways, including a slight reduction in the size of the construction program. When the antinuclear movement was small, government officials argued and probably believed it was the work of revolutionaries; when the movement expanded after the Messmer and Ornano plans, the government decided opposition arose from ignorance. As a result it tolerated the antinuclear movement while trying to reassure the public.

The Ornano Plan was itself an attempt to present the public with a fuller outline of the nuclear program, and the government followed it with other information on nuclear energy. In early 1975 the government began promising an assembly debate on energy policy, and this duly occurred in May. The government would also allow, Ornano had announced, debate in the regional assemblies. Yet it was clear to the public and to the opposition parties that these discussions were not designed to reassess government policy, but rather to inform the public and give the appearance of legitimate debate. Because the state fooled no one, opposition increased throughout 1975.

At the same time the state tried to appear open to debate, it took a small step to reduce the pace of nuclear deployment, largely due to the efforts of the Ministry of Finance.[25] Just as economists outside the government were shocked at the size of the nuclear program, so the ministry tried to apply the brakes within the government. It was powerful enough to have some effect. The ministry's arguments were the same made by cost-benefiters in Sweden and the United States: nuclear costs were high and uncertain, and demand would probably not continue to rise as rapidly as in the past. It took steps to reassert its control and curtail the

[25] The reassertion of control by the Ministry of Finance is detailed in Lucas (1979, 141–56).

nuclear program. It demanded *ex officio* seats on the PEON Commission for three of its most important directors; it developed its own energy model to counter the experts of EDF; and it insisted that the Planning Commission set up a special workgroup to examine EDF's claims about the use of electricity for space heating (Lucas 1979, 149–54).

The Ministry of Finance also supported the Agence pour les Economies d'Energie founded in 1974 to develop and encourage the use of energy-saving technologies and practices. This agency soon developed a strong economic case against the use of electricity for space heating, without which EDF would need far less nuclear capacity. The agency made no headway with EDF, whose organizational goals involved the penetration of electricity into new energy areas. But it did disconcert the agency charged with coordinating policies in different energy sectors, the Délégation Générale à l'Energie, whose members consisted of both engineers and economists.[26] According to Lucas (1979, 153), "Ever since, a certain sense of schizophrenia has prevailed in the DGE; a profound belief in the necessity of nuclear energy being challenged by the compelling economic logic that demonstrated the impediments to the introduction of this source into what was technically the most accessible market [home heating]."

In addition to the Agence pour les Economies d'Energie, other mechanisms were established in the years following the oil crisis to encourage rigorous forecasting and comparisons of energy sources and policies. The Ministry of Finance developed its own modest model of France's energy system, and INSEE, the National Institute for Statistics and Economic Studies, began to enlarge its economic model to include a better energy component. Gaz de France also developed an energy model, at least partly to resist EDF's encroachment on its territory of home heating. Finally, the Planning Commission began to use tools of economic analysis in its deliberations, although these were not influential until the development of the Ninth Plan in the early 1980s.[27] EDF's economic analyses of the French energy system continued to be the best ones, but they were no longer the only ones. Throughout the energy policy system

[26] The Délégation was part of the Ministry of Industry but during the 1970s had almost the status of a ministry itself. Usually headed by a member of the Corps des Mines, it typically had a policy style of technological enthusiasm. But the Ministry of Finance managed to have a finance inspector, Paul Mentré de Loye, appointed in this period, giving the agency greater concern for economic comparisons of energy sources. The technologists regained control and fired him in 1978, when André Giraud moved from the CEA to the Ministry of Industry. See Lucas (1979, 138, 164).

[27] Personal interview with Pierre Amouyel, in charge of the energy component of the plan, February 1985.

there was a proliferation of data and calculations that could be used for resisting EDF.

In the PEON Commission recommendations of December 1974, the Ministry of Finance had some effect. It wanted EDF to order 8,000 MWe of nuclear capacity in 1975, and to begin building 5,000 MWe in 1976 and the same in 1977. The technical representatives on PEON wanted 20,000 MWe ordered in 1975, including 7,000 MWe started in each of the following years. The ministry wanted to both curtail the program and delay it, in part so that better demand forecasts could be made. The PEON recommendation, also embodied in the Seventh Plan (1976–1980), was for 12,000 MWe to be ordered in 1975 (closer to the ministry position) and 6,000 MWe started in 1976 and in 1977 (halfway between the two positions). On 6 August 1975 the government officially adopted this compromise.[28]

One can only speculate why the Finance Ministry's recommendation was as large as it was—not really so far from that of the other PEON Commission members. The ministry seems to have accepted the inevitability of a large nuclear program and realized that recommendations far out of line would simply be rejected. The result was that it managed to curtail the program slightly. In addition, the ministry's real strength lies in its ability to slow down projects already under way by means of budgetary pressures, so it undoubtedly expected to do this.

The Finance Ministry's ability to curtail the French nuclear program is easy to explain. First, its proposed curtailment was minor. Second, a former finance minister, Valery Giscard d'Estaing, was now president. Although the controversy was settled at lower levels, the Finance Ministry would have been assured of at least a sympathetic hearing had it gone to the top. A sympathetic hearing is all, however, because the Messmer Plan fit nicely with one of Giscard's favorite goals: the integration of French industry with international industry. He was strongly in favor of the nuclear program, but this support was no doubt strengthened when he believed he was following the cautions of the Ministry of Finance.

At least one observer (Frears 1981, 146ff.) has argued that Giscard d'Estaing's personal support for nuclear energy helps explain France's strong nuclear program, but we must be careful in assessing Giscard's role. In this case the president's power was largely that of releasing or blocking certain bureaucratic initiatives. The most important nuclear policy decisions—to abandon the gas-graphite reactor and to launch the Messmer Plan—had been taken before Giscard became president, and continuing a policy is easier than launching a new one. At the same time,

[28] On these machinations, see Simonnot (1978, 274–78).

his own beliefs about nuclear energy mattered. His support was not the blind support of a technological enthusiast; he had been finance minister from 1969 to 1974, when his ministry curtailed EDF's nuclear construction. As president he also allowed the program to be scaled back slightly in response to Finance Ministry pressure. Finally, he could help the nuclear program in small ways, the most notorious being his plan to charge less for electricity in the vicinity of nuclear plants.

Giscard supported a large nuclear program partly because he had little choice and partly because it furthered (and symbolized) his own economic goals. He had little choice in that the French state and Giscard's own government had already committed themselves to the Messmer Plan, so that to back off in the face of public opposition would indicate weakness. In addition, industrial commitments had already been made that spurred technological development and promised to make French heavy industry competitive in world markets, two of Giscard's favorite goals. For him, nuclear energy symbolized more than an energy choice; it indicated the ability of the French state to act decisively, as well as the integration of French heavy industry into the world economy.

For Giscard d'Estaing and Pompidou, nuclear policy was a form of industrial policy, something dear to both men. Nuclear energy was a means of developing French heavy industry, making it competitive on world markets, and giving it access to foreign technology. The leaders of the PS and PCF also viewed nuclear energy as a form of industrial policy, but a bad one. For them it represented the domination of public policy by industrial profit-seeking. Industrial policy was another form of Left-Right conflict, so that in the hands of the major political parties, nuclear energy policy became another instance of class conflict.

France's nuclear construction program required unprecedented amounts of capital. But there were political constraints on raising electricity rates. Higher prices would discourage industrial development, and they would exacerbate the growing opposition to nuclear energy in France. Thus most of the construction had to be financed by the state or through loans. Figure 5.1 shows the sources of EDF revenues during this period, revealing the importance of loans, which provided more revenue from 1974 to 1983 than electricity sales. Public funds, which had provided most of the capital for EDF's hydroelectric program in the 1950s and 1960s, was under the control of the Ministry of Finance, so that their contribution to revenues in the 1970s was usually under 10 percent. The main source of investment funding was loans. What is more, roughly half the loans were from foreign investors, since France's capital markets were very limited.

EDF's credit in world markets was excellent, both because it was known to be well managed and because the French government stood behind it. Once the Ministry of Finance had capitulated on the Messmer Plan, it was willing to approve EDF's ventures in capital markets. EDF's borrowings jumped from 2 billion francs in 1973 to 5 billion in 1974, so that the majority of its revenues that year came from loans. American markets provided almost as much funds as French markets; in only three years, EDF borrowed almost $2 billion there, making it the third largest borrower on American markets in 1976, behind Ford and General Motors (Lucas 1979, 38).[29]

American funds may have flowed more freely to finance French rather than American nuclear construction. The main reason was that the French government was standing in between, guaranteeing EDF's loans and protecting EDF from the costly suspicions that American financial markets were developing of utilities. Even though the French government did not finance reactor construction directly to any great extent, it was instrumental in procuring private loans for EDF. The state both sanctioned the loans and also worked behind the scenes to encourage private French banks to help arrange them. In the United States, President Ford decided against any intervention, leaving American utilities to the judgments of the markets. In part EDF was shielded from these judgments, and in part the judgments were more favorable to it than to American utilities.

The French state faced the same debates between cost-benefiters and technological enthusiasts that occurred in Sweden and the United States. Despite the Finance Ministry's power, the cost-benefit position was suppressed. Technological enthusiasm fit the goals of Pompidou; Messmer; Giscard; and, most importantly, EDF. There were too many high engineers in the policymaking apparatus for the cost-benefit restraint to be taken altogether seriously. But at least debate was still possible, both in the government and outside it. The state tried to convince nuclear protestors, and it made some concessions to them and to Finances, as with the August 1975 cutbacks.

Conclusions

For perhaps the first time in French history, energy policy after the oil crisis was taken out of the hands of bureaucrats and engineers and decided by presidents and prime ministers. Their decision was to side with technological enthusiasts and begin a large nuclear construction program.

[29] EDF's extensive foreign borrowing blurs the model offered by John Zysman (1983). He contrasts financial systems in which industry relies on securities markets independent of the state (as in the United States) with systems in which the state plays a key role in allocating credit (as in France). In nuclear energy the French state intervened to help EDF enter U.S. bond markets.

They were able to do this because of a solid consensus at the technical bureaucracies such as EDF, the CEA, and the Ministry of Industry; the popularity of technical solutions and calls for French independence elsewhere in the French political elite; and the ability of the French state to assure loans for the construction. The disincentives were few, since public protest in 1974 was limited to groups easily dismissed as irrational or ignorant.

The Messmer Plan indicates a certain kind of flexibility on the part of the French political elite. This elite acted promptly and decisively, changing its energy policy sharply. But it did not have to set up new political and economic structures to carry out its new goal. It had established these in the years before the oil crisis. Now it only had to use them.

In the three years following the Messmer Plan, the government retreated only slightly, cutting the pace of construction. Most economic analysts outside EDF criticized the size of the construction program, but only the Finance Ministry had the power within the state to curtail it slightly. The size of the program, even after it was cut back, also alarmed many scientists, technicians, local politicians, and members of the public; but public protest had no effect on government policy. To affect it, the antinuclear movement would have had to convert the Socialist Party to its cause and help the Socialists win a general election. The Socialists were unwilling in 1976 to take a clear antinuclear stand, since many of its members and most of its official leaders valued the economic growth that nuclear energy promised. But the Socialists could easily have worked with the large moderate wing of the antinuclear movement. Unfortunately, even had the Socialists gone antinuclear, as they partially did in 1977, there were no presidential or parliamentary elections scheduled until 1978. Aside from changing the government, public protesters had few channels through which to affect public policy.

Nevertheless, the future looked bright for the antinuclear movement. It was growing in size and diversity, and its various segments tolerated, if not cooperated with, each other. It had attracted people who were not antinuclear in general but who opposed a nuclear commitment as large as the Messmer Plan. The movement had made an ally of the PSU and seemed capable of doing the same with the more important PS. If the Socialists demanded a curtailment or delay of the nuclear construction program, and if they won the 1978 elections, the antinuclear movement would be successful indeed. Many observers felt this outcome was possible. An EDF economist told me: "In that period most of us thought it unlikely that our full program would succeed. There were too many possibilities: Finances opposition, the scientists [in the antinuclear movement], a victory by the Left in 1978, hesitation on the part of the government. . . . [But part of our tactic] was never to speak those doubts. We had to be confident and not admit there was any alternative. There could be no choice about nuclear."[30]

Whatever the outcome of future battles, public debates had changed the nature of nuclear energy policymaking. Even though the Socialists, not to mention the Communists, did not embrace the antinuclear movement, they criticized the Messmer Plan in several ways, adding a new layer of partisan meanings to discussions of nuclear energy. Support for the Messmer Plan was support for a new industrial arrangement in which French and multinational corporations cooperated and made large profits. Giscard emphasized the cooperation, the Left emphasized the profits, but both were concerned with what nuclear energy involved for industry.

[30] Personal interview, anonymous, 1985. Many economists at EDF were critical of their organization, since their cost-benefit policy style was often overruled by technological enthusiasts. Had my informant been an engineer, he probably would have claimed that there simply *was* no alternative. Bupp and Derian (1978, 117) cite a similar opinion, although with a mistaken reference.

CHAPTER 10

Elite Discretion in Three Countries

FROM 1973 to 1976, the United States, Sweden, and France still had more similarities than differences in their nuclear energy debates and policies. The antinuclear movement and the oil crisis brought these policies out from behind closed doors. The structure, tactics, and social base of the movement; the ways in which the nuclear establishment dismissed it; and the lack of direct effects on nuclear policy characterize all three. So do the economic damage of the oil crisis, the short-term rhetoric, and the long-term development of rational energy comparisons and planning. The debates about whether to build or conserve that the oil crisis sparked were also parallel. But the cross-national differences that appeared would become important in driving nuclear policies in future years. When nuclear energy became a public, partisan issue; state structures, political parties, and politicians became important, and these differed in the three countries. Technological enthusiasm and the cost-benefit view were distributed differently in the political systems, so that the former won in France, and the latter in Sweden and the United States.

In all three countries the antinuclear movement grew from a combination of environmentalists, local citizens protecting a certain way of life, and occasional technicians and scientists questioning the safety of reactors. After several years they were joined by leftists opposing the capitalist state, but these remained a small minority in the movements. With different values and social bases, the groups formed only a loose coalition; they retained favorite tactics but were usually willing to cooperate with the other groups in unified demonstrations. Swedish and French groups often used information and arguments drawn from the United States, and some of the groups themselves had American models. The Friends of the Earth had an important affiliate in both France and Sweden. Diffusion from the United States and throughout Europe was important.

All three movements developed a moralist policy style, although it often accompanied cost-benefit arguments (nuclear energy costs too much) and technological enthusiasm (solar energy can solve our problems). The movements had links to radical ecological beliefs that laws of nature were being broken and would wreak havoc on society. In France there were some religious ties, for example the CFDT's Catholic heritage. The anger that participants felt when rebuffed by the normal political channels also developed into moral outrage and rhetoric. Ironically, the French antinuclear movement had the strongest technical and moderate wing capable of speaking the language of the technocrats, rather than a moralist language.

In all three movements, tactics evolved over time as the movements tried everything and stuck with what seemed most effective or promising. As activists became disillusioned with normal political channels, the evolution tended to be from peaceful and electoral routes to demonstrations and site occupations. Site occupations appeared in both France and the United States in 1976 and 1977, inspired by Whyl in West Germany. But a variety of tactics continued to be used, reflecting the coalition nature of the movement. The American movement spent more time on regulatory hearings, but all three movements brought lawsuits in the courts. The Swedish movement concentrated on electoral politics since the Center Party was open to it. But the other movements also devoted resources to lobbying elected officials, running for office, and holding referenda.

Herbert Kitschelt's (1986) structural account of antinuclear tactics claimed that the structure of the state in our three countries determined the tactics chosen. In his view the French moved to illegal occupations and violence because they were blocked from access to the state, while Swedish and American activists could use legal routes to influence nuclear policy. But all three political systems offered many routes of action. None of them provided victory, but this is as true of the United States as of France. The American and French movements moved to site occupations at roughly the same time (the summer of 1976), and the numbers of participants were roughly the same, taking into account the proximity of antinuclear populations. For example, Creys-Malville drew more protestors than Seabrook because it is a morning's drive from much of France, Germany, and Switzerland, while Seabrook could draw only on Boston and New England.

All three movements continued to try various tactics because nothing seemed to work, but everything looked as though it might *begin* to work. None of the movements stopped any reactors (except the early handful in the United States that simply changed sites), but it was reasonable to believe that as they continued to grow they would begin to have effects. In particular, with the victory of Fälldin and the Center Party, the Swedish antinuclear movement expected to have a large impact. The movements were at least having indirect effects: certain technical defects were brought to light; regulators tended to tighten standards in the face of growing controversy; and politicians and cost-benefiters began to scrutinize nuclear energy rather than simply accepting the assumptions that nuclear insiders fed them.

In all three countries the nuclear establishment and most politicians were rejecting the protestors as irrational or uninformed. But the possibility existed that as the antinuclear movements continued to grow, they would curb nuclear programs. In 1976 the battle raged fiercely in each country.

Like the antinuclear movement, the oil crisis hit the United States, France, and Sweden equally hard in 1973 and 1974, and their immediate responses were similar. In each there were a few short term measures to ease supply difficulties, a lot of strong language and posturing on the part of top political leaders, and stated intentions to increase energy independence through increased reliance on nuclear energy. Almost everyone in the three countries who was in a position to influence policy believed nuclear energy would expand. Each country had a significant nuclear program already, and the quadrupling of oil prices could only make the nuclear option more attractive relative to fossil fuels.[1]

By bringing energy to public attention as well as by raising its cost, the oil crisis inspired mechanisms for each country to consider its total energy picture and to develop an "energy policy," a new concept to replace a hodgepodge of oil policies, nuclear policies, and coal policies. These models and calculations would compare different energy sources and attempt to choose the most suitable combinations, whereas the traditional decisionmaking process had concerned itself with individual energy sources in isolation from the rest. Equally important was a shift from thinking solely in terms of the supply of each energy resource—what Leon Lindberg (1977) has dubbed the "Energy Syndrome"—to examining its most efficient uses, especially the degree to which each could be conserved. Each country improved its capacity for making rational energy comparisons, but the degree to which it used the new capacity differed. Technological enthusiasts saw energy policy as a problem of supply much longer than cost-benefiters or moralists did.

The new rationality broke the alliance of technological enthusiasts, who still favored rapid expansion, and cost-benefiters, who by then realized the evidence was inconclusive. Whether at the Council of Economic Advisors, the Ministry of Finance, or the Secretariat for Future Studies, cost-benefiters readily pointed to the new demand uncertainties that would result from large changes in the price of energy, and argued for caution and flexibility. They helped develop new techniques for modeling and forecasting electricity demand and for balancing energy sources more cost-effectively. Cost-benefiters had a difficult time encouraging utilities to view electricity demand differently, since technological enthu-

---

[1] Louis Puiseux (1977, 138–39) has argued that the oil crisis improved the case for nuclear the most in the United States, since its oil imports were increasing so rapidly, it had huge uranium resources, and it possessed the technology for the full fuel cycle. As for trade balances, the terms of trade figures reported in chapter 7 showed that France and the United States were hurt equally by higher oil prices, Sweden less so.

siasts in the industry revered the old growth patterns as natural laws. But new methods were eventually instituted. So while the oil crisis seemed to encourage nuclear expansion, it also created an apparatus that could be used to question nuclear energy.

Because elected officials now helped make energy policy, they were often the ones to choose how much to use the new economic rationality and how much to stay with the old technological impulse. At this point three differences in the French, Swedish, and American political systems became fateful for nuclear policymaking. One was the distribution of technological enthusiasts and cost-benefiters in the policymaking system. Another was the dominant partisan cleavage along which political parties competed. These affected the perspectives of state officials and elected politicians. Third was the power relations among different state bureaucracies and private companies. These cross-national differences in turn led to differences in the strategies and decisions made by top politicians, and in decisions concerning how to use the energy models being developed.

Technological enthusiasm tends to dominate policymaking circles because of the progrowth values widespread in Western societies. It especially dominates French technical policymaking, since the technical corps legitimate themselves on the grounds of technological progress. Cost-benefiters, usually but not exclusively economists, are typically present in their own organizations (the Council of Economic Advisors, the Ministry of Finance) rather than at large among politicians, and they are usually a minority. They have to be particularly persuasive to guide politicians along their policy paths. In the United States this happened because President Ford respected the Council of Economic Advisors and because Ford's free-market ideology supported the nonintervention advice the council was giving him. There were no structural constraints forcing him to listen to the council; he could as easily have ignored it. French politicians did ignore cost-benefiters, even though the Ministry of Finance had far more power than the U.S. Council of Economic Advisors. Prime Minister Messmer wanted a grand technological gesture, no matter what the cost.

Swedish policymakers were probably more attuned to technological arguments than to cost-benefit ones, but in this case they heard the specific recommendation to delay a decision. Delay appealed to the Social Democrats because it was their favorite way of dealing with a controversial decision—wait until it cools down. This had been their response to the weapons controversy in the 1950s, one of the reasons for Sweden's early civilian program. In addition, they were accustomed to being in power, so they expected to be responsible for implementing every piece of legislation. Delay might make a proposal more feasible. Hence the Social Democrats adopted the cost-benefiters' recommendation to wait.[2]

Technological enthusiasts and cost-benefiters did not influence policy just through their structural positions. If this had been true, the French Ministry of Finance would have had more effect. Instead, they won their policy battles by appealing to top political leaders, who had their own reasons for choosing one policy path over another. These reasons included political strategy as well as simply sharing the perspective of either technologists or cost-benefiters. Top leaders had more discretion once nuclear policy arose from public partisan politics instead of bureaucratic insider politics.

As nuclear policymaking went public, the dominant partisan cleavages of each country also began to shape nuclear policies. Partisan debates over nationalization and industrial profits left nuclear reactors themselves intact in France. The question of government intervention meant that a Republican administration in the United States was reluctant to intervene even when that was the only way to save new nuclear construction. Only in Sweden did the very existence of nuclear power become the issue for partisan politics. Nuclear energy was associated closely with the Social Democrats, while the main opposition party opposed it altogether. Thus the fate of Swedish nuclear energy would depend on the balance of power between the parties. As soon as political parties got into the action, each country faced a slightly different set of alternatives for its nuclear policies. Politicians called on to exercise their discretion faced different choices in the three countries.

Once nuclear policies were made, implementing them involved—as it had before the oil crisis—persuading utilities. Only Swedish politicians had full rein to consider a variety of energy options; conservation was seriously considered partly because Swedish utilities already had a large nuclear program. French politicians had to choose whether to constrain or unleash EDF's nuclear ambitions; they had less scope to develop a full alternative. American politicians had to choose whether to intervene actively, to encourage and protect nuclear development, in the face of utilities that were themselves uncertain. A pronuclear choice would have been more complex and difficult to implement in the United States, although it still could have been made. Utilities wanted somewhat different things: EDF wanted to build, while Swedish utilities were willing to wait. Constrained by PUCs and tight financial markets, many American utilities wanted to wait. These contrasting goals in turn result from differences in

---

[2] Note that delay did not interest the other parties, who, long out of power, were anxious to show they could govern effectively. Lundqvist (1980) calls the Social Democratic concern for deliberation and feasible policies the "Tortoise" pattern, as compared to the United States' "Hare" pattern of quick and ambitious policies that can later not be implemented.

the timing of nuclear construction programs; when the oil crisis struck, Swedish and American utilities had built heavily and were able to pause, while EDF had hardly begun.³

In addition to political differences shaping nuclear policies, our countries differed in two ways that had surprisingly little effect on policy choices. First were the three countries' financial systems. American utilities financed nuclear construction through securities markets, EDF had the support of the French state, and Swedish utilities combined state and private sources. In Sweden financial questions played no significant role in debates. In the United States utilities were being squeezed by skeptical financial markets, but Ford could have chosen to bail them out. Financial problems were the result of policy decisions and the rising costs of nuclear energy itself; they were not an independent factor. EDF had to turn to securities markets despite state support, although the latter improved EDF's credit rating and made borrowing cheaper. Again, financing was easier because of state policies and because of EDF's strong reputation and managerial tradition. Financing reflected these other differences.

A second difference that had little effect was the availability of fossil fuels in the three countries. We saw in part two that fossil fuel resources had no effect before the oil crisis. It also had no effect in the months following the oil crisis, when each country avowed its commitment to nuclear energy. It only began to influence policies when careful mechanisms for comparing energy mixes began to spread in the years following the oil crisis. U.S. coal reserves undoubtedly reassured American policymakers, and the lack of coal in Sweden and France upset their decisionmakers. But Swedish officials were willing to curtail nuclear energy despite a lack of fossil fuels. In all three countries the main alternative to nuclear was conservation, which French policymakers tended to ignore. France's Messmer Plan was based more on EDF's desire to expand its markets and technological enthusiasts' definition of energy as a supply problem than on France's lack of fossil fuels (although policymakers have used this lack to rationalize their large nuclear commitment). Fossil fuel resources had their clearest effect in the United States, where many coal burning plants were built instead of nuclear plants. Domestic energy resources began to play a role in policymaking, but a minor one. By the time systems for rational cost comparisons were set up, the major decisions in nuclear policy had already been taken.

To conclude, in the years following the oil crisis, French, Swedish, and

---

³ EDF did *not* need additional capacity to meet demand (none of the three electric industries did immediately after the oil crisis); it wanted the capacity to encroach on nonelectric markets.

American policymakers faced similar choices about supporting nuclear expansion. Should they listen to the technological enthusiasts and continue to build? Or should they listen to the cost-benefiters and pause to see if demand would require more nuclear plants? In all three countries nuclear policymaking had come out of the closet of bureaucratic infighting and into the media glare of partisan politics. In all three the antinuclear movements were growing but still too small to exert much pressure. Policymakers had real flexibility, genuine choices to make. What they did not realize was that those choices would set their countries on policy trajectories for decades to come. Swedish and American policymakers thought they were simply pausing in the deployment of nuclear energy, but by pausing they shut the door on future developments.

Politicians made different choices for several reasons. Foremost were their own ideologies, policy styles, and assumptions about how the world worked. Some better understood the cost-benefit view, others technological enthusiasm. Second were their own political strategies: upcoming elections, balances of power between one bloc or another. Third was the asymmetry of state power: what the state could do depended partly on what utilities wanted to do. The state could prohibit or release activities, but to create new ones it had to persuade utility managers. Whether nuclear energy looked like a good option partly depended on whether the utilities running it were competent. For all these structural, cultural, and biographical reasons, French politicians embraced nuclear energy, while Swedish and American politicians paused.

PART FOUR

The Structures Tighten: Policy Divergence and the Loss of Flexibility, 1976–1989

CHAPTER 11

High Costs and Decentralization of Control in the United States

AMERICAN, Swedish, and French nuclear energy policies, which had begun to diverge in the two years after the oil crisis, grew further apart in the late 1970s. What is more, they became rigid and relatively irreversible, especially in France and the United States. International events like the Three Mile Island (March 1979) and Chernobyl (April 1986) accidents merely reinforced the distinct policy trajectories, with each country interpreting them to fit its own policies. Elections and changes of government in all three countries in 1981 and 1982 brought no changes in nuclear energy policies, even though the new governments often had new views of nuclear energy. It appears that current policy paths are unlikely to change.

As early as 1976 new nuclear construction was a dead option in the United States, but few recognized this at the time. New orders disappeared and cancellations continued, but many observers considered this a temporary lull. In the years since the oil crisis one event after another has discredited the nuclear option and created obstacles to any revival. Continuing problems with utility management, high and unpredictable costs, and the recent decentralization of policy control are the main ones. The Three Mile Island accident in 1979 alerted the public and policymakers to the extent of these problems, but it merely cemented trends already well under way. Strong steps by the Carter administration might have saved nuclear energy, but at that time it was not yet clear that these were required.

This chapter traces developments in the late 1970s and 1980s, when the hidden weaknesses, in existence since the 1960s, became painfully apparent. In the realm of partisan politics, Carter's policies were similar to Ford's: a cost-benefit perspective discouraged intervention to support nuclear energy, which in any event was not yet recognized as moribund. After the dissolution of the JCAE, congressional control of nuclear policy was fragmented among a dozen committees, and the growing antinuclear movement was able to get a sympathetic hearing in several of them. In the realm of regulatory control, there was a decentralization of power not just in Congress but throughout the federal system, as state public utility commissions (PUCs) and legislatures nibbled away at federal authority. Most important for nuclear power's fate, however, were the increases in the cost of building each plant. The costs of nuclear reactor construction continued to rise rapidly, less from rising finance charges than from bad management and *post hoc* design changes (both the indirect results of early technological enthusiasm, premature deployment of reactors, and insufficient regulation in the 1960s and early 1970s). Changes in the political and regulatory arenas were moot, since the costs discouraged utilities from choosing nuclear energy in the first place.

Partisan Politics

By the late 1970s, politicians could do little to change the course of nuclear energy. First, Jimmy Carter's energy plan of 1977–1978, in spite of its concern with replacing foreign oil, detoured from nuclear energy. In part it was dominated by the same concern with oil and gas pricing that Ford's energy program had been; in part it reflected a new form of technological enthusiasm that looked to innovations in renewable energy sources to obviate the need for nuclear energy. Second, a growing number of politicians began to question the economics of nuclear energy, culminating in a sharply critical 1978 report in Congress. The familiar partisan cleavage reappeared—Republicans arguing that government interference had caused high costs, and Democrats blaming bad utility management. The debates had little effect on reactor orders. Finally, Ronald Reagan claimed to want to revive the American nuclear industry and made some efforts to do so. All have failed.

Carter's political style blends all three of our ideal types: a high moral tone, an attention to details of costs and benefits, and a confidence in technological progress to solve social problems. His energy plan of 1977 also combined these elements. Using moralist rhetoric, the plan identified energy problems as the "moral equivalent of war"; it exhorted people to make personal sacrifices in changing their behavior; and it provided penalties if the public consumed too much gasoline. In cost-benefit language, the plan was concerned with developing fair and realistic pricing throughout the energy system and with considering all possible energy options. With technological enthusiasm the plan hoped that development of nonconventional energy sources would provide increased supplies in the longer term. A partisan Democratic theme also ran through these concerns: "The government would correct market imbalances and inadequacies, and strengthen market forces that were inoperative or inefficient. The [energy plan] assumed it was the government's responsibility."[1]

---

[1] Katz (1984, 98) provides the best description and analysis of Carter's energy plan and

While Carter himself combined three policy styles, they were organizationally distinct in his administration. The development of the energy plan especially reflects a struggle between moralist and cost-benefit policy styles. Carter set up an energy team under James Schlesinger soon after his inauguration. Roughly a dozen people, supported by a staff of two dozen more, put together the Carter Energy Plan, a complicated package of legislative proposals, in only several months. At the same time, the Domestic Policy Staff, under Stuart Eizenstat, was also concerned with energy issues, partly out of a desire to prevent the proliferation of plutonium. The Domestic Policy Staff was strongly political, with a significant element of moralist rhetoric, while the Schlesinger team was economic and instrumental. In the end the moralists undermined the efforts of the cost-benefiters, largely because the latter ignored normal political processes of persuasion.

The roots of the two groups' policy styles are clear. The Domestic Policy Staff contained many people who had been active in Carter's campaign, including a large proportion of women and grass-roots activists. Many staff members hoped to curtail nuclear proliferation by stopping the Clinch River breeder project and nuclear fuel reprocessing, and the same people had encouraged Carter during his campaign to speak of nuclear energy as a "last resort." Many of the activists in this group were skilled in political manipulation and moralist rhetoric, and they used the latter in talking to the press (something they did freely and often).

The Schlesinger team consisted of economists, lawyers, and public administrators, and it was dominated by the thinking of its leader, the economist who had headed the AEC in 1971 and had told the nuclear industry it had to fight its own battles. In addition, the team was rather young, with a large representation from universities (Cochrane 1981, 553). In developing its plans and legislative proposals, the team avoided political ideology in favor of economic analysis. It favored incentives for conservation, since a changed price structure would change behavior in predictable ways. Team members saw little link between domestic nuclear energy and weapons proliferation and believed that the Domestic Policy Staff had little understanding of nuclear energy. Whether accurately or not, the Schlesinger team felt that many Domestic Policy staffers believed normal nuclear reactors could lead to proliferation. The energy team also opposed the Clinch River breeder, but on economic grounds.[2]

The energy team developed its plan in relative isolation, avoiding the consensus-building that would normally accompany a major legislative effort. In part the isolation resulted from the severe time constraints

---

its reception in Congress. See also Cochrane (1981), Goldstein (1978), Hill (1978), and Nivola (1980).

[2] This discussion of the Schlesinger team relies on interviews with John F. Ahearne, the team member responsible for nuclear energy.

190 · Chapter 11

Carter imposed, but it fit easily with a cost-benefit style. Schlesinger and Carter believed the hard facts of the international economy were already making themselves felt in the form of an energy shortage that was expected to continue indefinitely. Economic analyses to understand the constraints facing policy decisions, and price changes to deal with those constraints, seemed to carry their own obvious justification. The team expected the facts to speak for themselves, so it spent little time preparing for a political battle.

The Domestic Policy Staff, in contrast, was politically active while the energy plan was being developed. In conjunction with the State Department, it made an important announcement favoring nonproliferation just two weeks before the energy plan was to be unveiled. Staff members also described to Congress various pieces of the developing bill, so that Congress and the media already had ideas about the plan's elements and what they meant before the plan was finished. When the energy team finished and went to Congress with its proposals, it faced severe political problems. First, there was simple miscomprehension about the content and rationales of the bills. Second, even when senators and representatives developed new impressions of the proposals, it was often too late to change their public stands. Reacting to media coverage, organizations of their constituents had already voiced severe opposition.

The energy plan moved quickly through the House and was passed almost intact in August because House leaders bypassed the usual committee structure for considering legislation, setting up an ad hoc committee to oversee all the plan's proposals. By the time the bill reached the Senate, however, the momentum had faded and opponents had organized. Senate leaders were less cooperative. The lack of political preparation by the Schlesinger team had its effect, and the plan was picked apart over the next fourteen months. The version finally sent to Carter's desk in October 1978 contained most of the original proposals, but in reduced versions. The energy savings envisioned were barely half those of Carter's original plan, and critics thought even these were unlikely (Katz 1984, 111).[3]

Carter's energy program said little about nuclear energy *directly*. It did not support fuel reprocessing and recycling, and more importantly it did not support the breeder reactor at Clinch River. The moralists in the administration strongly opposed and the technological enthusiasts strongly favored the project; Schlesinger's cost-benefit calculations were

[3] Most accounts of why Carter's energy plan failed to pass (e.g., Rosenbaum 1981; Katz 1984; Sahr 1985) take a structural perspective: it was torn apart by special interest groups exploiting Congress's decentralized committee structure. This picture is accurate, but comparing the House and Senate reveals it was not inevitable. Much depended on choices made by congressional leaders.

decisive, and they showed Clinch River to be a bad investment.[4] Commercial nuclear reactors were not yet seen as being in danger, so no major steps were taken to support them financially. Carter's policies on this were similar to Ford's, partly because Schlesinger's economic perspective was similar to that of Ford's Council of Economic Advisors. The government had subsidized nuclear energy enough; there was no reason to continue doing so. Let the markets decide. By the late 1970s both moralists and cost-benefiters were rejecting nuclear energy.[5]

Carter's plan had a much larger *indirect* effect on nuclear reactors, since it encouraged the economic rationality in energy decisions that had been growing since the oil crisis. From home insulation to cogeneration, many of the proposals were designed to decrease the demand for new electric generating plants.[6] Even in its reduced version, the energy plan further undermined the already-weak rationale for building new reactors.

In summary, Carter's energy plan left nuclear energy on its own, but few realized yet that this path involved the death of nuclear construction. By 1977 cost-benefit rationality had fully shed the assumptions of the technological enthusiasts and adopted a few of the assumptions of the moralist opponents. This new perspective discouraged active support of nuclear energy.

Turning to Congress, by the time the Joint Committee on Atomic Energy was officially disbanded in early 1977, more than a dozen committees in the House and Senate had gained some jurisdiction over nuclear energy policy. Once this decentralization of authority had occurred, proposals to create a single House energy committee with concentrated authority were defeated (Katz 1984, 218). This proliferation of oversight is far more typical of the American political system than the centralized JCAE had been. With little discipline or financial support from their parties,[7] members of Congress are "political entrepreneurs" relatively free

[4] Cochrane (1981, 596) puts it similarly: "The debate takes place between technologists and environmentalists. But the real decision turns out to be based on economic feasibility."

[5] Interestingly, many technological enthusiasts were unwilling to actively support light water reactors. Breeders like Clinch River were more exciting, but so were new technologies like renewable energy sources and synthetic fuels. One interpretation of Carter energy policy is that it was as an attempt to skip a stage and move to even more advanced technologies without going through a stage of heavy reliance on nuclear power.

[6] One of the bills, the Public Utility Regulatory Policies Act of 1978, required utilities to buy electricity from small producers. Dodman (1988) reports that by 1985, 4 percent of U.S. electricity came from this source and that 10 percent of planned additions through 1995 would. At this rate of growth, nonutility producers could match production from nuclear reactors by the late 1990s.

[7] Party organization is concentrated at the state and local level (Janda 1980), so that candidates for national office rely on direct mailing and paid staffs rather than party activists (Hinckley 1981; Jacobson 1980; Epstein 1986).

to pursue their own agendas, with an eye to what will please their own constituents (or a national audience, if they aspire to the presidency).[8] Committee and subcommittee chairs are well positioned to pursue these agendas, and enough committees and subcommittees exist so that every senator and most representatives in the majority party can have one.[9] The committees and subcommittees with new authority over nuclear energy had various attitudes toward it, but their slants inevitably reflected the concerns of their chairpersons.

House chairpersons with oversight of nuclear energy have ranged from Tom Bevill (Appropriations Committee's Subcommittee on Energy and Water Development), who believed that the NRC was blocking nuclear deployment through incompetence and bureaucracy, to Edward Markey (Energy and Commerce Committee's Energy Conservation and Power Subcommittee) and Toby Moffett (Government Operations Committee's Environment, Energy and Natural Resources Subcommittee), both strong critics of nuclear energy. In between, Morris Udall's Subcommittee on Energy and the Environment (of the Committee on Interior and Insular Affairs) had perhaps the most influence on the NRC, since it was charged specifically with nuclear regulation; John Dingell's Energy and Power Subcommittee had wide powers over energy policy more generally.[10] Several subcommittees of the Science, Space, and Technology Committee dealt with nuclear energy, usually sympathetically. Not only did technological enthusiasts flock to this committee, but the subcommittee on Energy Research and Production was headed by Marilyn Lloyd, whose Tennessee district included the Oak Ridge Laboratories.[11] In the Senate, Gary Hart (chair of the Environment and Public Works Committee's Subcommittee on Nuclear Regulation) was very active in asking why the NRC had not done more after TMI.

While Hart, Udall, Markey, and Moffett were genuinely skeptical about nuclear energy, the latter two had an additional motive. They were young members of Congress who might be able to distinguish themselves

---

[8] Calling them entrepreneurs is more than a metaphor; by the mid 1970s each member of the House had eighteen staff members and a $250,000 payroll (Loomis 1979).

[9] There are roughly twenty-two standing committees and 130 subcommittees in the House, and fifteen standing committees and one hundred subcommittees in the Senate (Epstein 1986, 49).

[10] When Dingell became chairperson of the full Commerce Committee, he took his subcommittee with him, in a way, since the committee became the Energy and Commerce Committee. The entrepreneurial interests of individual members of Congress can change the structure of Congress itself.

[11] Lloyd's technological enthusiasm came through strongly in a speech she gave upon the publication of a major U.S. OTA report (1984) on nuclear energy (6 February 1984). She both espoused "inherently safe reactor concepts" and hoped "to reinvigorate the nuclear option."

by aggressively attacking nuclear energy as it became less and less popular, especially after TMI. Hearings in Markey's subcommittee often concentrated on the Pilgrim reactors in Markey's home state of Massachusetts (*New York Times* 1986).

The main effect of this growth in oversight committees and subcommittees was that the NRC received mixed messages. The main conflict was between those who felt the NRC should do more (Hart and Udall, whose subcommittees had the most direct oversight) and those who wanted it to do less (Bevill, whose subcommittee appropriated the money). In many ways the ambiguity of its directives allowed the NRC to choose its own priorities, but in major issues the appropriations subcommittee was most often the victor. If the money did not exist for a project, it could not be done.[12] Congressional influence on nuclear energy will never have the coherence (or the enthusiasm) it did under the JCAE. But the decentralization of authority arose because of nuclear energy's problems; it did not cause them.

Ronald Reagan's presidency revealed how hopeless nuclear energy had become in the United States. Reagan's rhetoric and his NRC appointments were strongly pronuclear, but they could not save nuclear energy. By 1981 there was little discretion left to policymakers, and almost no conceivable legislation could have revived reactor orders. As in France and Sweden, the state could not force utilities to buy nuclear plants they didn't want, and the costs of reactors in the United States were high enough that no utility wanted to order one. As with France in 1981 and Sweden in 1982, a change of government brought little change in nuclear energy policy in the United States.

The government support for conservation and renewable energy research that Carter had put into place was almost completely eliminated under Reagan. Department of Energy funds for research on conservation fell from a high of $779 million in 1980 to $22 million in 1983; those for solar and renewable energy from $751 million to $82 million. During the same period funds for nuclear energy fell slightly, but soared as a percentage of total DOE research: from 39 percent in 1980 to 86 percent in 1983. Nevertheless, rational energy comparisons had been institutionalized in the private sector, so that many utilities—especially when pressured by PUCs—still found conservation a more attractive option than nuclear reactors. Government policy could alter the cost-benefit calculations only slightly.[13]

---

[12] My discussion of the relationship between Congress and the NRC relies on interviews with former NRC Commissioner John F. Ahearne.

[13] Katz (1984, 163) reports the funding figures. Sawhill and Silverman (1983) describe numerous alternatives to building new generating capacity, including conservation, cogeneration, load management, renewable resources, and improvements in operating efficiency.

One of the biggest battles of recent years was the renewal of the Price-Anderson Act. The cap on liability was still under $700 million, with part of the money coming from a special fund to which all reactor operators would contribute in case of an accident. In 1988 Congress renewed Price-Anderson, but set a new cap at $7 billion. The vote reflected a much stronger antinuclear sentiment than the last renewal in 1975, and since private insurance stayed at $160 million per plant all the increase came from the operators' pool. But it was nevertheless considered a victory for the industry to receive a liability cap at all.

To sum up, after the oil crisis put energy policy on the agenda of national partisan politics in 1973, new politicians who were outsiders to the cozy world of nuclear policy began to clamor for influence. They gradually took away the JCAE's authority and spread it across a variety of other committees. As the antinuclear movement continued to grow, and as the dubious cost figures for nuclear plants became clearer in the late 1970s, a growing number of committee and subcommittee chairpersons expressed and pursued their skepticism about nuclear energy. In fact, they needed to do little except publicize nuclear energy's unfortunate position, which the industry and regulators had themselves created, independent of partisan politics. The economic and regulatory structures had tightened over the 1970s, so that by Reagan's inauguration in 1981 there was little a pronuclear president could do to revive the industry.

The Costs of Nuclear Energy

The main reason orders for nuclear plants disappeared in the United States was the high and unpredictable costs of building them, and no government policies could entice utilities back into nuclear energy in this cost situation. The average cost of building nuclear power plants increased 140 percent from 1971 to 1978 in constant dollars (Komanoff 1981); since then the rate of increase has probably been even greater. Worse, the increases have been unpredictable, with some reactors costing five or six times as much as others. The unpredictable cost increases are the main reason utilities have canceled orders for more than one hundred reactors and have ordered virtually no new ones since the oil crisis.

The high costs, in turn, were due to the hidden weaknesses we have traced to the 1960s, with bad management at the top of the list. Hands-off regulation merely allowed management problems to persist. Development costs continued to be paid as poor management combined with premature deployment to cause continual design changes. Inadequate

---

They criticize the traditional pro-build technological enthusiasm of most utility managers, but Flavin (1986) describes the huge commitments utilities are making to alternatives. Also see Allison and Carnesale (1983).

management and design changes led to many construction delays; regulatory delays were tiny in comparison. Delays combined with financial weaknesses to cause large interest charges. Increasing skepticism on the part of financial markets was confirmed by the oil crisis, after which new nuclear orders would have been harder to finance. Skeptical investors and soaring costs reinforced each other.

The period 1971 to 1978 is crucial for understanding why the nuclear industry in the United States has been paralyzed, because in 1971 nuclear energy still seemed to have in every way a promising future, and 1978 was the last year any utility ordered nuclear reactors. In 1971 the regulatory system was reorganized with the hope, on the part of regulators and politicians, that licensing would be speeded. And in the following three years reactor orders zoomed, peaking at 41 in 1973 (see figure 3.1). Yet all the reactors ordered in 1974 and later have been canceled, except for the two ordered in 1978 on which no work has been done (and which are probably kept on the books for symbolic reasons). By 1978 observers as different as *Nucleonics Week* (24 November 1977) and the House Committee on Government Operations (1978) could claim that the nuclear industry was dying. They did not recognize it was already dead.

Inadequate Management. Most observers now agree that poor management was the main cause of nuclear energy's high costs, from the capitalist tool *Forbes* magazine (Cook 1985) to NRC Commissioner Victor Gilinsky (1982) to nuclear opponents Bupp and Komanoff (1983). The best evidence is the extreme variation in construction costs by different utilities, ranging from 5 cents to over 20 cents per kwh in plants coming into service since TMI (Bupp and Komanoff 1983, 2). The best performers, such as Duke Power and Commonwealth Edison of Chicago, generally built reactors as cheaply as EDF in France, indicating that the problem for the rest of the industry lies with the utilities more than with manufacturers or the regulatory system (Bupp and Komanoff 1983). It is precisely those utilities with the greatest experience in building and operating nuclear plants that do the best job; they have overcome the challenges of the new technology. NRC Commissioner Gilinksy (1982, 5–6) says, "Some of the utilities that jumped on the nuclear bandwagon were ready to take on nuclear responsibilities, but some were not. . . . [The nuclear industry] is hostage to the worst performers in the industry."

The roots of the management problem are not hard to find. Most utility managers are engineers trained in coal and hydroelectric systems; fewer than one percent of chief executives and chief engineers have any nuclear training (Fenn 1980, 79). To treat nuclear reactors like coal plants, as "another way to boil water," is a sure recipe for construction cost overruns, high operating costs, and perhaps even accidents. Most conspicuously, utilities wanted the control rooms of nuclear reactors to look like those of coal plants. The result is that operators face a dangerous overload of information, only some of which is crucial (Wilson 1979, 16). Chapter 5 showed that the French utility EDF started off in the 1960s with these same wrongheaded attitudes but managed to change by the early 1970s. Many American utilities never learned their lesson, because the AEC and NRC were unable to teach it to them. Because of the ideology that they should not interfere in utility operations (the result of technological enthusiasm and American regulatory traditions) the AEC and NRC could only regulate from the outside, through retrofits and strict standards. What seemed like cumbersome regulation to the utilities was the only way to compensate for the fact that utilities were not treating nuclear energy properly.[14] Poor management increased nuclear costs by contributing to frequent design changes, construction delays, and operating costs for completed plants.

Development Costs and Design Changes. The nuclear plants that were built in the 1960s and that came into service before the oil crisis cost an average of 100 percent more than had originally been estimated. Part of the discrepancy was due to estimates built on extreme technological enthusiasm rather than actual experience. Part was also due to the normal costs of developing and commercializing a technology, which the technological enthusiasts were unequipped to see. Bupp and Komanoff (1983, 5) claim these cost overruns were typical development costs for new technologies of great complexity and novelty. They conclude that "those who designed and built the first Great Bandwagon Market plants did about as well as might have been expected, given the nature of the job they undertook." These costs, one might add, did not have to be borne so heavily by the French and Swedish nuclear programs several years later.

The surprise is that development costs continued to affect nuclear energy after the early 1970s. Utilities bought nuclear reactors when there were still many unsolved problems, so that throughout the 1970s the AEC

---

[14] Not only was the regulatory system unable to save utilities from bad management practices; it may have encouraged them. Having a set of precise standards to be met indicated to many utilities that they *only* had to meet those standards. A U.S. NRC study (1984) found that a key difference between utilities that succeeded with nuclear energy and those that didn't was that the former treated regulatory guidelines as a bare minimum rather than a satisfactory goal. It concluded that the worst performers' problems "had as their root cause shortcomings in corporate and project management." Because they had no control over managerial processes, the AEC and NRC couldn't instill better attitudes at the less successful utilities.

and NRC developed more elaborate and stringent standards for the design, construction, and materials of nuclear plants. The number of "regulatory guides" grew from twenty-one in 1971 to 143 in 1978 (Komanoff 1981, 24–25), and the amount of material and labor required in reactor construction more than doubled over the decade of the 1970s (Komanoff 1981). What is more, many of the changes in standards were applied retroactively to plants already under construction. Some observers believe these design changes were the main source of cost increases in the 1970s (Bupp and Derian 1978; Komanoff 1981; Bupp and Komanoff 1983; Zimmerman 1987), and the two years following the TMI accident in 1979 saw an even greater flood of changes.

Charles Komanoff claims that increased stringency in nuclear standards was partly the result of the AEC's and NRC's desire to hold total risks for society at a tolerable level while the number of reactors increased. Thus the risk per plant would have to decrease. He also argues (1981, 24) that knowledge about nuclear energy and its dangers increased continually, pointing out that "professional engineering societies developed new nuclear standards at an even faster rate [than AEC/NRC regulations grew] (often in anticipation of AEC/NRC regulations)." Bupp (1981) believes in addition that the antinuclear movement was able to pressure the regulators into increased stringency, perhaps through partisan politics.

The evidence seems to favor Komanoff slightly for the early 1970s, but Bupp after that, and most important of all is the interaction between increased knowledge and antinuclear pressures. The most important way in which nuclear opponents encouraged design changes was by encouraging the general growth and application of knowledge about nuclear energy. The ECCS hearings are the best example of technical risks coming to light because antinuclear organizations elicited them, but there are other cases on a smaller scale. Most whistleblowers, for example, make their case public only when supported by the antinuclear movement. The embarrassment to the regulators derived from having outsiders publicize design flaws may have been an important reason for many design changes, especially as Congress and the administration increased their scrutiny of the regulatory bureaucracy during the 1970s. The increased knowledge that Komanoff claims was behind tougher standards may have needed the encouragement of the antinuclear movement to become public and be applied.

Design changes play an important role in explaining the cost increases of nuclear construction over the last twenty years. But they are less useful for explaining why utilities began to cancel reactor orders at the time of the oil crisis, since the changes have been spread out over a much longer period. In fact, it was still widely believed in 1974 that the costs of nuclear construction would soon begin to fall. But new orders began dropping in 1974 and disappeared completely by 1978.

Standardization. Standardization of reactor and plant designs would have made nuclear energy cheaper, but during the great bandwagon markets it was not a reasonable option. Even though there were four reactor manufacturers, the American market was so large that each of them had enough orders to achieve economies of scale. By 1974 each had thirty or more orders, more than the entire French program at the time. More important than the reactors themselves were the other components of the nuclear plants, which represented the majority of construction costs and were customized for individual utilities. Once again, utility management made a difference. Well-run utilities, like Duke Power, standardized their nuclear plants, which they were able to build and operate almost as cheaply as French plants.

At first regulators refrained from stepping in to force standardization on utilities, less because they were reluctant to interfere than because standardization in the 1960s would have been premature. Light water technology was still rapidly evolving, and standardization would have frozen it. In 1973 and 1974 the AEC established incentives for reactor standardization, especially quicker licensing and fewer retrofits. Sixty percent of applications for construction licenses in 1975 and later used these standardization procedures (Campbell 1988, 43). But by then, nuclear orders were disappearing. There was nothing to standardize. Standardization is something of a red herring in cost discussions. Yes, it might have reduced costs, but it might also have discouraged further development. The underlying problem was the premature deployment of reactors.[15]

Construction and Licensing Delays. The time required to build and license nuclear reactors has risen steadily since the 1960s. Much of this increase is due simply to more stringent standards and greater amounts of materials and labor that go into the construction, but more is due to construction delays from retroactive design changes, delays in licensing, and bad utility management. Utilities blame much of nuclear energy's cost increases on delays caused by regulators and antinuclear intervenors. However, a wide consensus has developed that most delays are due to bad management on the part of utilities and builders.

Since 1959 the nuclear industry has regularly attacked "regulatory delays" due to licensing processes and demanded a "streamlining" of nuclear regulation (Lewis 1972, chap. 10). Chapter 3 showed that licensing

[15] See Jasper (1987b) for a comparison with France, and Campbell (1988, chap. 3) for a thorough discussion of standardization.

time increased in the early 1970s simply because of system overload from the enormous wave of applications (Anderson 1974). Increased staff and efficiency relieved the backlog, but accusations of undue delays continued into the 1980s. They reached a fevered pitch after TMI, when utilities argued that regulatory delays would cost up to $10 billion.[16] Similar concerns were voiced within the NRC itself in 1981 and again in 1984.[17] Attacks on regulatory delays are a normal part of the strategic rhetoric of the nuclear industry.

In spite of industry's claims, there is virtually no evidence that delays in granting licenses have led to delays in bringing power plants into service since the oil crisis, except in rare cases where there were major safety questions to be solved. Licensing delays could be caused by changes required by regulators, by public intervention through hearings and courts that result in no changes, or by intervention that caused regulators to mandate changes. Pure delay from intervention has not occurred, but, as argued above, public intervention has encouraged some retrofits and tightened standards. Linda Cohen (1979) argued that the small delays caused by licensing procedures represent vital processes of technological learning and evaluation, since nuclear technology is still evolving. Safety discussions rather than delay tactics lie behind regulatory delays. A study by the Federal Power Commission of twenty-eight reactors scheduled for completion in 1973 showed that changed requirements were more important than public legal challenges (table 11.1). A similar study in 1979 estimated that 81 percent of delays were due to private sector decisions and events completely unrelated to regulation.[18]

If the regulatory process was responsible for few delays, public participation in that process was responsible for almost none. As late as 1983 the NRC could tell Congress that "public hearings had never delayed the operation of a single reactor" (Union of Concerned Scientists 1985, 63; U.S. Congress 1981). Notorious cases like Diablo Canyon and Shoreham account for the legal delays of the entire industry. But these cases were unusual.[19]

---

[16] The Bevill Report for January 1981 (Washington, D.C., Office of Nuclear Reactor Regulation) and *Nucleonics Week* (February 19, 1981) describe the delays likely from regulatory processes. Testimony to the Senate Subcommittee on Nuclear Regulation is the source of the cost estimates (Union of Concerned Scientists 1985, 84).

[17] According to NRC Commissioner Peter Bradford (1982b, 5), "The great licensing panic [of 1981] has gone the way of its 1959, 1972, and 1977 predecessors, though its backlash still threatens to curtail some intervenor protections."

[18] One U.S. Department of Energy study (1980) found that only 19 percent of construction delays were due to NRC regulatory delays, forced retrofits, referenda, and court decisions combined. The rest was due to financing and management problems and decisions. See Zimmerman (1987) and the U.S. Congressional Budget Office (1979).

[19] One scholarly claim that regulation caused the delays in reactor lead times comes from

The reason that *licensing delays* have not been important since the oil crisis is that *construction delays* have overshadowed them. Table 11.1 shows that regulatory and legal challenges caused a total of thirty-two months of delay, while construction and management problems caused 220 months. Diablo Canyon may have been held up several months by legal intervenors, but it was held up several *years* because major supports to protect the reactors from earthquakes were installed backwards. A congressional witness (U.S. Congress 1978, 43) from the Council on Environmental Quality (of the Federal Council on Science and Technology) calculated that public intervention in construction permit hearings could have added an average of no more than five months to the proceedings, which he considered a small addition to the total lead time for nuclear plants (ninety-three months in 1975).

Construction delays have resulted partly from design changes and tightened regulations, but even more from financial problems, utility un-

TABLE 11.1
Construction Delays for American Nuclear Plants

Cause	Number of Plants Affected	Plant-Months of Delay
Poor Labor Productivity	16	84
Late Delivery of Major Equipment	9	68
Changes in Regulatory Requirements	8	23
Equipment Component Failures	6	15
Construction Labor Strikes	5	18
Labor Shortages	5	18
Legal Challenges	4	9
Factory Labor Strikes	4	5
Rescheduling of Other Facilities	1	12
Weather	1	9

Source: Davis (1976), citing a Federal Power Commission Study of twenty-three plants in 1973.

Golay et al. (1977), who compared the American regulatory system with the French. The former is more cumbersome, with adversarial relations and more channels for public participation. There have also been larger delays in the United States than in France. Golay et al., having shown that the delays *could* have occurred because of licensing, then conclude this is how they *did* occur. They provide almost no evidence of the mechanisms behind the raw correlation between regulatory systems and lead times. Even their best evidence, that the *Calvert Cliffs* case caused licensing delays, is limited, since the length of licensing times had been increasing steadily for several years before *Calvert Cliffs*. The large number of license applications simply overloaded the system. Golay et al. provide a good example of the pitfalls of formal structural analysis, especially when applied to France and the United States. Rycroft and Brenner (1981) make a similar logical leap in their analysis of how we can learn from foreign countries to make facility siting easier.

certainty about demand growth, and pure managerial incompetence.[20] Chapter 7 showed that the oil crisis eventually caused some utilities to defer construction projects to assess future trends; most PUCs enacted regulation to encourage such assessments. Energy modeling and forecasting as well as the existence of alternative fuels like coal played an increasing role in utility decisions to delay or cancel nuclear orders. In 1974 and 1975 a utility could not have seen how costly such delays might become, as they did after TMI because of both retrofits and even higher interest rates. The U.S. Department of Energy (1980) estimated that 27 percent of delays were conscious decisions by utilities to reconsider their demand decisions, and 19 percent were due to financing problems.

In addition to financing and demand reconsideration, 26 percent of delays were directly due to bad management of nuclear construction (U.S. Department of Energy 1980). NRC Commissioner Victor Gilinsky (1982, 4) claimed, "While plants are not being held up by lengthy proceedings, there is the possibility, if not likelihood, that plants which are largely finished will not go into operation on schedule because they have not been built right." The Marble Hill and South Texas plants were delayed for months because of large gaps in poured concrete. Rejection rates for welds and other mechanical features were around 38 percent at Nine Mile Point (Cook 1985, 90). These examples of poor quality control are typical and account for much nuclear construction delay.

In summary, poor management caused most (unwanted) construction delays, followed by design changes often required after construction had begun, followed distantly by public intervention. Design changes are also indirectly the result of bad management by utilities. But construction delays did not raise costs directly; they increased the interest charges utilities paid while they delayed.

Interest Charges. Some observers have pointed to interest payments as one reason for the high cost of nuclear reactors, and it is true that a handful of celebrated reactors have had enormous interest payments, up to $1 million a day for Diablo Canyon.[21] Interest charges are a function of interest rates and the length of time required to complete a plant. We have seen the causes of construction delays; what of interest rates? The high interest rates that utilities faced in the 1970s resulted from (1) generally high rates throughout the economy, (2) investor attitudes toward the utility industry, and (3) investor attitudes toward electricity generation from nuclear energy.

[20] These were also the conclusions of the U.S. Office of Technology Assessment (1984, 157).
[21] Interview with George Sarkisian of Pacific Gas and Electric, November 1985.

Before the oil crisis, high interest rates had little effect on utility decisions about nuclear reactor orders. Chapter 3 argued that high interest rates emerged in the late 1960s precisely when utility costs were beginning to rise, so that state regulators were slow to pass on cost increases to consumers. But this financial squeeze was slight, and it eased as PUCs increased their efficiency and allowed more funding of construction work in progress. Convinced of eternal demand growth, utilities saw no alternative to paying higher capital costs. If they didn't build nuclear plants, they would build coal or oil plants and still have to borrow large amounts of money. Nuclear plants did take longer to build (and hence involved greater interest charges), but Gandara's (1977, 33) careful analysis indicates that it was not until 1974 and 1975 that utilities "modified their construction programs so as to reduce construction interest charges and provide immediate cash-flow relief." As chapter 7 argued, the oil crisis revealed a no-build alternative, leading to the cancellation of many construction plans. Thus, after the oil crisis, high interest rates encouraged many utilities to consider conservation and cogeneration alternatives more seriously.

In spite of the reduction in construction plans after the oil crisis, investor attitudes toward utilities and toward nuclear energy continued to deteriorate during the 1970s. In 1970, 61 percent of utilities had bonds rated AA or better; by 1982 only 24 percent did; those rated BBB or worse grew from 8 to 37 percent (U.S. OTA 1984, 3). Both before and after the oil crisis, generally high interest rates were raised even higher for nuclear energy as investors expressed some cost-benefit skepticism. As early as 1968 investment banks showed concern over the cost of nuclear energy, indicating pockets of cost-benefiters who had not accepted the optimistic assumptions of the technological enthusiasts. The skeptics grew rapidly in the 1970s. The *Morgan Guaranty Survey* (1968, 5) expressed a common combination of technological enthusiasm and cost-benefit skepticism: "Since new industries generally offer greater scope for technical innovation than do "mature" ones, it is not unreasonable to think that nuclear power will stay competitive. But the rapid speed with which the capital costs of nuclear plants have been rising of late makes that judgment (and AEC forecasts as well) somewhat tentative."

The rising antinuclear movement may have played a part in raising investor doubts about nuclear energy. Costs were not just rising, but unpredictable. The antinuclear movement and the regulatory changes of the early 1970s increased the uncertainty, whether or not they directly increased the costs. The nuclear industry may have undermined its own financial position through its constant complaints about expensive licensing delays. The claims were largely unfounded, but they worried investors. Administrative changes at the AEC did little to satisfy antinuclear protestors, but they may have sent unintended signals to the financial community.

If the financial risk of nuclear energy was obvious to some investors in the early 1970s, by the early 1980s it was clear to all—even the industry magazine *Nucleonics Week* (25 November 1982). The Three Mile Island accident, discussed below, made all investors aware of the special risks of nuclear energy. Not only were cleanup costs higher than expected, but TMI's owner barely escaped bankruptcy.

Despite rising interest rates, interest payments remained a small part of the total costs of a nuclear plant. Komanoff (1981, 22) calculates that interest payments were roughly 6 to 10 percent of the capital costs of nuclear plants that were built in 1971 and 1978. Plants completed after 1978 had higher interest charges, since they took longer to finish and faced higher real interest rates. But these plants also bore enormous additional costs from design changes after the Three Mile Island accident, so that interest charges remained a small part of the total.

Interest charges were a significant but derivative cause of nuclear energy's high cost. They arose from the enormous construction delays that plagued the industry, and from the increasing skepticism of the investment community. That skepticism arose from public opposition and regulatory changes, and from the nuclear industry's own dire complaints about these. Even more, it was caused by the increasing and unpredictable costs the industry faced. Only a small fraction of these cost problems came from interest charges.[22]

Operating Costs. Once plants are completed and in operation, cost problems persist. Since interest charges, construction delays, and licensing play no role here, it is largely management practices that are at fault. The same utilities that have construction problems also typically have safety problems and low capacity factors when their reactors are operating. Operations and management costs have risen even faster in the 1970s than nuclear construction costs (Flavin 1983, 16), further evidence against utility management. American reactors also have low capacity factors—the percent of time over a year that a plant is able to produce

[22] Campbell (1988) and Schwartz (1986) make the withdrawal of financial support crucial to their explanation of why nuclear energy collapsed in the United States. Schwartz is concerned with describing the limiting role of banks in industrial development, while Campbell points out the U.S. government's inability to intervene in financial markets to guarantee funding to its favorite projects. Their structural accounts must be filled in with an explanation of why the cost-benefit perspective of the financial community became willing to abandon earlier enthusiastic assumptions about nuclear energy and a description of the economic rationality set in place after the oil crisis, which made no-build options more attractive.

power. Low capacity factors increase the cost of electricity since reactors out of service are not paying for themselves. An NRC report on the sixteen worst plants in the United States showed that they were unsafe when operating, but were rarely operating. At the time of the study (May 1986) only two were in operation.[23] As with construction costs, the best reactors performed as well as those in any country; but the worst did much worse than reactors abroad (Hansen et al. 1989).

In summary, nuclear energy's cost problems have their roots in the commercialization of the LWR in the 1960s. The "hidden weaknesses" (chapter 3) of hands-off regulation and a bandwagon that brought unprepared utilities into nuclear energy, both of which were fed by technological enthusiasm, combined to ignore various technical challenges to nuclear energy development. The development costs of the technology were extended over many years, requiring costly design changes. Along with the uncertainties created by the oil crisis, these design changes, which utilities often fought, led to delays in nuclear construction. These delays, along with investor doubts about utility competence, increased the interest charges associated with nuclear plant construction. Other minor factors have contributed to nuclear energy's high costs, but the most important cause is this chain stretching back to the original technological enthusiasm of the 1960s. Ironically, the same poor management that raises the costs of nuclear plants also makes them less safe when they are finished. On top of high construction costs there is thus an increased risk of a costly accident. Three Mile Island demonstrated how great that cost can be, making the investment community even less likely to support nuclear energy.

The American nuclear industry died from within over the course of the 1970s, as cost-benefit evaluations came to favor various alternatives to building reactors. Skepticism first arose in the investment community, but it was forced on utilities as economic rationality and energy comparisons spread after the oil crisis. Policies to save nuclear energy became more and more unlikely, since they would have had to change the internal workings of utilities and manufacturers. Policies can pull (restrain) more easily than they can push (initiate).

Decentralization of Authority

When the oil crisis and the growing antinuclear movement turned partisan political attention to energy issues around 1973, they set in motion a

[23] *New York Times* (1986). Ironically, these two were the Peach Bottom reactors, which the NRC closed in 1987 when operators were found sleeping during their night shifts.

gradual decentralization of the regulatory system for nuclear energy. The breakup of the AEC was a dramatic first step but had little effect. The replacement of the JCAE by a dozen congressional committees had more effect, especially since these committees rarely agreed on analysis or policy. In addition to these changes at the top, the government's federal structure finally began to have an effect. Even though the court system insisted that nuclear policy was the domain of the federal government, state laws and state PUCs encroached on this authority in the late 1970s and 1980s. But by then there were no new reactor orders to use this power against.

Many observers have claimed that a major cause of the demise of America's nuclear industry was the strength of the court system, which the antinuclear movement could use to derail the best intentions of Congress, the administration, and utilities. Opponents of nuclear energy have apparently also believed the courts could be used in this way, for they have devoted much time and money to lawsuits. They have little to show for it. Not only have lawsuits resulted in few significant delays, but the courts have resolutely left decisionmaking authority with the AEC/NRC. Most analyses point out that American courts could be used to challenge nuclear development, but they don't show that these challenges succeeded. (In fact, they did not.)[24]

Nuclear opponents' biggest court victory was *Calvert Cliffs* in 1971, when the D.C. Circuit Court criticized the AEC for making "a mockery" of the National Environmental Policy Act of 1970 (see chapter 3). *Calvert Cliffs*' effects were to require better impact statements and slow the licensing of several plants, but it stopped no construction, not even Calvert Cliffs itself. The way in which federal courts generally applied the decision was to require impact statements, but to allow construction to continue at the same time. In addition, the bite of the ruling was quickly undermined by congressional legislation allowing the AEC to grant "interim" operating licenses in special cases (Lewis 1972, 296).

Although *Calvert Cliffs* was not appealed to the Supreme Court, those cases that *were* appealed usually resulted in defeats for the antinuclear movement. Lettie Wenner (1981, 1982) studied fifty-eight federal cases over nuclear energy in the 1970s (most of them brought by opponents). Nuclear opponents won only thirteen, and seven of these were later re-

---

[24] Even the best analyses describe the broader role for courts in the United States, show that nuclear energy has been slowed there, and give no evidence that courts were the reason. For example, Bupp and Derian (1981, xx) say "Licensing decisions can be challenged in the courts in the United States and in the Federal Republic of Germany, but not in Canada, France, Japan, Sweden, or the United Kingdom." But their succeeding analysis ignores courts, except for a brief mention of *Calvert Cliffs*. Campbell (1988) and Kitschelt (1986) also exhibit this *post hoc ergo propter hoc* fallacy.

versed by the Supreme Court. The antinuclear victories concerned only minor points of procedure, inspiring Wenner's (1981, 92) understated conclusion that "the courts have not proved to be a very effective deterrent to nuclear power in the 1970s." Most of the cases questioned the exclusive right of the AEC and NRC to set standards and make technical decisions, but the Supreme Court resolutely defended that right on the grounds that in certain areas the federal government can preempt the ability of states to make laws. In several cases (*Northern States* in 1970; *Train v. Colorado PIRG*, 1976; *Vermont Yankee v. Natural Resources Defense Council*, 1978) the Supreme Court upheld the prerogative of the AEC/NRC and rebuked lower courts for intruding into the regulatory process (Wenner 1981, 86). In others (*Izaak Walton League v. AEC* in 1976; *Illinois v. NRC* in 1979) lower courts themselves carried out the Supreme Court's wishes. One exception to the court system's willingness to concentrate nuclear authority in the AEC/NRC is the group of state laws from the late 1970s that the courts have let stand.[25]

Even though the courts have clearly upheld the federal government's right of preemption, state governments have enacted a variety of statutes restricting or prohibiting commercial reactors. In 1975 and 1976 antinuclear activists concentrated their efforts on legislation and state referenda, and by February 1976 roughly fifty bills had been introduced in twenty-four state legislatures (Murphy and La Pierre 1976, 392). Only a handful of these passed, but antinuclear bills and referenda have continued to appear. By 1984, eleven states had laws restricting nuclear energy, four originating in referenda, one in the public utility commission, and the rest in legislatures (U.S. Office of Technology Assessment 1984, 216).

The most famous of these laws is California's, a model for many of the others.[26] The bill prohibits new reactor construction unless certain safety and liability conditions are met, most notably proof of a reliable technology for waste disposal. In April 1983 the Supreme Court allowed the law as an economic measure, protecting taxpayers and ratepayers from undue financial risk, rather than as a form of safety regulation, which remained the exclusive domain of the NRC. This decision opened the way for in-

---

[25] Raymond (1979) describes and Yellin (1981) criticizes the *Vermont Yankee* and other decisions. Stewart (1978), Byse (1978), and Breyer (1978) also debate the case. The underlying legal principle, federal preemption of state authority under certain circumstances, is grounded in the supremacy clause of article 6 of the U.S. Constitution. See Helman (1967), Holifield (1970), Hirsch (1972), *Columbia Law Review* (1975), Parenteau (1976), Owen (1981), and Woychik (1984) for the evolving discussion.

[26] California Assembly Bill No. 1579 (1975–1976 Regular Session). For discussions of the legal principles behind the bill, see Hays (1976), Parenteau (1976), and Tribe (1979). Murphy and La Pierre (1976) describe the bill's influence on other legislation, while Owen (1981) and Woychik (1984) discuss it in the context of nuclear waste debates.

creased intervention by state legislatures and PUCs, whose traditional function is to oversee the economic health of utilities as well as their ability to meet energy demand. In the Supreme Court review, no fewer than thirty-two states filed amicus briefs supporting the California law (Woychik 1984, 365). But no more nuclear reactors were being sited, so state laws inspired by California's had little practical effect. Thus the titles of two commentaries on the Supreme Court decision were both accurate: "Ruling May Have Driven Last Nail in [Nuclear] Industry Coffin" (*Los Angeles Times*, 21 April 1983, p. 1), and "High Court's California Decision Seen as Having Limited Impact" (*Nucleonics Week*, 28 April 1983, p. 5). By 1983 nuclear energy's coffin had so many nails that one more made little difference.

California's is the only statute that may have affected nuclear reactor construction, since it may have helped stop one plant, San Diego Gas and Electric's Sundesert plant. The Sundesert case, however, also demonstrates the fragility of state legislation, since the legislature in fact *exempted* the plant from the new procedures. The exemption depended on there being no alternative sources of energy to the plant. In 1978 the California Energy Commission claimed there were reasonable alternatives and that the predicted electricity demand might not materialize. Although it was never flatly denied permission to build the plant, San Diego Gas and Electric canceled its plans in May. The California statute may have contributed to this decision, although it was less important than declining demand forecasts and a PUC decision not allowing the utility to charge customers for the construction until it was completed. The skeptical attitudes of state bureaucrats played a larger role than formal legislation (which was quickly overruled).

After 1974 state laws restricting nuclear expansion were somewhat moot. Sundesert was canceled in part because it wasn't needed; *all* the plants in the United States ordered that year—1975—were eventually canceled. California's legislation slowed up the approval process long enough for the lower demand growth to become apparent. For the same reason, other states' laws approved after 1976 were not thoroughly tested in court. They were not a threat to nuclear reactors, because none was being sited. As Victor Gilinsky (1982, 3) said: ". . . all this modification of the licensing process applies to plants that are yet to be ordered. What good is it, if there are no new orders?"

Since the courts held that economic issues were valid territory for state regulation, PUCs have become more important in nuclear development. In twenty-four states PUCs (or other independent agencies) must validate the *need* for a power plant before it can be built,[27] and in Maryland and

[27] For a list of the laws, see the U.S. Office of Technology Assessment (1984, 152). Arti-

New York this right has possibly contributed to delaying several reactors (Cooper 1981). But in most cases the state is doing the utilities' dirty work, since many of these reactors would have been canceled anyway due to lack of demand. PUCs are also charged with setting utility rates, so that battles over who will pay the high costs of nuclear plants have arisen.[28] But the way the cost is split between rate payers and shareholders is an issue of the 1980s, and hence does not help explain why reactor orders disappeared in the 1970s.

Ever resourceful, the antinuclear movement tried many other avenues at the local and state levels. It has delayed at least two plants (Seabrook and Shoreham) by convincing local officeholders not to participate in evacuation plans. Since TMI the NRC has required detailed plans for evacuating people within ten miles of reactors, and the plans involved heavy state and local participation. When Seabrook and Shoreham applied for licenses to test at low power, the NRC could not grant them because state and local officials had not helped develop evacuation plans. The NRC then decided to change its own rules. Antinuclear activists pushed congressional legislation preventing the NRC from changing these rules, but the amendment lost in August 1987 by a vote of 261 to 160. Activists said they would have the legislation introduced again, but just after the November 1988 election, Ronald Reagan issued an executive order supporting the NRC. Nevertheless the licenses were delayed for months, and the New York State government will probably prevent the Shoreham reactor from opening, showing some decentralization of authority.

Other tactics: Antinuclear referenda continue to reach state ballots; for example, Maine voters have had three chances in the 1980s to close their one nuclear plant. Not until 1989 did a referendum to shut a nuclear plant succeed, closing the Rancho Seco plant near Sacramento, California. Washington State voters passed a 1981 initiative requiring public agencies to secure voter approval before issuing bonds to finance electric plants larger than 25 MWe capacity, further reducing the ability of the Washington Public Power System to complete its reactors. In 1981 the city of Austin, Texas, decided to sell its 16 percent share in the South Texas Nuclear Project; after several years searching for a buyer, it went so far as to appropriate $250,000 to convince its partners to cancel the project altogether.[29] A final example of innovative opposition was the

---

cles on specific state siting laws include Colton (1986) on Iowa and Weinberg (1980) on New York.

[28] PUC's have increasingly required utilities to absorb some of the costs of constructions overruns and plant cancellations. In 1989 the Supreme Court ruled that PUCs can force utilities to cover all the costs of cancellations even when the utilities have acted prudently (*Duquesne Light Co. and Pennsylvania Power Co. v. Barasch*).

[29] *The Nuclear Monitor*, 6 October 1986, p. 6.

New Hampshire drive for legislation preventing utilities from charging for construction in progress. Conservative Governor Meldrim Thomson vetoed such a bill, but his successor after the 1978 election, Hugh Gallen, signed it. This 1979 law helped drive the Public Service Company, which owns 36 percent of Seabrook, into bankruptcy.

The nuclear plants that have been canceled would have been canceled regardless of state regulations. This is clear if we examine the states with the best regulatory environments for reactor siting. States with "minimal power-plant siting legislation" (as defined by Ringleb 1986, 193) and "favorable climates" for rate relief (defined by Navarro 1986, 342) are Hawaii, Indiana, North Carolina, Texas, and Utah. Out of these five most favorable states, only North Carolina had nuclear reactors in operation in the mid-1980s. In Texas the Comanche Peak and South Texas plants began coming on line only in 1988. Indiana's only reactors, Marble Hill, were canceled in 1984; the best state regulatory environments could not save them from bad management. These five states had less than their share of nuclear energy in spite of favorable state regulations.

In summary, increased state authority over nuclear reactors arose because of the latter's problems; it did not cause them. States began to regulate nuclear reactors only after new orders had stopped. State authority could expand so easily because it supported and used the cost-benefit doubts that had already begun to undermine nuclear energy in national politics. When the states had any effect on nuclear plans, it was to delay construction long enough for economic rationality to catch up with the utilities and suggest canceling nuclear plants. While the federal courts did little to change the balance of power, they at least allowed some new state laws to stand. Most of these laws required full consideration of energy alternatives, including conservation (see, for example, Varanini 1981). PUCs became involved in nuclear regulation only because the costs of nuclear construction had skyrocketed. PUCs were called upon to solve the tricky problem of who would pay.

THREE MILE ISLAND AND SINCE

In the middle of the cost increases and decentralization of authority described above, the Three Mile Island (TMI) accident occurred. The March 1979 event made the public aware of the regulatory, safety, management, and cost problems of the nuclear industry. It boosted the antinuclear movement and turned American public opinion antinuclear for the first time. Just as the oil crisis had brought energy policy to the center of public attention, TMI put nuclear energy there. The drama of the days following the accident, when no one was certain what was happening, shocked all countries with nuclear energy. President Carter's visit to the plant only highlighted the event's importance.

In addition to an immediate agenda-setting influence, the accident had several longer term effects. It aggravated the many problems of nuclear energy; it helped observers recognize some of them for the first time; and it provided an urgent incentive for the nuclear industry to try to solve them. First, the accident showed that nuclear energy's survival depended on improving the way most utilities operated plants. Second, it indicated that nuclear regulation should be made tougher and more independent. Third, it told financial markets that nuclear energy was a clear risk, so that they charged even higher premiums for financing nuclear generation. In other words, TMI highlighted more clearly than ever the weaknesses that had been hidden in the 1960s and had begun to hurt nuclear energy in the 1970s.[30]

President Carter created the Kemeny Commission to investigate TMI and prepare recommendations for improving nuclear safety in the future. The commission's report placed the blame for TMI, and the main risk to the entire industry, squarely on "people problems" in both utilities and the NRC. Technological enthusiasm encouraged utility managers to believe that as long as they had sound physical technologies, there was no need to worry much about the people operating them. The most dangerous result of casual management was poor operator training. Kemeny (1981, 5) says: "We dug deeper and discovered that all theoretical background had essentially been removed from operator training programs. They were trained for button pushing—totally adequate for normal operating conditions—but they really had not been prepared for a serious emergency."[31] Utilities agreed that operator training needed improvement, without endorsing a broader critique of management attitudes. Within months of TMI a Nuclear Safety Analysis Center was established to study nuclear safety, and the Institute of Nuclear Power Operations was to develop better methods for training reactor operators. The NSAC had a staff of fifty and a 1979 budget of $3.5 million, and INPO had two hundred professional staff members and a 1980 budget of $11 million.[32]

Have these measures helped? A tentative answer is yes, but not

[30] The main reports deriving the lessons of TMI are the President's Commission's report (Kemeny 1979) and the Rogovin Report commissioned by the NRC. See Kemeny (1981) and Marrett (1981) for reflections on the Kemeny Commission work by two members. On the institutional effects of TMI, see the contributions to Moss and Sills (1981) and Sills et al. (1982). Wood and Schultz (1988) is a lengthy bibliography of materials on TMI.

[31] In a "Nova" program (29 March 1983), Michio Kaku asked, "Shouldn't we train nuclear power operators to be like airline pilots that make perhaps $100,000 a year only to make airplanes safe on the take-off and the landing, only just a few minutes out of every flight?" Most American operators have only a high school education.

[32] See Szalay (1981) and Marrett (1981) for details.

enough. INPO provides on-site inspections for its members; far more control room simulators are used in training than before TMI; operators now have regular shifts for retraining; workers' radiation exposure may be lower; various accident sequences are being studied carefully. At the same time the average number of unplanned reactor shutdowns has fallen, but remains high; capacity factors are still low; and operators lack much theoretical understanding of reactors. As of 1989, 80 percent of nuclear plants had not even made all the safety changes required after TMI.[33] Most of all, there is still a huge gap between the best- and worst-run nuclear plants. One suspects that the better-managed utilities have taken advantage of the information generated by TMI, while the problem utilities have not; it takes good management to recognize and utilize new resources. Thus in 1987 the NRC closed one of the country's "problem plants" because operators were habitually sleeping on the job. At least one utility had learned little from TMI.[34]

The NRC was strongly criticized for doing nothing to change utility management and attitudes and for having many of the same attitudes. Hands-off, formal regulation, an "emphasis on hardware," and technological confidence were all behind the NRC mindset that had to be changed. The Rogovin Report, created largely by NRC staff, concluded (p. 89), "The NRC, for its part, has virtually ignored the critical areas of operator training, human factors engineering, utility management, and technical qualification." The NRC was not even equipped to learn from operating experience at plants under its purview; a 1977 accident showed the possibility of a TMI-type accident and how to avoid it, but its lesson was not publicized.[35] The Kemeny Commission criticized the same weakness that other observers had for a decade: the NRC focused on licensing at the expense of ongoing safety of operations.

It is impossible to measure how much the NRC has changed since these reports appeared. It has probably improved, but probably not enough. Fines for utility violations increased dramatically in the months following TMI, but they may not have been sufficient to change utility operations

[33] *New York Times*, "Nuclear Safety Goals Are Not Met," 27 March 1989, p. D4.

[34] Philadelphia Electric's Peach Bottom plant was ordered shut on 31 March. Almost exactly eight years after TMI, the *New York Times* headline (3 April 1987) read, "Reactor Closing Shows Industry's People Problem." Philadelphia Electric was later fined $1.25 million by the NRC and sued for $500 million by the plant's three co-owners.

[35] At Davis Besse in 1977 and Beznau, Switzerland, in 1974, similar malfunctions had occurred, but the problems had been diagnosed and the proper valves closed before much damage was done. The Davis Besse incident was analyzed extensively, and two engineers pointed out how serious a risk (of a TMI-type accident) existed. The Kemeny Commission actually found nine incidents that could have provided information helpful in avoiding TMI. But the NRC did not use this information in strengthening internal utility operations. See also Budnitz (1981), Klein (1981), and Rogovin (1980, 94ff.).

(Fenn 1980, 53). Some of the Kemeny and Rogovin recommendations were followed, but not the most important ones, such as the creation of a single strong NRC head. Skeptics point to the fact that most of the reform was left to the NRC itself, and critics continue to attack the NRC for familiar faults.[36] A major thrust of the NRC's reaction has been to require new equipment at reactors, precisely the technological fixation that led it to ignore "people problems" in the first place.

The NRC is still regularly accused of being a captured agency. In 1987 members of Congress and some NRC employees accused the agency of relaxing safety regulation under pressures from industry. The chairs of several congressional subcommittees with nuclear oversight called for NRC Commissioner Thomas Roberts to resign because of improper ties with the nuclear industry.[37] NRC employees testified in Congress that a program to deal with drinking and drug abuse by reactor operators was abandoned because industry opposed it. Industry and the NRC still seemed reluctant to deal with all the people problems TMI had put in the spotlight.

In addition to revealing regulatory and managerial problems, TMI affected nuclear finances by making nuclear energy look riskier to the investor. By examining investor willingness to invest in nuclear as opposed to nonnuclear utilities, Hewlett (1984) found that the accident hurt nuclear financing both when the accident occurred in 1979 and when the estimates of enormous cleanup costs were announced in November 1980. General Public Utilities, the reactor's owner, claimed that cleanup costs would exceed $1 billion; since insurance totaled only $300 million, bankruptcy seemed a strong possibility (Hewlett 1984, 29). Investors demanded returns one to two percentage points higher from nuclear utilities than from nonnuclear utilities, a significant difference when the nominal rates were in the vicinity of 10 percent. Soon after TMI, thirty-five nuclear utilities formed Nuclear Electric Insurance Limited to obtain additional insurance coverage. Private insurers were unwilling to provide it, so the members of NEIL agreed to assess themselves millions of dollars each to cover claims in case of future accidents.

TMI also had effects outside the nuclear industry. American public opinion became predominantly antinuclear for the first time. Antinuclear organizations, still active at the time of TMI, moved quickly to take ad-

---

[36] See Marrett (1981), Klein (1981), and the Union of Concerned Scientists (1985). Temples (1982) additionally argues that structural reforms had little effect because they left the NRC's mindset unchanged.

[37] Charges included passing confidential documents to a utility under investigation and trying to thwart an internal NRC probe of the charges. Victor Stello, NRC Executive Director of Operations, was also accused of advising the TVA to stonewall against NRC requests for information.

vantage of the publicity. The spring of 1979 saw demonstrations in many cities, the largest of them drawing hundreds of thousands of protestors. Members of Congress noticed the changing sentiments. Before TMI there had been occasional skepticism, especially on the issue of costs (U.S. Congress 1978), but since TMI, most relevant committee heads have been relatively critical of utilities, the NRC, and reactor manufacturers. Nuclear energy had become an issue that could further the career of an entrepreneurial politician.

Since the TMI accident in 1979, further events have continued to batter nuclear energy, reinforcing the cost problems and decentralization of authority described earlier. Nineteen eighty saw the first cancellations of nuclear plants with substantial investments already made; by 1982 more than forty plants representing more than $50 million *each* had been scrapped (Hewlett 1984, 7). More than $30 billion in sunk costs were lost to canceled projects in the 1980s.[38] One of the most dramatic cancellations was the Washington Public Power Supply System in the summer of 1983, which dropped three of five planned reactors (worth $2 billion) and caused a default on municipal bonds.[39]

On 26 April 1986 an accident far worse than TMI occurred at the Chernobyl nuclear plant in the Soviet Union.[40] There were several similarities to TMI: the accident occurred early in the morning, operators had disabled some of the safety systems, and operators had a slim understanding of what they were doing. The nuclear systems in both countries were characterized by a complacency about the unlikelihood of reactor accidents.[41] American nuclear experts reacted in the same unhelpful way that French nucleocrats had reacted to TMI: "It couldn't happen here." Perhaps an identical accident could not, but many other, equally devastating accidents are possible and likely, since any reactor is a fragile and dangerous system.[42]

---

[38] *New York Times*, "Nuclear Plant Drain Put at $100 Billion for U.S." 1 February 1988, p. D1. This article estimates another $70 billion for the "extra cost of operating nuclear plants over that of plants using other technology."

[39] Leigland and Lamb (1986) document the WPPSS case, placing most of the blame on top management's lack of expertise with nuclear power and its lack of strategic or operational planning. See also Sugai (1987).

[40] See Ahearne (1986), Wilson (1986), and *The Observer* (1986).

[41] If casual managerial attitudes have been a problem, how did the Soviet Union react to the same problems American utilities demonstrated? The top six officials in charge of Chernobyl were sentenced to labor camp (three of them for ten years). *New York Times*, 30 July 1987.

[42] Charles Perrow (1984) has brilliantly shown that any complex and tightly coupled system like a nuclear reactor will inevitably have accidents. The relevant question is not how to avoid them but how to contain them, and whether society should take the risk to begin with. Although poor management is the cause of many of the United States' nuclear problems, even the best management could not make nuclear fission perfectly safe.

Chernobyl has not changed nuclear energy much in the United States because there is little left that can be changed. It made the public even more antinuclear, but it was already heavily antinuclear. Reactor manufacturers had already given up on the United States market. TMI had already taught the industry lessons about tighter operator training and accident sequence research. The NRC may decide to continue its research on the effects of radiation on humans, and Congress may ask it to oversee the military reactors owned by the Department of Energy (Ahearne 1986). Otherwise, nuclear energy policy seems unchanged.

In summary, the TMI accident let everyone know what problems American nuclear energy faced. Later incidents merely confirmed the death of the industry. TMI also suggested ways in which owners of reactors might improve their performance and safety, and many of the proper steps have been taken. But as with the entire history of commercial nuclear development, it is the well-managed utilities that have learned from TMI, while the problem utilities remain problems. TMI exacerbated some of nuclear's problems precisely by clarifying them (for example, scaring off investors), but it opened a small chance for solving some of them. As for saving future nuclear expansion, the post-TMI reforms were too little and too late.

LINGERING ISSUES

With America's last LWRs likely to come into service in the early 1990s, there remain two controversial issues related to the aftermath of nuclear energy. Who will pay for it? This involves not just remaining construction costs, but the costs of decommissioning plants and cleaning up particularly radioactive sites. Also, what will we do with the radioactive wastes? These questions have nothing to do with explaining past nuclear programs, but they are important to the possibility of any nuclear revival. With one or two exceptions, questions about which nuclear plants will be built and operated has been answered. But what then?

State PUCs decide how much of the cost of a nuclear plant is recoverable through rate increases. Although they have been more skeptical about the costs of canceled plants than completed ones, in most cases they have allowed almost all construction costs to be recovered.[43] They do not have to, as the 1989 Supreme Court decision in the *Duquesne Light v. Barasch* showed; even though it was ruled prudent for the company to have planned a nuclear plant as well as to have later canceled it,

[43] Two examples from 1987: Philadelphia Electric was required to refund $2.4 million spent to promote nuclear energy; Southern California Edison was allowed to recover 94 percent of its investment in San Onofre and 98 percent of Palo Verde. PUCs are squeezing only slightly.

the court ruled that the Pennsylvania PUC had the right not to grant rate increases to cover the canceled plant. PUCs have wide discretion in allocating costs. When the costs of decommissioning reactors—still largely unknown—come due, PUCs will probably pass them to consumers as well. In addition, American utilities owe $9 billion to the federal government for fuel enrichment fees that were undercollected; if these debts are not excused (they probably will be), ratepayers may be liable.

In 1988 the woeful state of nuclear facilities owned by the Department of Energy became known. Decades of accidents, leaks, and a general disregard for safe practices were documented. Although most of the facilities dealt with weapons production, some were research laboratories and other uranium enrichment plants. Costs for cleaning up these sites may reach $200 billion, another dark cloud over nuclear energy.

The longest-term challenge of nuclear energy is waste disposal. When Jimmy Carter refused to support the reprocessing of "spent" reactor fuel, and private industry also found it uneconomical, the fuel became simply a waste that would have to be stored safely for the thousands of years it takes to lose most radioactivity. The Department of Energy has spent over $700 million on a Waste Isolation Pilot Plant (WIPP) in New Mexico salt caverns that may never open because of shifts and cracks in the salt formation. For ten years state and federal officials struggled with each other over authority, but state officials have not managed to attain official veto power. One source of suspicion and problems has been the Department of Energy's tendency to make sudden decisions without consultation (Carter 1987, 176–91). Finally, Congress passed the Nuclear Waste Policy Act in 1982, setting up a process for choosing an underground site for high-level nuclear waste, and in 1987 Yucca Mountain, Nevada, was selected. But little progress has been made toward building a repository there. WIPP and Yucca Mountain show that political battles continue around authority over nuclear waste.

CONCLUSIONS

Why did nuclear energy collapse in the United States? The hidden weaknesses of the 1960s—utility financing, premature deployment, casual managerial attitudes, and hands-off regulation—began to have an effect in the early 1970s, and they overwhelmed nuclear deployment by the late 1970s. Yet nuclear energy was not widely seen as in trouble until the late 1970s. Only TMI in 1979 revealed clearly the nature of its problems. The hidden weaknesses contributed to nuclear energy's cost problems, which in turn caught the attention of cost-benefiters and contributed to the collapse of orders for nuclear plants. The technological enthusiasts who had controlled nuclear decisions at utilities and the government in the 1960s were blind to all the weaknesses. The outsiders who became concerned with nuclear energy because of the antinuclear movement and the oil crisis were better equipped to see them.

Utility financing was the first hidden weakness to surface, since there were obvious problems by the early 1970s. But before the oil crisis, the engineers in charge of electricity production saw no alternative to building generating capacity, and nuclear plant orders continued. The oil crisis encouraged cost-benefit thinking, first by policymakers, then PUCs, and then utilities; it became obvious that price increases were an alternative to building new plants. Only then did financial difficulties begin to hurt reactor deployment.

Problems in financing new construction, premature deployment, and casual managerial attitudes raised the costs of nuclear plants beyond the range dreamed possible. Management problems included a slim understanding of nuclear energy and how it differed from coal electricity, undue faith in the hardware and downplaying of people problems, and simple lack of control. Sloppy management and hands-off regulation were a disastrous combination. Plants were often not designed properly, so that costly retrofits had to be made. Part of the development process, in fact, continued after dozens of plants were operating or under construction. High costs resulted as well from every sort of managerial mistake, and the AEC and NRC were only equipped to deal with them after the fact. Since development processes were still occurring also meant that standardization would have been difficult in the late 1960s, when most plants were ordered. Finally, a few utilities—generally those that should not have entered the nuclear field—have especially sullied the reputation of the entire industry through spectacular mistakes in construction and operation.

"Cost problems" were words that policymakers and utility managers understood better than the moralist language of much of the antinuclear movement, which had otherwise made little dent in the policymaking arena. Costs became especially important in the period after the oil crisis, as economic rationality and cost-comparisons spread through energy policymaking and the utilities. Cost-benefiters could no longer accept the assumptions of the technological enthusiasts, as they had in the 1960s.

Paradoxically, America's early enthusiasm and confidence in nuclear energy helped kill it. Cost problems arose because we deployed reactors before they were truly viable commercially, because most utility managers did not see the special care they had to take in constructing and running them, and because regulators thought a hands-off approach would be sufficient given reactors' advanced technological designs. In the 1960s technological enthusiasm had resulted in a rush of orders for a technology that may not have been ready for commercial use, and many utilities built reactors without the expertise to operate them effectively. Technological enthusiasm blinded all participants to fundamental problems, and by the time these were recognized it was too late to save nuclear energy.

The political and regulatory system created new obstacles to nuclear energy in the late 1970s, but the decentralization of authority came in response to nuclear energy's problems; it was not their cause. State PUCs gained some authority because the financial situation of nuclear utilities was so bad. Courts also allowed state laws to protect citizens and ratepayers from undue financial strain. But the decentralization of authority did not proceed very far: when an obstacle like the lack of local and state cooperation on evacuation plans threatened Seabrook and Shoreham, the NRC could simply change the rules so as not to require local cooperation. Likewise TMI did not damage nuclear energy so much as reveal the problems it already had.

By the early 1980s political and economic structures had hardened enough to rob nuclear policymakers of much significant discretion: there was little anyone could do to revive America's nuclear program. Technological enthusiasts like Alvin Weinberg are pushing new reactor designs said to be "inherently safe."[44] Since this is what many believed about regular light water reactors thirty years ago, there is little chance policymakers or the public will believe the enthusiasts again very soon. On the other hand, the growing threat of the greenhouse effect and acid rain may encourage a rethinking of nuclear energy; 1988 bills by Senator Tim Wirth and others would have encouraged energy efficiency and the development of new reactor designs. But nuclear energy in the United States has dug itself a hole so deep that it will take many, many years to climb out, if it can at all.

[44] See Weinberg et al. (1984), Hannerz (1982), Marshall (1984), and Agnew and Johnson (1986). Two designs are the Modular High Temperature Gas-Cooled Reactor being developed in the United States and the Process Inherent Ultimately Safe Reactor from Sweden. These designs would shut themselves down without the aid of humans or electricity in the event of an accident.

CHAPTER 12

Political Paralysis and Antinuclear "Compromise" in Sweden

IF THE ECONOMICS of nuclear energy became decisive in the United States in the late 1970s, Swedish nuclear development continued to be dominated by partisan conflict. For several years nuclear energy was the most prominent symbol of differences between political parties, so its other meanings faded. The reason was the 1976 election of Thorbjörn Fälldin as prime minister, the only antinuclear moralist put at the head of a country during the period of public debate over nuclear energy. Although reluctant to compromise on the nuclear question, he was overwhelmed by the political system and the technical bureaucracy and forced to make a series of decisions allowing reactors to start. After two years in office he refused to compromise further and resigned. Antinuclear moralism and running a government, especially a system founded on the idea of compromise, proved incompatible.

The Liberal government that took office in October 1978 quickly moved to solve the nuclear issue by defining it in narrow technical terms rather than political ones. Its program was similar to that of the 1975 Energy Bill, with twelve instead of thirteen reactors and a large conservation effort. In spite of split attitudes in their own party and protest from the antinuclear parties and organizations, the Liberals were on the verge of succeeding with their energy plan when the Three Mile Island accident occurred. After TMI the Social Democrats, who dominated Swedish political life even out of office, found it prudent to call for a national referendum on nuclear energy. Here was a way of uncoupling party cleavages and the nuclear question.

Sweden's 1980 referendum results were interpreted as calling for an end to the nuclear program after twelve reactors and phasing out these reactors by 2010. Governments of the left and the right have clung to this policy out of desperate relief at removing nuclear power from the partisan political agenda. The Chernobyl accident in 1986 raised the level of antinuclear rhetoric but did not change energy policy. The nuclear industry and antinuclear activists remain quiet, but they foresee another round of debate in the 1990s.

A Moralist in Power

The bourgeois government formed in October 1976 consisted of the antinuclear Center Party and two pronuclear parties, the Liberals and the Conservatives, with Fälldin as prime minister and antinuclear Olof Johansson as energy minister. From the first day of the government to the last, the Center Party was torn between its uncompromising moral stance against nuclear energy and the need for compromise in running a coalition government. As soon as the election was over, Liberals and Conservatives ended their silence on nuclear energy, arguing that it would be irresponsible to abandon nuclear plants at advanced stages of construction—precisely what Fälldin had promised to do. Just as nuclear energy had been a lightning rod for differences between the Center Party and the Social Democrats, it became the focus of conflict within the coalition government and the controversy that destroyed it.

Political compromise won on the coalition's first day in office, when the government allowed fueling of the Barsebäck 2 reactor, contrary to Fälldin's campaign promises. In return for this concession, Fälldin insisted that the fuel loading be provisional on (1) the development within a year of satisfactory arrangements for the reactor's spent fuel, and (2) the creation of a commission to outline the options for the reconsideration of energy policy, already planned for 1978. By 30 December, the formal apparatus had been developed for assessing these and all future fuel arrangements, in the form of the Stipulation Act. An Energy Commission was also set up in late 1976.

The Nuclear Stipulation Act, passed in April 1977, required that operators show, prior to the loading of fuel in new reactors, how and where they would deal with the spent fuel "with absolute safety."[1] The fuel could be reprocessed or not, but there would in either case have to be a final resting place in hand. Sweden's five nuclear reactors operating by early 1976 were not covered by the Stipulation Act; Barsebäck 2 was allowed to operate under special status; the four reactors under construction and the three planned would be covered by the act.

Both the Center Party and the pronuclear parties agreed to this initial compromise because all interpreted the Stipulation Act according to their own particular policy styles and assumptions about technology. Because of their skepticism about technological development, Center Party leaders believed that no technology existed or could be developed to handle spent nuclear fuel with absolute safety, so that the Stipulation Act would eventually stop nuclear development (Vedung 1979). Liberal and Con-

---

[1] Johansson and Steen (1981) have described the Stipulation Act, including the conflict over its interpretation. Johansson and Steen (1979) is a useful summary.

servative leaders had enough technological enthusiasm to be confident that the terms of the Stipulation Act could easily be met (Johansson and Steen 1981, 7). Each contender was confident that reality was on her side, that technological development would unfold as she pictured it, and that the act would lead to the policy she favored. Yet the machinery designed to lead to determinant outcomes and avoid political strife would only cause further conflict. The question of whether nuclear energy was safe became that of whether the proposed waste technology was "absolutely safe." No scientific evidence could prove one policy style right and the other wrong.

The Stipulation Act prolonged the squabbling over nuclear energy instead of ending it, since each application sparked intragovernmental debate over whether the act's requirements had been met. In September 1977 Barsebäck 2's owners applied for a permanent license, and after intense debate the Center Party was again forced to compromise and grant the license. In December 1977 the State Power Board asked permission to load its Ringhals 3 plant, the first case under the normal Stipulation Act procedures. One side said the requirements had been met; the other said they had not. Months of reviews followed, but in September 1978 the government decided that the stringent requirements had not been met. In rejecting the application, however, politicians used narrow technical grounds, specifically that a satisfactory rock formation had not been found. While this tactic stopped the application, it meant that if such a formation *were* found the application would have to be accepted. The shift toward technical criteria was supported by the simultaneous decision to let the Nuclear Inspectorate (SKI) decide whether Stipulation Act criteria had been met in future cases. The Center Party had accepted the *theoretical* existence of an absolutely safe technology for spent fuel, since it felt certain the technology would never *empirically* exist.

Like the Stipulation Act, the Energy Commission that started operations in the winter of 1977 generated more conflict than consensus.[2] The governing coalition fought over who would fill the seats beyond those reserved for industry, labor, and the political parties. Especially important was who would chair the commission, since he or she would have great influence on the final outcome.[3] After haggling for three months,

[2] My discussion of the Energy Commission relies especially on interviews with six of its members: Ove Rainer, chairperson; Birgitta Dahl; Birgitta Hambraeus; Olof Hörmander; Per Kågeson; and Carl Tham.

[3] Steven Kelman (1981a, chap. 4) has described the dynamics of these commissions, arguing that they achieve consensus because in small groups the members come to respect each other, and because the Swedish tradition of deference helps the representatives persuade the groups they represent to follow the decisions taken by the commissions. Kelman's image of the commissions relies too heavily on their formal characteristics. In fact, accord-

the three coalition leaders chose Ove Rainer, a high civil servant who had chaired many committees, largely as a compromise because he was not associated with any particular position on the issue. Then they fought over what other representatives to include. The commission was much larger than most Swedish commissions, with fifteen members, a staff of 120, and a budget of 45 million crowns (roughly $8 million). The commission operated under intense media scrutiny, with members constantly badgered for comments by two dozen journalists who followed the meetings. Every disagreement was read as a sign of conflict within the governing coalition.

It became clear that Fälldin had failed to find antinuclear commissioners beyond the two representatives from the Center Party and one from the Communists. He had also misinterpreted Rainer as more antinuclear than he turned out. By January 1978 Rainer had developed, in the normal manner of Swedish commissions, a compromise plan for ten reactors, three fewer than were currently being planned. This program would have curtailed Sweden's deployment after those reactors then operating or on the verge of operation. According to Rainer, if Fälldin had accepted the compromise, the nuclear question in Swedish politics would have ended. For two months his coalition partners tried to convince Fälldin to accept the compromise, but in the end he refused. The commission's report of June 1978 recommended twelve reactors, but by a divided vote of twelve to three. Like the Stipulation Act, the Energy Commission buried its politically sensitive calculations in a mass of technical details.

The Center Party had seen its antinuclear position eroded decision-by-decision for almost two years. With each defeat, a few young party members resigned and several newspapers denounced Fälldin for breaking campaign promises. Only rarely did Fälldin consciously decide to compromise; more often he was defeated by the system of bureaucratic, technical decisionmaking. Olof Johansson, the Center Party minister of energy, felt isolated at the ministry, surrounded by civil servants who had been chosen by the Social Democrats and were generally pronuclear. It was hard for the Center Party to find antinuclear experts to help in the ministry or to place on the Energy Commission staff (Abrahamson 1979, 34). It even had to hire an American, Dean Abrahamson, as an energy

---

ing to Bjurulf and Swahn (1980), one out of four Swedish commissions does not reach a consensus, while one out of three consists of an individual person. This implies that of those commissions where disagreement is possible, three out of eight fail to reach consensus. The dynamics of the group may be less important than what ideas and goals the members bring to it. In labor-capital issues (Kelman studied workplace safety) noncompromising moral politics will play a smaller role than in new issues like nuclear energy. The chairperson's agenda is also crucial, because it may range from isolating a minority position to playing up the disagreement.

adviser. Once Fälldin had made the early mistake of letting the nuclear issue be defined—at least potentially—in technical terms, he had lost several of the battles to come.

The ecology movement was unable to provide much support, partly because it had split over leadership issues in 1976. It could neither help pressure the other political parties nor organize engineers or scientists who might have questioned some aspects of the Swedish nuclear program. The Center Party had to rely on its own resources, but it had no energy experts respected by members of the Swedish energy establishment. According to Abrahamson (1979, 34): "Fälldin had few visible allies outside his own party, and no organized allies. The unified pronuclear forces applied enormous pressure from their side but there was virtually no organized counterbalancing pressure from the other."

After a series of demoralizing little defeats that allowed further reactors to be built and come into operation, Center Party leaders began to resist. Refusing the Energy Commission's compromise plan was the first instance. The second came over funding for an eleventh reactor, not yet under construction and not clearly needed for its electricity. The main argument for the Forsmark 3 reactor was to keep the nuclear industry alive, something Fälldin could not agree to. Olof Johansson, closer to the women's and young people's segments of the party than Fälldin, interpreted an energy proposal in a way unacceptable to the Liberals and Conservatives, and at a Center Party congress in October 1978 he received support for his refusal to compromise. Buoyed by a party united, and drained by so many compromises, Fälldin resigned, saying, "All parties in a coalition must be able to compromise, but no coalition party should demand of another to extinguish its soul."[4] Many Center Party leaders felt they had at last done the right thing. Party activists at the October congress seemed to support them, and opinion polls showed their popularity increased after Fälldin resigned. Their final no began to restore their tarnished moral image. Only a party of moralists could feel triumphant in the collapse of their government.

THE LIBERALS FINESSE THE CONTROVERSY

After Fälldin resigned, the Liberals refused to form a government with the Conservatives, whom they considered somewhat extreme. With the tolerance of the Social Democrats, still by far the largest party in the Riksdag, the Liberals (with only thirty-nine out of 349 seats) formed their own government. Tacitly, they had one task to accomplish while waiting

---

[4] Abrams (1979, 39). My discussion relies on Vedung's excellent and detailed account (1979), and on an interview with Olof Johansson.

for the elections the following September: to solve the energy question. They quickly took steps to do this, and a coalition with the Conservatives would have complicated their efforts.[5]

Energy Minister Carl Tham worked closely with several Social Democrats in developing an energy bill. One reason was that, along with Prime Minister Ola Ullsten, Tham belonged to the party's left wing, which was close to the Social Democrats ideologically. Another reason was Tham's concern at having a bill that would be passed by the Riksdag, so he also consulted with the Conservatives. Tham set up another commission and a study group, and he tried to dissuade OKG, a private company, from building an additional reactor, but for the most part the bill he presented in March 1979 followed the 1978 recommendations of the Energy Commission. The bill proposed that nuclear construction be stopped after twelve reactors—the number recommended by the Energy Commission and a compromise between the thirteen favored by the Social Democrats and the eleven by the Liberals. The bill was scheduled to be considered in June.[6]

While formulating its long-term plan, the Liberal government also moved decisively to solve the specific decisions that had split the previous coalition. With the support of the Conservatives and the Social Democrats, it approved funding for Forsmark 3, which would become Sweden's eleventh reactor. There remained the difficult question of whether the Stipulation Act had been satisfied so that the two waiting reactors could be started. In February 1979 nuclear industry researchers presented SKI with new drilling results to show that a safe repository for spent fuel was now available. Yet seven out of eight geologists in a review group decided that the new drilling site still did not meet the tough standards of the Stipulation Act. On 27 March the SKI Board of Directors chose to ignore this review and approve the application, on the grounds that it had not been proven there was *not* a satisfactory repository at the new site.[7] With this suppression of technical findings for political ends, there was little doubt the Liberal government would approve fueling the

---

[5] This section is based on interviews with Carl Tham, Liberal energy minister; Hadar Cars, Liberal MP; and Peter Steen. On the Liberals and the formation of their government see Sahr (1985, 99ff.), Vedung (1979, chap. 5), Petersson (1979a; 1979b, chap. 7), and Abrahamson (1979). Tham (1981) has also presented his own views.

[6] The state commission report was published in 1980 as *Kärnkraftens Avfall: Organisation och Finansiering* SOU 1980:14 (Stockholm: Industry Ministry). The "yellow energy bill" was *Proposition 1978/79:115*.

[7] The board was composed of four politicians, one academic, and three people who had academic posts but were closely tied to the nuclear industry. The academic, Thomas Johansson, and the Center Party representative voted against the application. I have relied on Abrahamson (1979, 35–36) and a personal interview with Per Unckel, the board's Conservative representative.

two plants. Interpretations of the Stipulation Act depended on one's confidence in technological progress, and the Liberals believed that nuclear waste technology was or would be quite safe.

The Liberal government's strong moves to solve the energy controversy are remarkable because the party's own membership was sharply divided over the nuclear issue.[8] Like the Social Democrats, the Liberals had a visible antinuclear organization within the party, and unlike the Social Democrats and Conservatives, the Liberals lacked close ties to trade unions or industry organizations that would have pressured them to take a pronuclear stand. The symbolic message of calling for eleven rather than thirteen reactors was that the liberals were willing to compromise. But party leaders were still pronuclear; they believed that nuclear technology was relatively benign, so they concluded that the spirit of the Stipulation Act had been met.

It was precisely the split in party opinion on nuclear energy and the party's lack of strong ties to economic organizations that let the Liberal leaders redefine the nuclear question as one of governance. Conflict over nuclear energy had paralyzed the Swedish state machinery for two years; the Liberal government was determined to get it working again. The party believed itself well-suited to the task. It prided itself on representing neutral expertise and abstract principles rather than a particular group in the class struggle. Party leaders were able to hide divisions over nuclear energy by making energy policy a test of their ability to "get things done." They had wide discretion in solving the nuclear issue, so their own worldviews played a key role.

In six months the Liberals succeeded in stabilizing Sweden's energy plans, even if they stretched the meaning of the Stipulation Act to do it. There would be twelve nuclear reactors, but also a research program aimed at renewable resources and conservation. After two years of constant bickering between coalition partners, the lines of the 1975 energy bill would be renewed, minus one nuclear reactor. Yet the party's formidable success was for naught, since the day after the SKI board approved fuel loading, the Three Mile Island Accident (TMI) occurred in the United States.

Fallout from Three Mile Island

Media coverage of the TMI accident on 28 March 1979 was as heavy in Sweden as in other industrial countries, but there it led to a major change of position by the Social Democrats. On 4 April Olof Palme announced

---

[8] The urban middle class base of the party contains both business owners, who tend to be pronuclear, and professionals and middle white collar employees, who tend not to be. The party was split almost evenly on the nuclear issue.

that his party now favored a national referendum to decide Swedish nuclear energy policy. Before the referendum, state commissions would investigate nuclear safety and the implications of a complete shutdown, and until the vote all nuclear decisions would be postponed. Within hours the Conservatives and Liberals had agreed. The environmental groups and Center and Communist parties had been calling for a referendum for years and had increased their efforts in early 1979 as the Liberal government tried to close discussion on energy policy. The "People's Campaign" formed from these antinuclear groups and parties—roughly forty in all—had gathered 520,000 proreferendum signatures in only six weeks, the equivalent of 10 percent of the electorate (Little 1982, 20).[9]

Critics said Palme's call for a referendum was an opportunistic—albeit clever—move to heal the split within the Social Democratic Party over nuclear energy in time for the September 1979 elections. Liberals especially felt betrayed, believing they had solved the energy problem and that the referendum merely prolonged the controversy a year. For them TMI did not provide enough new information about nuclear energy to warrant a change of policy. SAFE—the organization of antinuclear Social Democrats—was small but very visible because of heavy media coverage,[10] and two days before Palme's announcement the president of the Young Socialists had called for a shutdown of nuclear plants pending further research (Little 1982, 4). Palme (and the small group of advisers he consulted) must have been aware that the referendum could defuse these persistent critics. In addition, by calling for a referendum to be held *after* the September 1979 elections, Palme could avoid the issue he thought had hurt him in 1976 by separating the nuclear issue from partisan politics.

Social Democrats claimed that supporting a referendum was compatible with their earlier policies. They had always tried to remain flexible, they said, especially when confronted with new information. And they had recently shifted their program from thirteen reactors, reiterated at their 1978 congress, to twelve, in line with the Liberals' pending energy bill. Palme's call for a referendum was probably not solely a political stratagem, because as such it was quite risky. Referenda always have the potential to shift partisan alignments, and the largest parties usually have the most to lose (Valen and Martinussen 1977). Palme hoped to let party

[9] This section is based on interviews with Birgitta Dahl, Lennart Daléus, Tor Ragnar Gerholm, Per Kågeson, Lars Liljegren, Måns Lönnroth, Carl Tham, and Per Unckel. Holmberg and Asp (1984) is an excellent study of the referendum campaign. Sahr (1985, 101–23), Zetterberg (1980), and Little (1982) are also useful sources.

[10] Holmberg and Asp (1984) exhaustively discuss the effects of the media. By presenting the reaction of the "other side" to every energy pronouncement, the media portrayed the party as more divided than it probably was.

supporters vote for or against nuclear energy directly, rather than using Riksdag elections for that purpose. But the risk was that once someone voted against her party on one issue, she would continue to do so in future elections. The referendum was a risky wager, and the Social Democrats would probably not have undertaken it if they had not been genuinely willing to reconsider their energy policies.

Both genuine flexibility and political opportunism probably played a role in Palme's decision, but the media universally inferred the latter. It is surprising that both nuclear proponents and prominent newspapers criticized Palme for bowing to public opinion rather than sticking to his pronuclear policies, with one paper proclaiming, "Palme Afraid of the Voters" (Sahr 1985, 105). The idea, described in chapter 8, was that Swedish politicians should lead, not follow. It is considered bad form for a politician to appeal to voter dissatisfaction.[11] In contrast it is hard to imagine American politicians being criticized for following the desires of the public, or for opening up decisionmaking to public pressure.

Comparing Social Democratic and Liberal strategies for dealing with antinuclear dissent within their parties shows the different connotations nuclear power had for the two parties. The Social Democrats saw nuclear energy as the main reason they were out of office. In spite of their best efforts to shift attention from nuclear reactors to total energy policy—especially conservation—nuclear energy symbolized many of the differences between the Social Democrats and the Center Party. It lined up perfectly with the dominant partisan cleavage in Swedish politics. Compare the United States election of 1976: nuclear energy did not appear as an issue because it provided neither party any leverage against the other. The Swedish Social Democrats saw TMI as a chance to end nuclear's life as a partisan issue, in spite of the other risks of referenda.

Nuclear energy was not such an important part of the Liberal Party's public identity, so Liberal leaders could play on another of the issue's connotations, that of government paralysis. According to Carl Tham, the Liberals would have preferred their policy of aggressively settling the energy issue, even after TMI.[12] The literal policies of the two parties concerning nuclear reactors were almost identical, but the connotations of those policies encouraged different political strategies.

The Liberal government followed all of Palme's suggestions. It set up two state commissions, both scheduled to produce reports by Novem-

[11] See Kelman (1976, 117). Olof Ruin (1982, 149) writes, "Strong government is taken to mean an ability to take unpopular decisions, an ability to distance itself from special interests, and an ability to formulate long-term plans."

[12] Personal interview with Carl Tham, June 1985.

ber.[13] In June the Riksdag passed the Liberal energy bill, minus the sections on nuclear energy, and the "Respite" law delaying the operation of any new nuclear reactors until after the referendum. The government would reimburse the owners for part of the cost of this delay. The planned referendum served the purpose of removing nuclear energy from the agenda of the 1979 election campaign, although this separation didn't allow the Social Democrats to win enough seats to control the Riksdag. After the September elections another coalition of the three bourgeois parties was set up, again with Fälldin as the prime minister.

That autumn the details of the referendum, scheduled for March 1980, were worked out. The antinuclear parties and organizations had proposed the wording for their ballot line in the spring: that no more reactors be built and those currently in use be phased out in ten years. The development of a pronuclear line was more complicated. Social Democrats and especially trade union representatives took the lead in developing this option, which called for completing the twelve reactors and using them until the end of their working lives. After that Sweden would be nonnuclear.

The pro- and antinuclear positions were remarkably similar; neither favored continued expansion of nuclear energy, and both foresaw a gradual phaseout, although at different speeds. Palme said to the antinuclear forces, "You have already won." More importantly, Sweden's normal channels for resolving conflicts, namely negotiations between representatives of parties and interest groups, had broken down. The strong norms that encourage political compromise and consensus in Sweden could not succeed against politicians and activists who were moralists, since the essence of a moral position is that it should not be compromised. The antinuclear moralists changed their position very little from 1973 to 1979, while the "pronuclear" position in Sweden had changed greatly since the early 1970s, when continual expansion of nuclear energy had been forecast. In any other country, a moratorium on new construction and an eventual phaseout would have been considered clearly antinuclear, but industrial representatives and the party closest to them, the Conservatives, didn't want another bourgeois coalition to collapse over

[13] The Reactor Safety Commission, chaired by Hans Löwbeer, submitted its report November 19: *Säker Kärnkraft?* SOU 1979:86 (Stockholm: Industry Department). The Consequences Commission, chaired by Lennart Sandgren, submitted its report November 23: *Om Vi Avvecklar Kärnkraften.* SOU 1979:83 (Stockholm, Industry Department). Like all commissions on nuclear energy, these were controversial, and like all the Liberal government's energy efforts, these were widely charged with having a pronuclear slant. Several antinuclear members of the Consequences Commission resigned in protest in October 1978, and a relatively objective member of this commission admits that nonnuclear energy options were not given a fair hearing (Peter Steen, personal interview).

nuclear energy, and they were afraid of even more antinuclear policies if voters were faced with only extreme choices. In a system that insists on compromise (as opposed to stalemate, as the United States tolerates), moralists often win.

Just before the Riksdag was to debate the referendum bill, the pronuclear parties decided to offer two lines instead of one. The main reason was that the Social Democrats did not want to support the same line as the Conservatives. When this had happened in neighboring Norway in the 1972 vote on common market membership, many Social Democratic voters had continued to vote Conservative after the election.[14] Norwegian Labor politicians themselves warned the Swedish Social Democrats that such an alliance could be disastrous. The Swedish Social Democrats therefore encouraged the Conservatives to present a separate line, especially by insisting on a clause calling for public ownership of all nuclear plants, an affront to the Conservatives' ideology. This clause would appear on the back of the Social Democratic line, while the fronts of the two ballots would be identical. The result: the Conservatives supported Line 1, the Social Democrats and Liberals Line 2, and the Center Party and Communists Line 3.[15]

---

[14] Valen (1972, 1973) and Valen and Martinussen (1977) discuss the fascinating effects of the EEC vote in Norway.

[15] All three ballots began with a statement that Sweden currently had six reactors in service, four ready for service, and two under construction. The first paragraph of Lines 1 and 2 read: "Existing nuclear power plants are to be closed down at a rate consistent with the need for electricity to maintain employment and welfare. In order to, among other things, reduce dependence on oil, and pending the availability of renewable energy sources, only the twelve reactors now operating, completed, or under construction will be used. There will be no further expansion of nuclear energy. Safety considerations will determine the order in which reactors are taken out of service."

Line 2 added, on the reverse side of the ballot: "Energy conservation will be vigorously promoted and further stimulated. The weakest groups in the community will be protected. Steps will be taken to steer electricity consumption, partly to prevent the heating of new, permanent buildings by electricity. Research and development concerning renewable energy resources will be accelerated under government auspices. Environmental and safety improvements will be made at nuclear plants. A special safety study will be made for each reactor. To inform the general public, a safety committee drawn from local inhabitants will be set up for each nuclear plant. Electricity production from oil- and coal-plants will be avoided. The main responsibility for producing and distributing electricity must be in public hands. Nuclear plants and any other future installations of any importance for electricity production must be owned by the state and the municipalities. Excess profits from hydroelectric production will be absorbed by taxation."

Line 3 read: "NO to the continued expansion of nuclear power. The phasing out of the six reactors currently in operation within at most ten years. A conservation plan to reduce our dependence on oil will be carried out on the basis of (1) continued and intensified saving of energy, (2) substantially increased investment in renewable sources of energy. The safety requirements for the operating reactors will be tightened. Reactors not yet charged will never be put into operation. The mining of uranium will not be permitted in our country.

The construction of the referendum was already a small victory for the same bureaucrats who had transformed Fälldin's political and moral attack on nuclear energy into a narrow technical question about rock formations. Instead of a simple yes or no vote, the referendum results were almost bound to be ambiguous, since it was unlikely that one line would get more votes than the other two combined. Not *too* ambiguous, since all three lines called for phasing out nuclear energy, but politicians and bureaucrats would have a chance to influence the interpretation of how quickly the phaseout would occur.

When the Riksdag passed legislation setting up the referendum and funding the three campaigns for the lines, it faced a similar dilemma. Either the three lines were distinct, or they were two pronuclear and one antinuclear lines. If they were distinct, they should each get the same funding, but after the referendum the votes for the two pronuclear lines could not be added together to support a policy of a longer phaseout. If the government admitted that two of the lines represented the same position, it would have to give as much funding to Line 3 alone as to the other two lines together. The government chose this option, even though in debates the lines were treated as three separate choices (with equal time).

Further differences emerged between the three lines as the highly visible and emotional campaign progressed from December to March. The campaigns for Lines 1 and 2 were run like regular election campaigns, and many of the staffers and volunteers were in fact party activists. Line 3 depended instead on a vast network of grassroots activists and organizations willing to go door to door; on artists who performed free; and, in the final weeks, on large bus tours of the country. An enormous 25 percent of those supporting Line 3 worked actively for its success, compared to 3 percent of those supporting Line 2 (Sahr 1985, 112). In one poll 35 percent of Line 3 supporters reported a "very strong" conviction on the issue, while only 15 percent of Line 2 voters and 29 percent of Line 1 voters did (Zetterberg 1980, 49).[16] In spite of their formal similarity, a distinction between Lines 1 and 2 also emerged, with the latter position-

---

If current of forthcoming safety analyses so require, this proposal naturally means that nuclear plants must be immediately shut down. Work to prevent nuclear weapons proliferation and atomic weapons must be intensified. No reprocessing of nuclear fuel will be permitted and the export of reactors and reactor technology will be discontinued. Employment will be increased by alternative energy production, more effective energy conservation, and the increased processing of raw materials."

[16] According to Zetterberg (1980, 50), Line 3 volunteers visited one in five homes in big cities and one in ten homes elsewhere in Sweden. He chides them for not reaching their goal of visiting every home, but this is probably the widest penetration of any antinuclear movement in the world. His criticism indicates what strong expectations of political participation Swedes have.

ing itself as the middle way, a compromise, and yet genuinely committed to a "reasonable" phasing out of nuclear energy. Supporters of Line 1, while formally backing a phaseout, seemed to many people to hope for some sort of continuation. To underline its sincerity, the Line 2 campaign produced a program for alternatives to nuclear energy.

American information and public figures remained important in the Swedish debate after TMI, just as they had been since public controversy began in the early 1970s. TMI itself and the American reactions to it—such as the Kemeny and Rogovin reports—were central, but there were additional imports from the United States. Three members of the Union of Concerned Scientists visited Sweden in March 1980, and their statements received heavy press coverage. Coverage was also given to an Oak Ridge report on Zhores Medvedev's (1979) *Nuclear Disaster in the Urals*, which described a large Soviet nuclear accident and contamination in the late 1950s. Line 3 sponsored a series of presentations and discussions at the Stockholm Technical Institute: Gordon MacLeod, Pennsylvania's secretary of health at the time of TMI, claimed that mental damage among newborns increased after the accident; a Colorado health official said that cancer rates were unusually high around the Rocky Flats plant in his state; at a third meeting, Hannes Alfvén spoke about TMI, as well as accusing Edward Teller of promoting nuclear war. An American documentary on weapons testing, "Paul Jacobs and the Nuclear Gang," having been rejected by Swedish television, was shown at theaters throughout Sweden. Whatever the scientific status of the claims presented, they received wide publicity in Swedish newspapers and television.[17]

The campaign began with opinion polls showing antinuclear Line 3 ahead with roughly 35 percent support, compared to roughly 30 percent for Line 2 and 25 percent for Line 1. Lines 2 and 3 both gained ground steadily at the expense of Line 1 and the "not sure" respondents. A poll taken days before the vote was the first one to show Line 2 ahead of Line 3, and that was also the result of the referendum. The vote: 39.1 percent for Line 2, 38.7 percent for Line 3, and 18.9 percent for Line 1. The main reason that Line 2 came from behind to win was the Social Democrats' mobilization of their supporters behind it. Whereas only 50 percent of their adherents were found to support Line 2 in January, 67 percent did by March. The party took a great risk by pushing hard for a position many of its members did not support, but it was aided by its ability to portray Line 2 as the "middle way."[18]

[17] These examples of American information used in the debates before the referendum are taken from Mills (1980).

[18] These results are taken from the analysis by Hans Zetterberg (1980), head of SIFO, Sweden's most prestigious polling organization. SIFO's accurate prediction several days before the referendum of a Line 2 victory surprised many people.

Sweden breathed a collective sigh of relief after the referendum, with leaders from all parties happily proclaiming that the people had spoken clearly and that there were no more energy conflicts to resolve.[19] Activists on both sides were simply exhausted. Within weeks the bourgeois government, although headed by Fälldin, approved fuel loading for four reactors completed but not yet operating: the two approved but pending since the Liberal government, and two new ones. Sweden would then have ten reactors in operation. In June the Riksdag approved brief legislation to the effect that Sweden would phase out its nuclear power by 2010. The Social Democrats had first mentioned the year during the referendum campaign, as their interpretation of how long the planned reactors would last, and it has since become "a holy year." In 1981 the government introduced more specific measures for reducing Sweden's dependence on both oil and nuclear energy, and in 1984—after another commission report—the Riksdag changed the basic nuclear legislation, eliminating the Stipulation Act.

Quiet Tension in the 1980s

Since the 1980 referendum and the resulting legislation, there has been almost no discussion of nuclear energy policy in Sweden. Both the bourgeois government and the Social Democratic one that followed it in 1982 have said they were following the 1981 legislation committed to abolishing nuclear energy by 2010. In the meantime, both governments increased the liability for nuclear accidents in the Swedish Nuclear Liability Act of 1983, submitted by the old government and passed by the new (Jacobsson 1985).

The Social Democratic minister of energy, Birgitta Dahl, confidently predicted that Sweden's extensive research into renewable energy, especially wind power, would provide adequate alternatives to nuclear and oil. As early as 1977, 39 percent of government funds for energy research, development, and demonstration were devoted to conservation, and 11 percent to renewable sources (OECD IEA 1978). Since then the latter figure has increased. Sweden has even been willing to fund research like that of Thomas Gold, who believes that deep underground there exists methane gas created when the earth was formed (rather than from later biological decay). Since 1986 the State Power Board has been drilling the deep hole necessary to test the theory, with mixed evidence so far. Conservation research has already had some results, giving Sweden an important energy "source." Research on renewable sources may

---

[19] Although legally the referendum was only advisory, politicians gladly treated it as binding.

do the same. Policies can create energy resources, as several members of the Secretariat for Futures Studies recognized (Lönnroth, Johansson, and Steen 1980, 563):

> Measures to create a solar industry therefore threaten the nuclear industry. And conversely, measures to protect the nuclear industry and the demand for nuclear-generated electricity may very well create an insurmountable barrier to a nascent solar industry. . . . A flexible policy will require a balance of interests, which means more control by the national government than is required for clear-cut implementation of either the solar or the nuclear option. But that may be the price that has to be paid in order to have freedom of choice.

The accident at Chernobyl in April 1986 strengthened the Social Democrats' commitment to phasing out nuclear energy. Sweden was the first Western country to notice the fallout from Chernobyl, since the wind was blowing directly from the Ukraine. In the months following, large amounts of food had to be destroyed—a total economic loss of over $2 hundred million. Social Democratic Prime Minister Ingvar Carlsson announced to a rally of the huge LO trade union: "Nuclear power is one of the greatest threats to our environment. . . . Nuclear power must be gotten rid of."[20] A state commission was formed to explore the effects of advancing the 2010 deadline, although it ultimately discouraged such a plan (Swedish Government 1986). Tentative gropings toward reopening the debate have been stilled by the accident, but they may reappear. Carl-Erik Wikdahl of OKG said (Berman 1986, 5): "After Chernobyl, I don't know. But in politics, everything is possible." In *Swedish* politics everything is possible.

Both pro- and antinuclear activists wait and listen carefully to what is said in public. They agree only in their skepticism when Birgitta Dahl, the Social Democratic Energy Minister, says that renewable energy research and development is proceeding very well. Nuclear opponents claim the government is purposely placing Sweden in a position where it will have no energy choices but nuclear energy in 2010. Proponents don't believe this strategy is purposive; they feel there are no feasible alternatives to nuclear energy. ASEA-ATOM has developed a new, small reactor especially well adapted to heating needs, PIUS. It claims that the reactor, unlike current light water technology, has an "inherently safe" design. It would be hard to fathom this development work if ASEA-ATOM believed it would continue to be prohibited from selling any reactors at home or abroad. American nuclear proponents have embraced the new design (Weinberg et al. 1984), in an ironic reversal of the days when Swedish

---

[20] "Swedish Chief Assails Nuclear Power," *New York Times*, 18 August 1986.

technological enthusiasts eagerly swallowed America's new nuclear technologies.

Even though in private many politicians, antinuclear activists, and members of the nuclear industry express doubts that the phaseout will occur as planned, few politicians will do so publicly. They could only lose, becoming lightning rods for controversy. When one Social Democrat criticized the phaseout plan in 1985 his party leaders immediately forced him to back down.[21] The current generation of politicians is still tender about the nuclear question, so it is unlikely to touch it. But everyone is aware that nuclear energy may resurface as a political issue in the 1990s, when many or most politicians have been replaced by younger politicians who did not endure the controversy of the 1970s. By the end of 1986, all the party leaders of the 1970s had already been replaced, taking with them much of the bad blood and personal animosities of the nuclear conflict.[22] Most politicians say with relief that they do not want to tie the hands of future generations. Whatever their motives, Swedish politicians have retained far more discretion for directing future energy policy than their counterparts in France or the United States.

If research into renewable energy sources does not yield the results hoped for, then nuclear energy could look better to many politicians. The moralists in the Center and Communist parties will not change their minds, and a new Environmental Party won representation in the Riksdag in 1988.[23] But most politicians acquiesced in Sweden's nonnuclear plans for pragmatic reasons. Most simply wished to find a compromise that would take nuclear energy off the immediate political agenda. Many had more of a cost-benefit perspective and were skeptical that Sweden needed further nuclear plants. The eleventh and twelfth reactors were much more expensive than the others, so the cost advantages were not clear.[24] If the balance of costs swings in nuclear's direction in the future, many of these pragmatists will be willing to reconsider the current plan. If any party can raise the issue of nuclear energy in the 1990s, it will

---

[21] Nils-Erik Wåäg had only suggested that the date of 2010 was a goal rather than an absolute law. See *Dagens Nyheter*, 31 May 1985. The paralysis of politicians is parallel to oligopolistic behavior in firms, where innovative behavior will be harmful unless the competitors follow along.

[22] Prime Minister Olof Palme was assassinated in February 1986, and Karin Söder replaced Thorbjörn Fälldin at the head of the Center Party in June. Palme and Fälldin were the political leaders who had developed the greatest personal bitterness during the nuclear energy debates.

[23] On the origins of the new "Miljöpartiet," see Vedung (1988) and Miljöpartiet (1982).

[24] For example, Oskarshamn 3 cost 11 billion Swedish crowns, compared to 800 million for Oskarshamn 2, built ten years earlier. Even controlling for inflation and its larger size, Oskarshamn 3 was almost three times as expensive per MWe. I thank Carl-Erik Wikdahl of OKG for providing these figures.

certainly be the Social Democrats. Says Conservative Per Unckel, "If the Social Democrats aren't prepared for a turnaround, there won't be a turnaround."[25] As always, the Social Democrats will propose, and the opposition will attack.

In the meantime, Sweden's twelve reactors are a success in terms of cost and safety. The first ten were roughly as cheap as French reactors, although the last two were more expensive because of high interest rates, delays, additional safety features, and uncertain demand. All twelve, however, are extremely safe, perhaps the safest collection of any country. A study to assess plant safety by observing how well plants were run during the dangerous hours of the early morning ranked Sweden high, with France and then the United States further down the list.[26] Reactor availability has also been consistently high, at least for the nine boiling water reactors. Only recently has France's performance caught up. Good management and training are the key reasons for the high safety and low costs. In addition, Sweden's reactor industry has been very successful at refining the boiling water reactor (far more than Framatome reworked the pressurized water reactor in France) and developing the new PIUS design. Yet all these examples of competence and technical success, in the end, had no influence on Sweden's nuclear energy policies, which belonged to the realm of politics.

Sweden's technical success does show that a large series of standardized reactors—as France has—is not necessary for cost containment, safety, or availability. Neither is centralized control or ownership. Even the State Power Board owns outright only four reactors, and Sydkraft owns two. The others are operated by consortia of producers. Serious management has come from close cooperation with SKI and from the high technical skills of operators and managers. The structural factors often said to explain France's successful nuclear program do not exist for Sweden's even more successful one.

Sweden's cautious approach to nuclear waste, technically robust and sophisticated, has aroused little controversy. Spent reactor fuel will be stored thirty to forty years in caverns on the coast at the Oskarshamn nuclear plant; by then a permanent repository deep in Swedish granite is expected to have been found. The fuel will be brought to Oskarshamn by sea on a ship especially designed for this, the *Sigyn*, avoiding risky and controversial land transport. Low- and middle-level wastes will be permanently stored in rock and concrete under the Baltic, offshore from the Forsmark nuclear plant. Like Sweden's nuclear plants, these waste re-

[25] Personal interview with Per Unckel, May 1985.

[26] I am indebted to Måns Lönnroth for information about this study, conducted by Ed Schmidts, an Austrian GE employee who visited various plants early in the morning.

positories seem well designed. After extensive study of radioactive waste policies throughout the world, Luther Carter (1987, 306) concludes, "No nation has done better in the past than the Swedes, or seems likely to do better in the future, to honor the political imperative to contain radioactivity in nuclear operations."[27]

CONCLUSIONS

How did Sweden come to have a moderately antinuclear energy policy? Despite the bitter partisan debates of the late 1970s, Sweden's 1981 policy of stopping after twelve reactors differed only slightly from the 1975 Energy Bill, so that explaining nuclear energy policy involves three successive questions. How did nuclear energy become the most debated political issue of the 1970s? Why was Fälldin unable to change nuclear policy more than he did as Prime Minister? Finally, why had the 1975 Energy Bill already been so moderate?

We saw the conjunction of chance events that propelled nuclear energy into such prominence in Swedish policymaking. With another prime minister, another bourgeois party in the lead, or a less tight electoral balance in 1976, nuclear energy policy would not have been so visible. But it is not accidental that a moralist prime minister took power at the height of the nuclear controversy. The composition, ideology, and history of the Center Party predisposed it to moral ecologism and to leaders like Fälldin, and it was the party's antinuclear, ecological stand that had made it the largest of the three coalition partners. What makes Sweden different from some countries is the spectrum of parties that made one available for capture by strong antinuclear attitudes. France and the United States had moralist voters and politicians too, but they lacked a political party that these groups could flock to. The Swedish Center Party provided a home unavailable in the other countries. Had there been fewer parties, or if all parties had been decisively tied to clear class organizations and ideological positions on the left-right spectrum, the Center Party wouldn't have been this haven.

But energy policy in the 1980s was not very different from the 1975 Energy Bill. Thirteen reactors became twelve, and the commitment not to build more reactors was made. The story from 1976 to 1980 is how Sweden's policymaking system fended off attempts by the Center Party to change existing policies radically. Politicians and bureaucrats worked to interpret the Stipulation Act differently from their prime minister so that the nuclear program could continue. After the pragmatists had

[27] On Swedish radwaste, see Carter (1987, chap. 9); for a critique of the *Sigyn*, Anér (1982).

forced the moralist to resign, the bureaucrats cooperated with the Liberal government to quickly reach the compromise position that had existed before Fälldin. Even though the 1980 referendum ruined this work, it was constructed so that it would probably yield the same results: twelve reactors to be eventually phased out.

Chapter 8 argued that Sweden's policy system did a good job of incorporating cost-benefit skepticism about future energy growth so that policy avoided a large additional commitment to nuclear reactors. Decisionmakers were aided because Sweden already had so many nuclear reactors, in turn the result of the early victory of light water technology there. Both Sweden and the United States made large and early commitments to nuclear reactors. America's commitment was accompanied by lax management that eventually undermined and destroyed the program. Sweden's program was better managed, and it allowed the Swedish political system room for negotiations and discussions. While French and American political and economic systems have become rigid, blocking many energy policy options, Sweden has retained a full range of options to discuss and debate in the future.

CHAPTER 13

Political Repression and Low Costs in France

AFTER SOME hesitations, internal debates, and toleration of the antinuclear movement in the three years after the oil crisis, the French state moved decisively in 1976 and 1977 to silence criticism and eliminate alternatives to the nuclear program. Critics outside the state could do little except organize for elections, and they helped the Socialists come to power in 1981. When the Socialists retained most of the nuclear plans of their predecessors, the antinuclear movement finally disappeared. The Socialists did more to support internal critics of nuclear policy, finally taking conservation more seriously and curtailing the nuclear program as it became less economically efficient. The French public resigned itself to nuclear energy and even grew proud of its technical successes. The result is a program too big from a cost-benefit perspective, but with reactors that are cheap and perhaps relatively safe. The program's uneconomical size reveals that France has not installed rigorous cost comparisons of energy options to the same extent as the United States and Sweden, and that the original Messmer Plan was not driven solely by economic considerations.

THE STRONG ARM OF THE STATE

In 1976 the growing antinuclear movement began to focus on the Super-Phoenix breeder reactor proposed for Creys-Malville—a good target since it was connected to weapons production, was being built for research purposes instead of commercial electricity, and was to occupy a beautiful site on the Rhône. A peaceful "antinuclear fête" in July attracted two or three thousand people as well as the first major repression by the CRS. (The CRS is a vivid and all-too-material symbol of the repressive capacity of the French state, a highly professional and usually brutal force trained to intimidate students, striking workers, and other enemies of the state.) When many participants remained on the site several days after the protest, mostly camping and enjoying the sun, the CRS decided to clear them out with tear gas and clubs. The brutality backfired, since local residents sheltered and sympathized with the protesters, and the departmental Conseil Général was moved to hold hearings about nuclear energy.[1] N. J. D. Lucas (1979, 197) observes of the incident, "The indiscriminate brutality of the forces of law and order in France on occasions like this does have to be seen to be believed; it was, in this particular case, in striking contrast to the restraint of the demonstrators." Unfortunately, the attack was only a warm-up for the summer of 1977.

The French antinuclear movement reached its peak strength, popularity, and visibility in the spring and summer of 1977. Public opinion turned antinuclear for the first time that year (Jasper 1988); ecology candidates did well in the municipal elections in March, winning 10 percent of the first round vote in Paris and several other urban districts;[2] and large demonstrations were held in the spring (including one drawing ten thousand people at Nogent-sur-Seine, organized by the Amis, the PSU, and the CFDT). Many local elected officials, especially those of the left, began to oppose nuclear plants.

The movement continued to focus on Creys-Malville. Almost two thousand scientists at nearby nuclear research facilities in the Alps and Switzerland signed petitions against the project. But the main event of the opposition was to be a large occupation of the construction site on July 30. The choice of tactic indicated that the radical wing of the antinuclear movement had the momentum, but the event had the support of all elements of the coalition.

Demonstrators began arriving near Malville on the twenty-ninth, but cold rain and a prohibition on the march by the departmental prefect confused the weekend plans, which had been for a day of education on the thirtieth and a march to the site the thirty-first. Little happened the thirtieth, although both the CRS and the violent fringe among protesters were ready for warfare; one side circled in helicopters, the other prepared Molotov cocktails. The war arrived on the thirty-first, when there were tens of thousands of protesters in the area.[3] The CRS moved to clear the site with brutal efficiency, using tear gas and concussion grenades.

---

[1] This unusual move had little effect; the hearings were published as *Le Dernier Mot?* (Grenoble: Presses Universitaires de Grenoble, 1977).

[2] The myth has crept into the English-language literature on French nuclear energy that the ecologists won 10 percent of the vote *nationwide* in the 1977 elections. The error seems to have appeared first in Bupp and Derian (1978, 116) and to have been blindly repeated since then. These results would be astounding if true, since the ecologists ran candidates only in selected districts (although they did well in those districts). To set the record straight: they won over 12 percent in three of the eighteen districts in Paris, 10–12 percent in six of them, 8–10 percent in seven, and 6–8 percent in two (*Le Monde*, 15 March 1977, p. 5). Not a bad showing, but they emerged from the second round with no representatives.

[3] Estimates range from 20,000 (by the police) to 60,000 (by the organizers). Lucas (1979, 195–207), who was there and has written the best account of the event, thinks the number was closer to the higher estimate.

One person was killed and five were seriously wounded. More than one hundred suffered gashes or other minor injuries. The CRS pursued the retreating protesters, and went through the nearby town street-by-street to remove them (by clubbing them). Lucas (1979, 205) describes the effects of the repression: "The main effect of the reaction of the State is to criminalize the antinuclear movement. This effect depends in turn on three main features: the provocation of the violent elements [among the protesters], the intimidation of the peaceful participants and the implication of the latter in the deeds of the former."

The repression at Malville had additional effects. The antinuclear movement shed its violent fringe and moved definitively away from site occupations as a tactic.[4] It also undermined the legitimacy of the nuclear program for many French; the grenading and clubbing of protesters, the vast majority of whom were peaceful, seemed to support protesters' claims of a connection between nuclear energy and a police state. But for most politicians and state bureaucrats, the antinuclear movement consisted of criminals, just as Louis Puiseux, an EDF economist, was accused of "treason" when he published an antinuclear book in 1977. More and more after 1977 the conservative government tried to attach the label of "les Marginaux" to the antinuclear movement, but they could not succeed as long as the opposition parties felt some sympathy with the movement.

As a result, support for France's nuclear program became equivalent to supporting the state and the current government. The enemies were traitors and criminals. The Socialist Party found an opportunity to attack the government for its arrogant and undemocratic way of handling the protest and the whole decisionmaking process, and in October 1977 it called for a two-year moratorium on new reactor orders and Super-Phoenix construction. Sympathy for the protesters and opposition to the government were closely linked in polls; supporters of the majority strongly felt that the protesters had been wrong, and the left felt they had been right.[5]

It was difficult for the Socialist Party to adopt a clear stance on nuclear energy, since it was racked by conflict over the issue, as described in chapter 9. The party spoke with one voice when it could assimilate nuclear questions into issues of labor versus capital, the dominant partisan cleavage in French politics; the Socialists attacked private ownership of

[4] The violent fringe hardly disappeared: occasional bombings and sabotage continued for years. Many were carried out by the ultraleft, many by farmers, and one probably by Israeli agents (the 1979 destruction of the reactor core to be shipped to Iraq).

[5] Lucas (1979, 209) reports that 73 percent of Communist and far left supporters felt that the antinuclear protesters at Malville had been right, 54 percent of socialist supporters did, but only 19 percent of majority supporters felt that way.

reactor production and fuel exploration, as well as secret business profit-making. It was also united in attacking the conservative majority for pushing through energy policies without proper democratic debate. But the Socialists were bitterly divided when nuclear policy was seen as a question of ecological balance versus economic growth.

For the upcoming assembly elections in the spring of 1978, the Socialists were willing to reach out to the moderate wing of the antinuclear movement, especially the CFDT, the union closest to the party. It was not clear how much a Socialist government would have curtailed the nuclear program, but the ecologists continued to run their own candidates. After winning 611,000 votes, or 2.2 percent, in the first round, they generally supported the Socialists in the second round of voting. When the right stayed in power, there was widespread frustration and demobilization in the antinuclear movement. Realistic hopes for a major change in France's nuclear plans disappeared; by the time the next elections came in 1981, the program would be mostly in place. Antinuclear activities once again concentrated on specific sites: court challenges, organized responses to public inquiries, and pressure on local officials. None of these strategies was successful very often. The trajectory of protest in France partly parallels that in the United States—from local protest in the early 1970s, to national protest in the mid 1970s, and back to local action in the late 1970s.

The French antinuclear movement's experience in the court system is also more similar to the American case than most observers have admitted. The stereotype is that American courts have hurt nuclear energy because they will consider substantive issues, while French courts only deal with formal procedures.[6] But in two cases French opponents managed to win court decisions against EDF for procedural errors, although one had little impact and the Council of State overturned the other.[7] The same thing happened in the United States; occasional victories in lower courts, always over procedural issues, were generally overruled by the Supreme Court. In neither country has the judiciary constrained nuclear development.

In summary, the French antinuclear movement began to dwindle because there was little for it to do. Defeated in elections and pummeled

[6] Campbell (1988) and Bupp and Derian (1978) both mention the importance of courts in the American system.

[7] In the case of Flamanville in 1979, the permit to build into the sea was not obtained before the construction permit, the result of poor coordination between the ministries involved. The latter had to be applied for again, but the delay of four weeks did not actually stop work at the plant, but merely shifted it. In the case of Belleville the tribunal administratif ruled that there was not enough environmental impact work in the files, but the Council of State overturned this decision.

by police in site occupations, activists were barred from regulatory participation like that possible in the United States. Activity decreased because the movement had been beaten, not because it was losing popularity. A 1978 poll showed more French to be antinuclear than pronuclear; French opinion was more antinuclear than American opinion at the time (Jasper 1988). But social movements require projects, activities, and a sense of possibility. The French antinuclear movement was losing these, as the centralized structure of the state excluded it. The conservative government was actively using the strong arm of the French state to marginalize the antinuclear movement.

Nuclear Policy as Usual

There were few changes in French nuclear policy in the late 1970s, although immediately after Malville, President Giscard d'Estaing announced he would create a Conseil de l'Information Electronucléaire to "relieve confusion and anxiety." State policymakers still assumed that opposition arose from ignorance. The president seemed to take the council seriously by naming Simone Veil (then minister of health and social security and a respected politician) to head it and by appointing to it local officials, academics, and representatives of nature protection groups. But the purpose of the council was to convince the public that nuclear energy was safe, not to create a serious debate about it. For two years Veil complained about a lack of cooperation, and the council's first report did not appear until May 1979. It recommended greater information and participation for the public.[8]

Doubts about the economics of the French nuclear commitment continued, and the Ministry of Finance continued its efforts to tighten the reins on EDF. But the fact that the Council of Ministers itself decided the number of reactors to be ordered each year gave EDF immense leverage in resisting the Ministry of Finance. In November 1977 the Assembly Finance Committee, in preparing the 1978 budget, strongly criticized the size of the nuclear program, nuclear energy's cost compared to coal, the debt required, and the composition of the PEON Commission; but the committee was impotent against EDF and the Ministry of Industry.[9] The

[8] Two earlier attempts (in 1975) by the state to inform the public and increase debate had been blocked by the technical bureaucrats even more completely than this council was (Lucas 1979, 186).

[9] *Rapport fait au nom de la Commission des Finances.* Annexe no. 23. Assemblée Nationale No. 3131, 15 November 1977. Lucas (1979, 105) discovered that many EDF officials had not even heard of the report, it was so unimportant to them. It parallels the report by the U.S. House Committee on Government Operations (1978) critical of nuclear energy on cost grounds.

so-called Schloesing report was ignored. Energy conservation, the natural policy implication of the cost-benefit perspective, was also resisted. Throughout 1977 the Agence pour les Economies d'Energie pushed for various sensible conservation measures, including one to slow down the penetration of electric heating that directly conflicted with EDF's nuclear plans. The proposals were consistently blocked in the Council of Ministers, although the pressures may have come from French industry as well as from EDF (Lucas 1979, 152–56).

EDF and the state continued their efforts to make local residents like their reactors, but they stopped trying to change people's beliefs through increased information about nuclear energy. Instead they emphasized economic payoffs. The benefits to local governments included tax payments on the payrolls during construction, temporary housing that would later revert to the communes, and in several notorious cases the construction of public amenities like swimming pools. These benefits are often said to have made nuclear energy more popular in France, but in their effects they differed little from steps taken in the United States and Sweden. Local officials in those countries were also anxious to get the economic advantages of large construction projects and were typically in favor of siting reactors in their localities. In all three countries, when local populations turned against the projects, local officials often did too. No matter how many swimming pools might be built, elected officials would be replaced if they took the wrong stand on such a disputed issue as nuclear energy.

In 1982 EDF did go further in dispersing economic benefits than Swedish or American companies could, promising to make outright payments of roughly $1 million a year to the Midi-Pyrenées Regional Council for the operating life of its Golfech plant (Fagnani and Moatti 1984b, 266). The council then dropped its opposition to the plant. Although EDF claimed this was an exceptional measure, the state soon set up guidelines for future deals of the same kind. It is inconceivable that similar payoffs would be made in Sweden, with officialdom's clear sense of public image, or the United States, with its deep suspicion of government-business cooperation.

Giscard d'Estaing also wished to help, by reducing electric rates near reactors. EDF was said to oppose the plan, presumably because it indicated that nuclear plants really were risky and not just a nuisance during construction. Giscard's reductions went into effect in 1980, averaging 15 percent for residents of the three hundred affected communes, but in 1985 the Council of State ruled them unconstitutional. This is one of the few instances in which the highest courts in Sweden, France, or the United States overturned government policies on nuclear energy.

By excluding the antinuclear movement from any role in the deployment of reactors, the state could continue its strongly pronuclear policies. Yet no one enjoys being hated, so EDF and Giscard d'Estaing continued to court public opinion. They had some successes and some failures in their tactics. Opinion since 1977 has gradually become more pronuclear, less from the positive efforts of the state than from the disappearance of an antinuclear movement that could provide an alternative viewpoint. Just as externally the state excluded the antinuclear point of view, internally it resisted the cost-benefit point of view by refusing a full hearing to nonnuclear options.

"It Can't Happen Here"

The state maintained its advantage over the lethargic antinuclear movement even when the accident at Three Mile Island occurred in March 1979. President Giscard d'Estaing and Prime Minister Barre both appeared on television declaring that a similar accident could not happen in France; Industry Minister André Giraud declared that France had "no serious alternative to nuclear energy"; and CEA head Michel Pecqueur labeled TMI "a very large safety experiment."[10] The government, as though to prove its point, soon announced a slight acceleration in the pace of EDF's reactor orders. French reaction to the TMI accident contrasts strikingly with American and Swedish reactions: Jimmy Carter visited TMI and established several panels to examine its causes; Swedish politicians quickly agreed to a national referendum on nuclear energy. French policymakers had little discretion left by 1979; the political and industrial structures were in place, and most of the reactors envisioned in the Messmer Plan were already under construction.

The government's attempt to paint the Americans as incapable of handling complex technologies that the French had mastered was largely successful. Public opinion continued to grow more favorable to nuclear energy after the TMI accident (Jasper 1988). This shift occurred in spite of the fact that most French took the accident seriously; they believed a similar accident could happen in France and that their government would lie to them about it.[11] The French have a long tradition of sensitivity about their technological capacity in comparison with America's, so they embraced nuclear energy as a technology they could handle better

---

[10] Pierre Tanguy, head of IPSN, was more frank, saying "a 'similar' accident, i.e. an accident with the same consequences, could certainly take place in France." Quoted in Vallet (1986, 162).

[11] Only 20 percent of the population surveyed believed they had been told the whole truth about TMI (Fourgous et al. 1980, 178); only 13 percent believed a similar accident could *not* happen in France (Mitchell 1980, 8); and 97 percent of the population had heard about the accident (Mitchell 1980, 8).

than the Americans could. Since 1977, public opinion has grown ever more favorable to nuclear energy in France, partly because of the repression of the antinuclear movement and partly because there have been many technical successes.

Nevertheless, a group of activists inspired by Alain Touraine began to circulate a national antinuclear petition after TMI (Chafer 1982, 212–14). The moderate antinuclear groups, especially the CFDT, continued their antinuclear work, and succeeded in convincing PS leaders—many of them reluctant—to sign the petition in June. The coalition of groups backing the petition included the Amis de la Terre,[12] the CFDT, the PSU, GSIEN, the Mouvement des Radicaux de Gauche, and the national consumers' organization (UFC-Que Choisir) as well as the Socialists. Although the petition drive spurred some grassroots organizing, there was not much remobilization. The radical wing dissolved, accusing the CFDT of selling out by subordinating environmental goals to political ones. Most moderate activists blamed the violent fringe for attracting violent repression, as well as for using moral rather than cost-benefit arguments against the nuclear program.[13]

When tiny cracks were discovered in tubes supporting the cores in several reactors (autumn 1979), it was the CFDT that first made the problem public, and a strike that spurred a serious investigation.[14] Even the Conseil de l'Information Electronucléaire complained that it had received inadequate information concerning this important matter. Many observers felt that American and probably Swedish regulators would have required temporary shutdowns if faced with similar cracks (Bupp and Komanoff 1983, 13), but no shutdowns occurred in France. In 1982 a similar dispute, with the same outcome, arose over ruptured pins in the reactor cores (Vallet 1986, 169). The pattern seems to be that CEA safety engineers often wanted to force safety changes on EDF, but EDF could stop them in the Ministry of Industry.

---

[12] Les Amis de la Terre were now active in the form of the Réseau des Amis de la Terre (RAT), or "network" of the Friends. RAT grouped together the various local organizations and emphasized the decentralized nature of the membership. The idea for a national petition was first raised publicly at a RAT Congress at the end of April.

[13] Pierre Samuel, personal interview, March 1985.

[14] The cracks in the vital steel tubes supporting the reactor cores were discovered at Tricastin and Gravelines in September 1979 by members of the CFDT. Several weeks of safety inspection and a strike by the CFDT and CGT unions delayed fuel loading, but in the end the Ministry of Industry agreed to monitor the progress of the cracks *after* the plants were put into operation. Not only would this make repairs much harder, but the robotics technology to repair them had not even been developed yet. Technological enthusiasts at the ministry assumed it would become available before the cracks became dangerous. See Vallet (1986, 167–69) as well as the *Financial Times*, 5, 11, and 26 October 1979, and the *Economist*, 6 October 1979.

In 1980 the head of the government's own interministerial committee on nuclear safety, Jean Servant, resigned because his work had been hindered by other government agencies. Hardly antinuclear, Servant is another example of an engineer frustrated when prevented from doing his assigned task of improving the details of nuclear safety. He is similar to the members of the CFDT itself, who were committed to technological development except when they thought it was forced in an unsafe way, to the engineers involved in the ECCS hearings in the United States, and to the American engineers who resigned from GE and the NRC in 1976 to protest a lack of attention to safety issues. Like these other cases, Servant got some support from the moderate antinuclear movement in coming forward with his complaints.[15]

The CFDT continued to monitor the reprocessing facility at la Hague, staging a large strike after a bad fire occurred in January 1981. The CFDT's militant stance at la Hague can hardly be seen as a ploy to gain members, since it has in fact lost ground to Force Ouvrière; as mentioned earlier, the decentralized structure of the union allows its members to follow their own policy styles and ideological concerns. These concerns outweighed material and organizational interests in this case. Even more dramatic was local resistance to a proposed reactor at Plogoff, on the beautiful bluffs of the western coast of Brittany. Like Fessenheim before it, this conflict was fueled by regional hatred for Paris; the violence was so intense that the documents for the public inquiry were brought in vans (labeled "city hall annex") and guarded by the CRS. In early 1980 there were almost daily clashes with the police, and one demonstration had a turnout of twenty thousand people and fifteen sheep (reminiscent of Sweden's antinuclear sheep rancher Fälldin). The town of Plogoff was covered with slogans like "Yes to sheep, no to neutrons," "No Harrisburg at Plogoff," "No to capitalist profit," and "Plogoff is not for sale."[16] Yet the official permits continued to flow, including those from the regional assembly and the departmental council.

In short, the technological enthusiasts within Giscard d'Estaing's government managed to keep the cost-benefiters at bay, and they hardly needed to worry about the external critics any longer. No matter how controversial the proposed site, EDF received the necessary permissions from the state. No matter how dubious economically, fuel reprocessing and breeder technology were still being developed. Giscard d'Estaing

[15] Interviews with Jean-Paul Schapira, Bénédicte Vallet, Monique Sené, and Jean-Claude Zerbib in the winter and spring of 1985. See the *Financial Times*, 17 December 1980 and the *Wall Street Journal*, 9 January 1981.

[16] Eric Stemmelen provided me with a list of slogans he had collected at Plogoff. In descending order of popularity, they dealt with democracy, death, capitalism and markets, local lifestyles, and various other themes.

made a few efforts to sooth the antinuclear movement, but since he viewed them as simply ignorant he had little success. In spite of his 1974 claim that "Nuclear power stations will not be imposed on populations which refuse them," he continued to do just that (Lucas 1979, 161).

THE SOCIALIST GOVERNMENT

The antinuclear movement had its last chance to curtail French nuclear energy in the 1981 elections for president and parliament. CFDT tactics veered toward reminding the Socialists of the petition they had signed calling for a moratorium, while the Amis and other ecologists supported distinct ecology candidates, especially Brice Lalonde running for president. No group was officially affiliated with a political party, and many feared that electoral politicking for regular parties could disrupt the organizations' main agendas. One CFDT activist told me: "We had to twist their arms to get them [Socialist party leaders] to sign the petition; we knew most of them favored nuclear. But the only realistic alternative was the Right, and the nuclear program was theirs to begin with. . . . What could we do? We hoped their public stand for a moratorium would shame them into cutting the program, and that they would open up public participation so this kind of thing couldn't happen again."[17] Many antinuclear activists and voters must have felt this way, choosing the lesser of two evils. Lalonde received 3.9 percent of the first round presidential vote, while two other candidates sympathetic to the antinuclear movement (Bouchardeau of the PSU and Crépeau of the Mouvement des Radicaux de Gauche) gained another 3.3 percent. Most of these voters supported Mitterand in the second round, easily providing the edge for him to beat Giscard d'Estaing by 51.8 to 48.2 percent. France's high threshold for representation prevented the ecologists from gaining any assembly seats.

Once in power, the Socialists had the assembly votes and bureaucratic power to carry out whatever energy policy they wished. Among their first actions were the suspension of construction at five sites, the cancellation of Plogoff, and the commissioning of several reports to lead to a fall debate in the assembly.[18] While all observers saw the actions as payoffs for ecologist support in the election, some still hoped for a genuine public debate and perhaps the severe curtailment of new construction. This would not have been unreasonable since forecasts of electricity demand

[17] Personal interview with a CFDT activist now employed by the AFME. Spring 1985.

[18] The suspended sites were Chooz, Civaux, Cattenom (reactors 3 and 4), Golfech, and le Pellerin. In addition, a commission to gather information on Nogent-sur-Seine was appointed. Six of the nine reactors suspended were later resumed; le Pellerin was the only site to be dropped altogether (Sorin 1981b, 475).

growth were shrinking and the program already under way would provide most of France's electricity needs.

Emerging Socialist energy policy involved the promise of public participation and an explosion of reports and committees designed to gather information and opinions. Decisions about particular sites would go through three tiers of approval, if needed: first municipal councils, then regional ones, and finally the National Assembly itself. Success at any one of the levels meant approval of the project. Second, there would be a genuine debate on energy policy in the assembly, as opposed to the short and sparsely attended debate in 1975. In the following years the Socialists also brought new representatives onto the energy planning commissions, for example consumers and small builders, and (in May 1983) passed legislation to expand the scope of the public inquiries, opening at least the possibility that the public could question officials and that more documents would be public. These formal changes had little effect, however. As two close observers (Fagnani and Moatti 1984b, 265) said, they indicated "a genuine attempt by the government to break away from the formality and arbitrariness of past practices. Yet it would be premature to attempt to evaluate the impact of the changes, and to foresee in what manner the government will respond to the criticisms formulated. The traditional procedures are deeply entrenched in a system that is prone to inertia." This skepticism was well grounded.

The reports the government commissioned also promised to open genuine debates on energy policy. Edmond Hervé, energy minister, requested one report on the "institutional and procedural aspects" of energy policy from Maurice Bourjol, an academic at Tours, and one more generally on energy policy options through 1990, compiled within the Ministry of Industry and labeled the Hugon Report. The Bourjol Report strongly condemned existing procedures, saying the current form of the public inquiry "cannot exist in a Democracy."[19] The Hugon Report reflected the goals of the PS leaders. These were somewhat competing goals: increased use of coal, nuclear energy, and natural gas, and the general conservation of energy. The report forecast a whopping 5 percent annual growth rate to justify these expanded energy supplies.

In addition, the prime minister asked the Economic and Social Council to report on energy needs in the coming years, and the assembly set up its own mission for energy information. The latter was headed by Paul Quilès, who had been the Socialists' main energy spokesman before the elections. The seven-person mission (four of them Socialists, one a Communist, and two conservatives) wrote a report recommending that only

[19] As an indication of how seriously the Bourjol Report was taken by the nuclear industry, the article describing it in the *Revue Générale Nucléaire* misspelled his name (Sorin 1981b).

four reactors be purchased at that time, that the treatment facilities at La Hague not be expanded, and that the breeder reactor program be cut back. This proposal was quite at odds with the intentions of most PS leaders.

The Socialists remained bitterly divided on the question of nuclear energy. In the years before the election, Quilès had been allowed to take the lead, and in 1981 the party had published his report critical of the nuclear program, *Energie: l'autre politique*, as the party's official position. Among grassroots activists in the party there was probably strong support for his position. Most party leaders, however—who were older and favored economic growth as a primary goal—were hostile to the position and had seen it largely as an electoral strategy.[20] As a result, Quilès was not given the position of energy minister in the new government. But to the annoyance of PS leaders, he insisted on defending his new report in the assembly. It was not clear which side would win in the weeks leading to the assembly debate on October 6.

Socialist Party leaders, in order to assure that the Hugon Report's more pronuclear goals were adopted, called a party meeting before the assembly debate at which deputies sympathetic to the antinuclear movement were strongly brought into line. In particular, prime minister Pierre Mauroy "exchanged heated words" with Paul Quilès and, it is said, threatened to resign if he were opposed. Despite two days of genuine debate in the assembly, the Socialist energy program (the Hugon Report) passed easily, restoring four of the five suspended sites, ordering six reactors for 1982 and 1983, expanding la Hague, continuing the breeder project, and generally following Giscard's nuclear policies.[21] The party's antinuclear positions before the elections were simply dropped. More emphasis was placed on nationalizing CGE (which owned Alsthom-Atlantique) and Paribas (a parent company of Framatome) and on creating a board to run Framatome that gave the CEA increased sway.[22] Chapter 9 showed that Framatome's profits had long been a favorite target of the

---

[20] Bell and Criddle (1984, 38–39) cite the case of environmental sympathizer Alain Bombard, who was given an important post on the environment, but was fired months after the election. He claimed in a *Paris Match* interview (28 May 1982) that Mitterrand privately dismissed ecology as a youthful preoccupation.

[21] The vote was 331 in favor (the Socialists and Communists), 67 against (mostly UDF), and 87 abstaining (the RPR). Not a single Socialist deputy voted against the bill (*Revue Générale Nucléaire*, Sept.–Oct. 1981, p. 484).

[22] Paribas was a major shareholder in Creusot-Loire, which in turn owned most of Framatome, which produced France's nuclear reactors; Alsthom-Atlantique produced the turbines for EDF's nuclear plants. The Socialists eventually did change Framatome's ownership, first increasing the CEA's share and later allowing EDF and CGE (which was itself nationalized) to buy shares. They were aided by Creusot-Loire's bankruptcy in 1985.

Left. The Socialists could agree on nuclear energy as a problem of capitalist profits, but not as an ecological issue.[23]

Although it may have been a minority of Socialist deputies who had ecologist sympathies, it was a large minority to ignore altogether.[24] Party leaders could do this partly because the Socialists are an electoral party rather than a party of mass organizations and grassroots activists (like the Swedish Social Democrats), and they would face no national elections for five years. The progrowth technological enthusiasm of party leaders was not to be diluted, especially at a time when there were good tactical reasons for stressing growth: unemployment was high, and the Communists in the government had to be placated. In addition, a Socialist nuclear program was more palatable to many nuclear opponents than a conservative one had been. EDF was a state enterprise, and there were plans to nationalize several of the private companies involved. Those for whom nuclear energy had been a tool for attacking the Giscard government were satisfied: the dominant ideological cleavage erased the nuclear issue for many voters and activists.[25] Finally, many of the CFDT activists who had been vocal and respected critics of nuclear energy became government employees at the Agence Française pour la Maîtrise de l'Energie. They turned their attention to more technical issues involving conservation in the hopes of having practical effects.[26]

For years the conservatives had been trying to identify the antinuclear movement as marginals, but they had succeeded only partially. When the Socialists rejected and marginalized the ecologists within their own party, the label could stick. The movement was finished.

Of all the mechanisms the Socialists established in 1981 to spur public debate but later quietly ignored, the Castaing Commission was one of the most long-lived. It was to examine scientific and technical, but not economic, aspects of the back end of the fuel cycle, and its members were chosen to represent a range of opinions. The only way for a commission composed largely of academics to be taken seriously by state

---

[23] On the assembly debate and the Socialist divisions, see Sorin (1981b), Fagnani and Moatti (1984b), and the *Revue Générale Nucléaire*, Sept.–Oct. 1981, pp. 482–95.

[24] Although no one knows precisely how many Socialist deputies were antinuclear, Monique Sené of GSIEN guesses that approximately one third were (personal interview, March 1985).

[25] Fagnani and Moatti (1984b, 272) report that before the elections, teachers and students had been more antinuclear than the general population—but not after the elections. Their interpretation is that nuclear energy had "constituted a charged political issue as well as a means for some groups to express discontent with the former administration."

[26] Many ecologists felt it was a mistake for these activists to "sell out," but most of the CFDT activists were engineers by training. After years of working on nuclear energy, they were relieved to build and develop new technologies that were also in line with their political ideals of a decentralized and nonnuclear society.

technocrats of the technical corps is for it to be headed by someone who—like Castaing or Bernard Gregory—spans the two worlds.[27] Three years of work brought about three reports, which were increasingly critical of the French program for fuel reprocessing and waste disposal and which were increasingly ignored by the state. The final one was signed only by Castaing; the state did not officially accept it but only congratulated the commission on finishing it.

Five years of Socialist rule did little to change France's nuclear commitment. Slightly more discussion, information-gathering, and perhaps public participation left the breeder program and fuel reprocessing intact, and it cut the light water program only slightly. The cuts were made because of economic pressures more than political resistance, and a conservative government might have made similar cuts. Of course, the Socialists' stronger commitment to energy efficiencies made the economic pressure stronger, but this effect was small. PS leaders made their choices, and the structure of the party and the state prevented serious challenges.

A SUCCESSFUL PROGRAM?

Even though the Socialists resisted political pressures to de-emphasize nuclear energy, they soon ran into economic ones: cost-benefiters who during the 1970s had said France was overbuilding turned out to be right. Even without strong programs to encourage it, energy efficiency had increased since the oil crisis (due largely to industry's response to higher prices, as economists had predicted). Economic growth was far below the 5 percent Hugon predicted and EDF endorsed; electricity demand grew by 1.2 percent in 1982 and 2.5 percent in 1983. By almost any measure, France now has more nuclear plants than it needs (Gerondeau 1984). From the three plants a year (for 1982 and 1983) that the government allowed EDF to order in its 1981 decision, the Socialists (and then the Conservatives) have allowed only one reactor order per year since 1985. And this token rate was merely to keep the industry alive, since the exports Framatome hoped for did not materialize.

What does EDF do with its excess nuclear electricity? It has rapidly retired oil- and coal-fired generating plants. It has also pressured the state into allowing it to market aggressively within France, and the electric heating so long resisted by energy efficiency experts is increasing rapidly. In October 1984 EDF and the state signed a contract envisioning

[27] Bernard Gregory, director of the Centre National de la Recherche Scientifique (CNRS), chaired a thirty-seven-member commission, established in 1975 after the "Appel des 400," to evaluate France's energy policy. Its two reports recommended greater use of conservation.

a 50 percent increase in sales by 1990. Third, French net electricity exports have grown steadily, currently to about 10 percent of production. A new transmission line to Britain was opened in 1986 and is capable of sending the electricity of two nuclear reactors across the channel. But most foreign countries are reluctant to become too dependent on imports, and in at least one case national laws have forbidden special arrangements.[28] Finally, EDF singlehandedly stopped plans for commercial breeder reactors that were uneconomical and unnecessary.

All these measures have still left EDF in a tight position. By the time construction projects began to taper off in 1984, EDF had accumulated a debt of over 200 billion francs (around $30 billion), roughly half of it in foreign currencies. Payments on interest and principal represent more than one quarter of EDF's annual sales.[29] EDF's endebtedness and overcapacity could be countered by selling portions of its reactor capacities to foreign countries, and it persuaded Belgium to buy a 25 percent stake in a reactor at Chooz. Beyond this, EDF has had little success with this new tactic. More likely, French nuclear plants will be operated at less than full capacity, a practice inefficient both technically and economically. Brady (1984, 106) estimated that in 1990 France would have seven to ten reactors more than it needs for domestic purposes.

Although France has more reactors than it needs, its reactors are cheap to build and to operate, and they are probably safer than those in most countries. A French LWR plant costs between $1 billion and $1.5 billion to build, the equivalent of just over one thousand dollars per kWe of capacity. This is roughly the same as the cheapest American plant, half the cost of the average American plant, and far less than the $5 billion for the more expensive American plants. Swedish plants are closer to the French costs. French electric costs also compare well to France's European neighbors; they are 20 percent cheaper than German, 30 percent cheaper than British, and 40 percent cheaper than Dutch electricity (Brady 1984, 110). The costs of nuclear electricity in France also seem to compare favorably to those of coal-fired electricity if the nuclear reactors operate at reasonable capacity. EDF learns quickly, and reactor availability has increased in France to around 80 percent of capacity (Hansen et al. 1989). But if overcapacity forced reactors to cut back to only three thousand or four thousand hours a year, then the costs of nuclear and coal power might be about the same.[30]

[28] Because German law requires utilities to purchase expensive German coal, a German utility managed to prevent a direct arrangement between EDF and a large German chemical company.

[29] David Marsh (1984a) provides the estimate of 26 percent of sales. The figures on EDF debt come from a table entitled "Evolution de l'endettement à long et moyen terme en MF," provided by Jean-Michel Fauve of EDF Public Relations.

[30] For the *Revue générale nucléaire* (1984, no. 3, 273) the break-even point of coal and

How have the French managed to build cheap nuclear reactors?[31] We can begin by reversing most of the factors that have made American reactors expensive. In place of the bad management of many American utilities, we have EDF, one of the world's best-run companies. After having the same casual attitude toward nuclear energy of American utilities in the 1960s, EDF took tight control over the construction of its plants. Through its extensive experience it has standardized construction processes as well as the plants and control rooms that surround the nuclear reactors; these are important since their costs exceed that of the reactors themselves. Good management has helped EDF avoid construction delays and the retrofits that result from shoddy work; in addition it has simply resisted some retrofits it probably *should* have made.

EDF benefits strongly from its bargaining with Framatome, the monopsonistic buyer overpowers the monopolistic reactor producer. While all contracts are secret, it is widely thought (even by EDF officials themselves) that EDF pays lower prices than other Framatome customers. In a one-on-one negotiation, industrial structure tells us little about who will benefit; EDF gains because of its wide contacts and lobbying within the state, its mastery of nuclear technology, and the generally high quality of its negotiators. Yet EDF still complained of Framatome secrecy (Picard et al. 1985, 211), until in 1985 it obtained 10 percent of Framatome's stock and thus access to its books.

Standardization, the word used by most commentators to explain the low cost of French nuclear power, can now be understood more clearly. Manufacture of reactors that are more or less identical is only part of the explanation. Standardization of control rooms and the rest of the plants, and of construction processes and contracts are also important. Bargaining power lowers the reactor prices even more. But the Swedes, and many American utilities, have built nuclear plants as cheaply without such standardization. Standardization was not an unalloyed blessing, especially in the early 1970s. "To put all your eggs in one basket" in choosing a single reactor design was very risky; one major design flaw could destroy the whole program. This is one reason American regulators were reluctant to encourage standardization in the 1960s, when our great bandwagon market began. By 1974, when the Messmer Plan was launched, more was known about the LWR. France benefited from beginning its great surge of nuclear orders at a time when America's surge was ending. The United States bore most of the development costs, a shock

---

nuclear comes when each plant operates only twenty-three hundred hours a year, or when the nuclear plant operates four thousand hours and the coal plant almost full time. With capacity factors averaging seventy or higher, French reactors' actual operating time has averaged around six thousand hours.

[31] I have dealt with this question in Jasper (1987b).

from which American nuclear energy has never recovered. Besides, the French standardized in 1974 in order to build reactors quickly as much as to build them cheaply. Even in 1974 they took a great risk in standardizing so completely.[32]

One reason individual nuclear plants are cheaper in France is that they are built in less time and hence entail lower interest charges. But "lead times" were longer in the United States largely so that utilities could reconsider their demand forecasts. Had French decisionmakers done this, they might have built fewer nuclear reactors, so the total cost to French society would have been lower. Short lead times may have lowered per unit costs but raised the total costs of France's nuclear electric system. EDF's profits grew at the expense of the rest of society.

By embracing nuclear energy almost ten years after the American boom began, the French could avoid some of our hidden flaws. Through their own debates over reactor lines in the 1960s, they learned more about the problems of the LWR. Added to what they learned from the experience of the United States and to the managerial skills of EDF, this information kept the French from deploying nuclear reactors in a premature form. Following a more natural path of development than the American boom market, Framatome has gradually increased the size of its reactors, from 900 MWe after 1970 to 1,300 MWe in 1976 and to 1,500 MWe in 1987.

It is harder to assess the safety of a nuclear program than its cost, but French reactors are probably more safe than American and less safe than Swedish reactors. EDF's resistance to retrofits and even to prompt inspection (as with the cracks) as well as to outside interference (in the case of Servant's safety committee) is balanced by its tight management and careful training of operators. A corps of safety experts with eighteen months of training has been developed, especially after TMI.[33] Six or eight people at each plant are completely outside the normal operations hierarchy, studying safety procedures and waiting to take command in emergencies. Some are young engineers on their way up through EDF and hence have broader interests in nuclear energy, while others are former plant operators who have been retrained and have 10 or 20 years of experience at EDF. They are far better equipped to protect nuclear safety than are the high school graduates running American reactors. Management, not hardware, is the key to safety.[34]

[32] See Bupp and Komanoff (1983). Thomas and Surrey (1981) argue that the performance of light water technology throughout the world has been so disappointing that it should not be crystallized by standardizing it in its current form.

[33] My discussion relies on Vallet (1986, 100–103).

[34] Aaron Wildavsky (1988) has even argued that the complicated safety mechanisms of

French technocrats have long claimed that the reprocessing of spent fuel would decrease the cost of nuclear energy, but this claim is doubtful. Reprocessing facilities have been built and expanded largely to satisfy technological enthusiasts and foreign contracts, not for sound economic reasons. In fact, Cogema has not and may never make money from its la Hague facility. EDF plans to use recovered plutonium in some of its LWRs in place of enriched uranium, but it has backed away from the use of breeder reactors for commercial electric production. Cogema is not nearly so well run as EDF, but it has at least received high marks for its process to vitrify nuclear wastes (Carter 1987, 322). But no place has been found to put the waste once it is vitrified. As with the reactors, however, the French state will find a place for it. Although ahead of the United States in solving the waste problem, France is not taking as careful precautions for interim storage as Sweden.[35]

The very strengths of France's nuclear energy program are also its weaknesses. The state's ability to build virtually as many plants as it wished revealed the technological enthusiasm that caused it to build far more plants than it needed. This overcapacity is the main threat to EDF's otherwise fine cost figures. EDF's excellent managerial record is what allowed it to build cheap nuclear plants as well as to keep external interference to a minimum, but occasional interference might have made the plants even safer. France concentrated on one reactor design because it could build many of them quickly and cheaply after the oil crisis, but the whole system was left vulnerable to hidden design flaws. EDF and the French state did not adopt nuclear energy because it was cheap, but it became cheap because of France's strong policy commitment.

Conclusions

Technological enthusiasm triumphed in France in 1974, fully ten years after it won in the United States. Whereas in the United States it was competition between two private companies that encouraged it, in France competition between two government organizations (EDF and the CEA) had to be ended before it could triumph. Top French politicians released the enthusiasm of the nuclear establishment first in 1974, hesitated slightly in 1975–1976, and finally moved to quash public debate in 1977. They simultaneously discouraged the cost-benefit perspective, which might have saved EDF from overbuilding. The triumph of technological enthusiasm had certain costs in France as well as in the United States. France was able to build as many reactors as it wished, and it

---

American reactors reduce overall safety, a claim not far from Perrow's (1984) idea that any complex and tightly coupled system is bound to have occasional failures.

[35] See Carter (1987, chap. 10) and Campbell (1988, 172–73).

clearly ended up with too many. French energy policy became less flexible than that of either the United States or Sweden. Its confident pursuit of nuclear energy did leave France with cheap and comparatively safe reactors. EDF has learned to manage nuclear plants effectively, a lesson not all American utilities have gained. Thus technological enthusiasm did less damage when it conquered France than when it conquered the United States.

The technocrats of France are fond of saying they have won their "nuclear wager." They usually mean that it was risky to concentrate on one family of reactor designs, so that a single fatal flaw could have killed the whole program. While true, this risk was hardly the only one they took, but they have rewritten history to deny the others. Policymakers say that France had no choice, that without other energy resources it had to adopt nuclear fission in a big way. They make the decision sound far more simple and rational than it was.[36] They have also continued their efforts to paint the antinuclear movement as an odd assortment of extremists and "marginals," to the extent that they now deny it was ever a threat to EDF's nuclear program. These decisionmakers make it sound as though they knew from the start how cheap nuclear power would be. They claim that all political parties accepted and even embraced nuclear energy, since France lacked other options. Decisionmakers often deny their own power, claiming that they are simply reading off the best available data and have no discretion of their own. In France the state elite conveniently forgets the internal and external debates, the alternatives raised and statements made, and the severe uncertainties behind the Messmer Plan of 1974. The rest of us should not.

[36] Listen to EDF's Jean Guilhamon (1983, 3) describing the rationale for France's nuclear program: "Indeed, for us, *there were few choices*: France had no oil or gas, and the coal from its own fields was too expensive and insufficient in quantity. New forms of energy (solar, wind, tidal, etc.) would, then and now, require many years of research and development, *with no certainty as to the result*. Only nuclear power, with the mastery acquired by the French Atomic Energy Commission (CEA) at the end of the 1950s, held out any promise of a dependable supply of adequate quantities of energy *sheltered from the fluctuations of the international market*." (My italics.) All the themes are there: French policymakers had no discretion of their own; they merely reacted to the "givens" of the situation, the resource distribution, and the state of technology (described for new energy forms as beyond our control). Of course even here the final phrase alludes to a universe of possibilities and shows how much discretion was in fact involved.

CHAPTER 14

Structures and Flexibility in Three Countries

THE THREE distinct paths of nuclear policy adopted by the United States, Sweden, and France in the two years after the oil crisis were confirmed and strengthened in the late 1970s, and by the 1980s each path had taken on an aura of inevitability. Flexible responses to the oil crisis grew into inflexible structures that few politicians could question or change. No one could revive nuclear construction in the United States, while no one could question it in France. The regulation and economics of nuclear energy were transformed in both countries to reinforce each trajectory. Swedish policy was more flexible, but it was consistent with the lines first laid down in 1975. National nuclear policies that had been very similar at the beginning of the 1970s could hardly have been more different at the end.

Since the late 1970s American, Swedish, and French nuclear energy programs have faced similar public debates and have been shaped by the same international developments. In all three countries slow economic growth and high energy prices have caused much slower growth in the demand for electricity than had been forecast. All three experienced intensive media coverage of the accident at Three Mile Island and had to ask themselves what it meant for them. Likewise the debates between moralists, cost-benefiters, and technological enthusiasts remained a shouting match of the deaf. In all three countries the antinuclear movement gained some influence over elected politicians but none over the bureaucratic insiders who still heavily influenced energy policymaking. Among insiders, the debate between cost-benefiters and enthusiasts took similar lines in all three countries, although it was resolved differently. Parallel public debates continued in France, Sweden, and the United States, not least because they influenced one another. By the 1980s it was the policies themselves that shaped, sustained or killed the public debates, rather than the reverse.

Our three national states reacted quite differently to the pressures from the oil crisis, slackening electric demand, and the antinuclear movement. One key difference was the way in which conflict between cost-benefiters and technological enthusiasts evolved, according to which of them had more power within the state. This conflict in turn was influenced by the degree to which nuclear energy became grounds for contention between major political parties. How important nuclear energy became as a source of partisan conflict in turn depended on legislative structures and the balance of power between existing parties. These factors led to three divergent policy paths.

In Sweden and the United States, cost-benefiters quickly gained an edge over enthusiasts after the oil crisis. In Sweden cost-benefit warnings of slower demand growth fit well with Social Democrats' reluctance to commit themselves to irreversible policies that would preclude compromise. After 1976 Center Party moralists were outmaneuvered and failed to alter the flexible policy of combining conservation with nuclear. In the United States the prominence of the Council of Economic Advisors and the economic thinking of Schlesinger's energy team made Ford and Carter aware of alternatives to nuclear energy construction. American utilities were reluctant to accept precise methods for comparing energy options, but state PUC's have widely forced them to adopt this rationality. By contrast, French politicians have continued to allow technological development and new generating capacity to be EDF's main response to energy shortages. Conservation, demand management, and renewable sources have been embraced less than in Sweden or even the United States.[1]

The attitudes of top political leaders helped determine whether the cost-benefit perspective was allowed to undermine nuclear programs. In France, Giscard d'Estaing had an unshakable faith in EDF's nuclear program, and the French political structure allowed him to indulge this faith by ignoring and suppressing all doubts. Not facing reelection from 1974 to 1981, and with his party in control of the assembly, he could remove all barriers to nuclear deployment. When the Socialists came to power in 1981, Mitterand and other top leaders actively silenced their party's own antinuclear sympathizers. Carter and Ford were also pronuclear, but they had more confidence in economic analysis, which told them not to intervene to help nuclear energy. Fälldin was as fervently antinuclear as Giscard d'Estaing was pronuclear, but Fälldin and his party could not take advantage of the cost-benefit doubts to stop the last two Swedish reactors. His insistence on moral arguments and his party's relative isolation from intellectuals and academics prevented them from using the strongest evidence against nuclear energy.

Competition between political parties was as important as, and interacted with, the attitudes of particular leaders. Leaders' goals and deci-

---

[1] Energy research spending in our three countries indicates clear differences in emphasis. In 1977 Sweden was spending 39 percent of government research funds on conservation technologies and 11 percent on renewables. The United States was spending 2 and 8 percent on these (IEA 1978). Although we lack comparable figures for France, Lucas (1979, 105) provides a proxy in the form of 1978 Ministry of Industry spending on energy: 1 percent on energy efficiency and 0.6 percent on new energy sources.

sions were influenced by awareness of electoral support, partisan tactics, and practical constraints on their power. In the United States nuclear energy represented a partisan cleavage over government intervention, so Ford refused to intervene with help even though most Republicans favored nuclear energy. In Sweden, the positions on the very existence of nuclear energy distinguished the Center Party (the main party of the Right) from the Social Democrats (the main party of the Left), so it became the heart of partisan debates. In France the Right and the Left disagreed over the issue of private profits more than the existence of nuclear energy. Thus dominant party cleavages influenced the degree to which nuclear energy programs were up for grabs in political discussions.

If political goals arose from policy styles, leaders' preferences, and dominant partisan cleavages, political structure and the electoral balance between political parties affected who could achieve their goals. In Sweden the antinuclear movement captured the soul of a major party, which happened to become the largest party in a new governing coalition. The Center Party had ample opportunity to curtail nuclear energy, but it was not politically effective. Without ties to intellectuals and with its uncompromising moral style, it could not capitalize on its structural position. Antinuclear movements had less success in the other countries. In the United States some members of Congress questioned nuclear energy and were well positioned to curtail it—once administrative authority had been decentralized. This decentralization did not occur until nuclear energy was already dying. In France the antinuclear movement justifiably expected some benefits after it helped the Socialists come to power in 1981, but it was outmaneuvered by the technological enthusiasts at the top of the party. But in Sweden and the United States, elections and party politics mattered greatly.

These struggles between contrasting policy styles, individuals, and political parties were influenced by the economic and political structures in which they took place. In France and Sweden top politicians directly decided the number of reactors to be built, while in the United States they discussed policies for supporting services, financing, and regulation. But many other structural differences either had little effect or reflected conscious choices rather than unyielding structures. The structure of the French state allowed officials to reject the antinuclear movement more thoroughly—and violently—than in Sweden or the United States, but the movement had little effect in those countries either. The French state took a more active role in financing nuclear expansion, but the others could have done the same had they chosen to.

Political and economic structures have become more rigid in the last twelve years. At the height of struggles over nuclear energy in the mid 1970s, politicians and bureaucrats had great flexibility to create and change the administrative structures for deploying nuclear reactors. By the 1980s, their flexibility had shrunk. Technological choices were already made, production processes stabilized so that costs were known more accurately, regulatory systems developed stable practices, and even public opinion eventually stopped changing. Nuclear programs were no longer an unknown venture and could be judged more accurately.

Nuclear energy policies began to diverge because the three political systems responded to similar challenges and similar debates in different ways, but once the policies diverged they took many other factors along with them. Policies influenced many variables that are often thought instead to influence policies, such as the costs of nuclear energy, alternative energy resources, public attitudes toward nuclear energy, and the strength of the antinuclear movement. Even the policymaking structures themselves began to change as a result of the policies adopted.

The costs of nuclear energy have been heavily influenced by government policies. The strong French commitment has allowed economies of scale, standardization of reactors, lower costs of fuel processing and reprocessing, and priority in construction resources. Many of the costs have simply been shifted from EDF to the state, but this shift has made nuclear construction cheaper to EDF. In the United States, regulators' reluctance to intervene to improve utility management eventually led to huge cost increases. While utility management explains most of the cost differences between the two countries, government policy not only interacted with that management but influenced costs directly. Sweden falls somewhere in between, and the costs of the two reactors built under the storm of the controversy were higher partly because of the withdrawal of full state support.

To emphasize the impact of government policies on nuclear costs is not to deny the other factors that have emerged in this book. American costs were high mostly because of bad management and early technological enthusiasm, hidden weaknesses that hands-off regulators could not correct. French costs were low largely because of good management by EDF, its strong bargaining position with Framatome, and its ability to block many regulatory safety changes. Swedish costs were low because of good management, but due to strong regulation they were not as low as in France. Yet on top of these factors, state policies also had an impact.

As the costs of nuclear energy diverged in the late 1970s, and as they became more accurately measured, they fed back into the policymaking system. EDF continued to win its arguments with the Ministry of Finance partly because costs were low, and low costs helped make nuclear energy popular in France. In the United States, high costs discouraged both cheap financing and political support. Costs were determined by commitments made in the 1960s and early 1970s, when costs were poorly known. As costs became better understood in the late 1970s, they reinforced policy paths already taken.

The influence of state policy on the safety of nuclear plants is similar. The most important factor is again management attitudes and competence, but state policy influenced these. In Sweden and France, good management is reinforced and given technical support by the close relations between industry and regulatory bodies. In the United States bad management cannot be corrected by the formalistic regulatory style of the NRC. Likewise the United States lacks mechanisms for screening out utilities that should probably not have nuclear plants.

In addition to regulatory style, state support for nuclear energy may affect safety directly. The sharpest contrast is again between France and the United States. Because nuclear energy is still controversial in the United States, the nuclear industry pretends for strategic reasons that nuclear energy is a perfectly safe energy source, continually underestimating the potential for serious accidents. The French nuclear industry has full government support, which allows it to admit there are serious risks. Thus French operations aim to keep minor incidents from developing into major accidents, hence their safety teams at each plant. In contrast American managers try to keep incidents from occurring, but are always surprised when they do.[2] Given the huge difference between incidents, which occur frequently, and major accidents, which can be catastrophes, the French approach may produce greater public safety. EDF officials are so confident in their system that many say they almost wish an accident would occur so that the public would see that nuclear energy is simply a risky industry like many others rather than one which must be absolutely perfect.[3]

State policies have also shaped the perceived alternatives to nuclear energy. The cost-benefit comparisons of energy sources instituted after the oil crisis in the United States kept alternatives in the foreground of energy debates so that American energy wealth could be seen as an alternative to nuclear. (Before the oil crisis, abundant fossil alternatives had not influenced nuclear policy.) In France less attention was paid to alternatives, in part because of policymakers' enthusiasm for nuclear energy and in part because EDF could expand its market only through nuclear energy. Swedish policymakers have long considered a range of energy alternatives, but since deciding to phase out nuclear energy, Sweden has renewed its commitment to new energy sources, from wind

---

[2] I am indebted to Jean-Paul Moatti for this insight.

[3] In fact EDF has had several serious accidents, although most were on older gas-graphite reactors (Carter 1987, 307).

power to increased efficiency to methane deep in the earth. The French commitment to nuclear energy makes such development unnecessary and unlikely, since excess electric capacity already exists in France. Contrasting patterns of research funding also link energy policies to the creation of alternative energy supplies.

Electricity demand should ultimately determine how much nuclear capacity is built, but even demand was shaped by nuclear policies. Curtailing reactor deployment led the United States and Sweden to try to reduce demand, especially through conservation and energy efficiency. In France pronuclear policies actually discouraged full consideration of techniques for encouraging conservation. EDF was even allowed to expand demand by actively marketing electricity.

Public policies have also affected public attitudes toward nuclear energy. In another work (Jasper 1988) I have shown that opinion began to diverge in France, Sweden, and the United States in the late 1970s, with each public coming to support its country's de facto policies. I attempt to summarize this in figure 14.1 with an "index of acceptability" for each country. Following Rankin et al. (1981), this is the number of pronuclear responses divided by pronuclear plus antinuclear ones, so that a figure above fifty indicates a majority of pronuclear responses, below fifty a majority of antinuclear ones. French public opinion has grown steadily more pronuclear since 1978, while American opinion has grown more antinuclear. Sweden, where the direction of nuclear policy is still unclear, has retained a roughly even split of opinion. In contrast to recent years, public opinion in all three countries was fairly evenly split on nuclear energy

FIGURE 14.1 Acceptability of Nuclear Energy in American, Swedish, and French Public Opinion

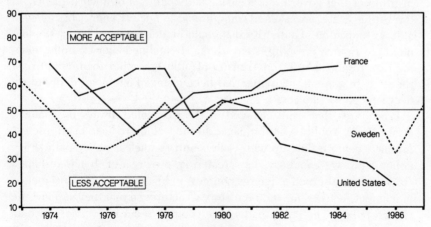

Source: Calculated from data and estimates in Jasper (1988).

in the late 1970s (except that American opinion was more pronuclear), when current policy paths were being chosen. Public opinion was influenced by state policies more than it influenced them.

Although it is often thought that the strength and tactics of the antinuclear movements affected nuclear policies, those policies determined the fate of the antinuclear movements. In France the movement was unable to gain any permanent access to policymaking, and its supporters were ignored if not physically beaten by the state. It disappeared. This outcome is predictable from the insulated structure of the French state; what is surprising is how little impact the movements had with more open structures in Sweden and the United States. In the United States, continued uncertainty over the fate of nuclear energy has allowed the movement to survive in two forms: local watchdog groups and several national organizations like the Union of Concerned Scientists. Both wings have a role to play, testifying at NRC and PUC hearings, lobbying legislators, and monitoring existing nuclear plants. Their existence also influences public opinion; antinuclear commentators are on the news after an accident such as TMI or Chernobyl, whereas in France only government spokespersons appear to reassure the public. In Sweden antinuclear activists have moved into other arenas, including the new environmental party, since the country is too small to support permanent national antinuclear groups. But they would be ready to fight other battles should any emerge.

Finally, nuclear energy policies have reshaped the systems of regulation and political oversight that are used to control or encourage nuclear deployment. In each country a regular flow of nuclear legislation has allowed current policy preferences to become embodied in the rules and institutions that govern the licensing and operation of nuclear plants. In the United States a weakened commitment to nuclear energy has allowed a decentralization of authority that could hinder potential future deployment. In France procedures have been streamlined—for example, making EDF responsible to the Ministry of Industry rather than its old rival the CEA. In general, legislation has been changed often and easily to reflect new policy choices.

Procedures that were changed in the past could always be changed again. But it would be harder, since all the factors we have been discussing have come to line up with each country's nuclear policy path. Each policy path, we could say, has created its own reality. It has reshaped costs, safety, alternative energy sources, public opinion, the antinuclear movement, legislation, and regulatory structures so that they support the chosen policy. The policy paths have come to seem inevitable, since all these factors that were determined by policy can be reinterpreted in retrospect as the *causes* of those policies. While this rewriting of history is most noticeable in France, it has occurred in the United States as well.[4]

The direction of causation between policies and these other variables is not strictly one-directional. Rather there is a reinforcing feedback loop. Nuclear policies had begun to influence costs, safety, and the rest by the mid 1970s, when those policies were beginning to be laid down. But these effects became known only in the late 1970s, at which time they could mostly reinforce policy paths already taken. But they have performed their reinforcement functions well.

What does the future hold? How much discretion do elites have today? Political and economic structures have closed in around nuclear policy choices to different degrees and in different ways in France, Sweden, and the United States. It would be almost impossible for France to alter its extensive commitment to nuclear energy, especially since secondary variables like costs and public opinion have come to support it. The United States has less explicit policies about nuclear energy, but the economic and political structures will probably prevent any new construction for a long time. In both countries the discretion and flexibility of the early 1970s has vanished; decisions taken then launched the two countries on policy trajectories from which they cannot escape.

Only Swedish policymakers retain the flexibility they had in the early 1970s, even though they pretend not to. Although today's politicians say they are committed to phasing out nuclear energy, tomorrow's politicians could easily reverse that policy. The reason for such flexibility is that nuclear energy policy has arisen in recent years from an explicitly political process in Sweden. Swedish policymakers are aware of costs, safety, and alternative energy sources, but they do not pretend that these are guiding their decisions. This is the reason they have not rewritten history to portray nuclear choices as inevitable. In fact, they may have rewritten history to highlight the political process, ignoring the economic reasons for the twelve or thirteen reactor plan as early as 1975. After the 1980 referendum politicians could say, "The people have spoken," ignoring the previous stage at which the bureaucrats had spoken. But when a policy is said to be grounded in political preferences, it can be changed if those preferences change.

Little was known about the exact costs and benefits of nuclear energy in the early and mid-1970s, when the three countries' energy policies were beginning to gel. As a result, political positions were adopted on the basis of policy styles, ideologies, and assumptions about how tech-

[4] From the perspective of the late 1980s it is easy to think that the antinuclear movement curtailed nuclear energy in the United States, since the plants most often in the news (Seabrook and Shoreham) have been delayed partly by opponents. But more than one hundred reactors have come into service, rarely slowed by opponents.

nologies develop. By the early 1980s the costs and benefits were clearer, but by then the policy choices had been made, often irrevocably. This is a paradox of policy choice: a policy option can be fully evaluated only when it has been adopted. But in that case it may be too late to change it. Policies must always be chosen on the basis of incomplete—often terribly incomplete—information. Hence cultural grids play an important role.

Evidence that energy policies became less flexible in the late 1970s comes from the Three Mile Island accident. The same accident caused all three countries to ask themselves what it meant for them. The answers they gave were completely determined by their current policy trajectories. The French claimed it meant nothing, and they accelerated their construction program. The United States was devastated, adding further delays, safety mechanisms, and costs to reactors under construction as well as discouraging any future construction. Swedish policymakers decided to abandon the compromise policy that was emerging under the Liberals and throw the question open to a national referendum. Each country interpreted the TMI accident according to its own policies.

Finally, each country had a change of government in the early 1980s: Reagan replaced Carter in 1981; the Socialists replaced the Conservatives in 1981; and the Social Democrats replaced Fälldin in 1982. Each new government had different attitudes toward nuclear energy than its predecessor, but none changed nuclear outcomes at all. Reagan tried hard to revive nuclear construction, but there was no flexibility left. The structures had hardened.

With a controversial issue like nuclear energy, the decisionmakers in each country were anxious to claim they had little choice of policy. This became true in the late 1970s in France and the United States, although in the mid-1970s all three elites had exercised a great deal of discretion. Policy choices taken after the oil crisis developed into inevitable trajectories, as they pulled associated factors along with them. Public opinion, plant costs and safety, alternative energy sources, the strength of the antinuclear movement, legislation, and regulation were all transformed to fit chosen policy paths.

PART FIVE

Conclusions

CHAPTER 15

What Have We Learned?

AFTER RECAPPING the reasons that nuclear policies in the United States, Sweden, and France diverged, I want to draw two types of conclusions from this study. The first concerns how political systems work and how we should study them. The second group concerns practical political arrangements for promoting sound technological choices. How generalizable are these conclusions? The theoretical conclusions deal with how three political systems typically make policies, rather than nuclear policy specifically, and most of them are applicable to political systems in general. Many point to the limits of purely structural theories of the state and provide counterexamples to the general claims of these theories. The practical conclusions, on the other hand, concern either energy policy or technology policy, rather than policymaking in general. Other policy areas probably have similar practical challenges, but recommendations that were too abstract would not be useful.

WHY NUCLEAR POLICIES DIVERGED

At the time of the 1973 oil crisis, American, Swedish, and French nuclear energy policies were converging, in spite of the contrasting origins of nuclear energy in the three countries. All three had settled on light water technology; all had set up centralized and insulated regulatory systems to facilitate reactor construction and licensing; and all planned widespread deployment of reactors. A combination of technological enthusiasm and the cost-benefit perspective had captured the small circles of bureaucrats and managers who made energy policy. By raising the question of energy independence as well as quadrupling the price of oil, the oil crisis seemed to assure nuclear's future, and leaders in all three countries quickly reiterated their commitments.

The oil crisis and the rising antinuclear movement put energy policy on the agenda of politicians and the media. At the same time cost-benefiters began to question many of the assumptions fed to them by technological enthusiasts. In all countries a debate emerged: was the proper response to the oil crisis to build more reactors or to let price increases lower demand to avoid the necessity for much new capacity? One of the indirect results of the oil crisis was the development of energy models and rational cost comparisons that bolstered the cost-benefit case. The antinuclear movement of the mid-1970s had more effect on the public debates than on the eventual outcomes.

In the three years following the oil crisis, politicians and bureaucrats in the United States, Sweden, and France made different choices without realizing they were choosing the policy trajectories for decades to come. Cost-benefit arguments against actively supporting nuclear construction appealed to Gerald Ford's free-market ideology. The previously irrelevant financial weakness of utilities now resulted in the cancellation of orders, since a no-build option had appeared. The Swedish Social Democrats were attracted by the idea of delaying a difficult decision for several years, so they also refused to push for additional nuclear construction. Much to their surprise, they were out of office when the time came to reassess energy policy. French policymakers, in contrast, chose a bold technological option that promised national independence; they unleashed EDF's nuclear ambitions.

Only French leaders knew they were embarking on a policy path from which they could not turn back. Swedish and American leaders thought that nuclear deployment was only pausing so that the technology could catch up with itself. Swedes did not see how controversial an issue nuclear would become, freezing their political machinery for two years after Fälldin's election. Americans did not realize that their most fundamental hidden weakness, bad utility management, was only beginning to show its effects. In the following years their pauses became the end of nuclear construction.

Policy choices made in the mid-1970s began to reshape their own contexts in the late 1970s. The French nuclear commitment lowered reactor costs, prevented alternatives from being recognized, crushed the antinuclear movement, and eventually made nuclear energy popular with the public. Swedish and American misgivings about nuclear had the opposite effects. In particular, alternatives were exploited more fully: coal in the United States and conservation and renewable sources in Sweden. In the United States costs and safety risks have remained high for many utilities, and public opinion has grown steadily more antinuclear. Whereas France's nuclear program allowed reactor production to be standardized, the United States stopped ordering new reactors soon after standardization became possible. By the time enough information was available to make careful decisions about nuclear energy, those decisions had already been made.

The view that American nuclear energy was strangled by our decentralized system of decisionmaking and the veto power of interest groups and movements (Campbell 1988; Kitschelt 1986) is not completely accurate. Weaknesses, largely resulting from the enthusiasm with which we embraced nuclear energy in the 1960s, went unrecognized and hence uncorrected for too long. Only after they had appeared during the 1970s did we decentralize authority over nuclear policy. But by then new nuclear construction had already disappeared. The decentralization affected who would pay for cost overruns, but not whether reactors would be built. Decentralized political oversight is an effect, not a cause, of nuclear's problems in the United States.

France, Sweden, and the United States all had popular and visible antinuclear movements in the mid and late 1970s. Technical, economic, and moral arguments spread between them, as did tactical innovations. But none of the movements had a strong effect on nuclear power in its country. The French movement was beaten by riot police and later ignored by the Socialist government. The American movement was active longer, but it was not the reason nuclear energy collapsed in the United States. The Swedish movement actually influenced the heights of government, thanks in part to Fälldin's and the Center Party's moralist policy style, but even at the top antinuclear politicians were outmaneuvered by bureaucrats and coalition partners. In each country one of the most vibrant social movements of the 1970s had little effect on state policies.

In the last ten years the flexibility of policymakers in all three countries has shrunk; the political, legal, and economic structures they created have hardened around them. In the 1960s state bureaucrats had to create institutions for persuading manufacturers and especially utilities that nuclear energy would be the cost-effective way of the future. After the oil crisis the discretion shifted to elected politicians, who faced genuine choices of building or conserving. Even in the United States, many of the emerging weaknesses of nuclear energy could have been solved with active government policies. Not so by the late 1970s. New nuclear construction was dead in the United States. Nuclear energy had paralyzed Swedish politics, so politicians avoided new options for reviving it. French energy policy was the least flexible of all; a huge nuclear system was in place that covered all parts of the fuel cycle. Because it was too large for French electric needs, France was forced to find new outlets. Flexibility in energy policy had allowed decisionmakers to create their own institutional structures, but those structures took on lives of their own.

The tightening of these structures provides the confidence for several predictions. There is little chance of a nuclear revival in the United States in this century given public opinion, the decentralization of authority, and recent cost figures. France, at the other extreme, could not possibly give up on nuclear development; even when EDF has a major accident it cannot hide—as it inevitably will—France will continue developing new designs and replacing old reactors. Sweden is less predictable since its structures are less firm. Smaller, safer reactors will probably be debated seriously in the 1990s.

THEORETICAL CONCLUSIONS

Most of the implications of this study explain politics and policymaking rather than suggesting how to conduct them. Several concern the limits of a structural approach restricted to variables like formal position in the state, centralization, and state ownership. Several others suggest how one must supplement structural factors with policy styles, cultural meanings, and personal biographies. There are also several implications to explain what happens when an issue moves from bureaucratic arenas to the glare of public debate and partisan conflict. Finally, the study shows what can happen after clear policy paths are chosen—how history can be rewritten and how the policy itself becomes a causal factor. What follows is a list of conclusions, with examples to stir the memory.

1. *The formal power and positions of groups and organizations do not fully explain their successes and failures and hence poorly explain political outcomes.* Identifying an organization's position in a political structure helps explain its success or failure in achieving its goals. But a prior step is to explain why it has the goals it does, a project that often requires analysis of ideologies and policy styles. Succeeding steps must examine informal skills, luck, and the discretion that individual policymakers exercise on the basis of their own goals and desires. All these steps are needed to explain policy outcomes.
    a. *Policy positions and outcomes are strongly influenced by struggles within the state and within each state organization.* States almost never act as unified actors, and the struggles between different segments of the state can influence outcomes as much as struggles in civil society. But because these conflicts are unofficial, their outcomes often depend on informal persuasion that is hard to study. To explain the influence of the Council of Economic Advisors on Gerald Ford or the relative success of the CEA and EDF at persuading de Gaulle or Pompidou, the mental worlds of the leaders are as important as the structural positions of the bureaucracies.
    b. *Individual discretion and decisions are important factors, and they can never be completely predicted from position in the state or an organization.* Political leaders do make important and fateful decisions, especially when different segments of the state disagree. Their decisions can never be predicted fully in advance, since, like all humans, leaders have complex mental grids that influence their decisions. Not only longstanding policy styles and ideologies but concerns

of the moment may loom large: how to defuse an angry mob, how to respond to an Arab embargo, how to inspire the public in an election year. De Gaulle favored the gas-graphite reactor; Pompidou the light water design. Nixon might have favored a dramatic technological fix; Ford hesitated to intervene in economic markets.

c. *Personal skills such as the power to persuade others, which partly depend on shared cultural meanings, are often as important as formal power.* This follows from the importance of bureaucratic fighting within the state and the huge discretion of individual policymakers. EDF won the day in France in part because it could refuse to do certain things, but also because it lobbied continuously in one-on-one conversations and in committee meetings. Cost-benefiters in Sweden and the United States were persuasive in large part because they had the precise rhetoric of economics behind them and because they presented data that cast doubt on the costs of nuclear programs. Some, like the U.S. Council of Economic Advisors and the Swedish Secretariat for Future Studies, had no formal power; they were effective only through persuasion.

d. *Centralization and size, concepts often used to describe state structures, are clumsy variables commonly hiding several other factors.* In the United States, a centralized and determined AEC could not impose its will on a myriad of electric utilities; in France a centralized CEA had no better luck with a centralized EDF in the 1960s. Centralization does not improve nuclear operations, good management does, and the two are not always correlated. Size is also dubious; small Swedish firms and a handful of American utilities could do as well as EDF. National size plays a role in the case of Sweden, however; the interpersonal connections of the small elite (both in politics and the nuclear industry) facilitated cooperation and communication. But the key was a shared policy style; their cooperation did not extend to moralists.

e. *Public versus private ownership, like centralization, seems important but turns out to obscure more specific processes and factors.* EDF succeeded with its nuclear program because of careful, tight management, not because it is a state enterprise. EDF and CEA officials cooperate easily because they share goals, worldviews, and Polytechnique training. They also both answer to the Minister of Industry. Without these shared cultures, their common boss would only keep conflict buried at a lower level, not eliminate it. In Sweden and the United States there are no significant differences in how private and public utilities handle nuclear power or in how they are regulated. Good management at utilities matters more than size or ownership.

f. *At many levels, legal and political structures change continuously in ways that structural approaches are often slow to recognize and for reasons that may be external to the structures themselves.* When a policy is developed that is incompatible with existing institutional structures, the latter can often be changed. Certain basic political structures are enduring, such as federalism in the United States. But the institutions that directly affect nuclear policies are far more malleable. Legal structures changed often in all three countries. The United States provides the best example of how easily regulatory and administrative structures can be changed too. In the early stages of nuclear development, a tight, centralized structure was created quite unlike that of most American traditions. When decisions were made unfavorable to nuclear energy in the late 1970s, new structural constraints appeared that were local and could restrict future nuclear expansion.

2. *Explanations of policy outcomes can be improved by supplementing structural factors with cultural meanings. Two sets of meanings that proved useful are dominant partisan cleavages and policy styles.* Groups and individuals form agendas out of beliefs about the world, and people's beliefs do not always line up with their material interests. While individuals' material and economic interests are easily predicted from their places in social and political structures, their mental grids may be affected by their biographies as well as their current positions. Because many different factors influence worldviews, no one's mental world is fully predictable; individuals always have some discretion. At the same time, we can identify salient points in the perspectives of key individuals as well as clearly defined groups.

   a. *The beliefs of individuals are influenced by particular cultural meanings and styles as well as biographical details, and these styles are often important for predicting actions.* In many cases nuclear policy has been shaped by the decisions of top political leaders; a structural balance of power may have given them discretion to choose, but structural factors alone cannot explain why they made the choices they did. Because of his free-market ideology, Ford chose to follow the noninterventionist advice of economists rather than that of technological enthusiasts. Because of his moralist policy style, Fälldin became interested in nuclear energy as a moral issue and led his party to oppose it. Because of their engineering backgrounds, utility managers listened to optimistic claims about the promise of nuclear energy and ignored cost-benefit doubts.

   b. *Beliefs about the world occasionally lead individuals and groups to act in ways not predictable from (or are even contrary to) their material or organizational interests.* Although this is rare, given the tendency for interests and beliefs to influence each other, it does occur. Many engineers spoke out because of artisanal concerns for safety even though they knew they would lose their jobs. The CFDT pursued its antinuclear activities even though it lost members at La Hague as a result. Many EDF employees questioned their company's nuclear program even though that program was key to EDF's expansion into new markets. Many motivations drive people's actions.

c. *But in general, cultural meanings interact with political and economic structures so that they reinforce each other. A purely structural or purely cultural account of state policies will be incomplete.* People try to enter occupations or organizations in line with their cultural expectations; if there is conflict, one or the other is likely to change. In the 1960s technological enthusiasts created institutions and nuclear programs in line with their expectations. After the oil crisis politicians and bureaucrats again reshaped many structures to fit their own preferences. Conversely, certain institutional structures encouraged technological enthusiasm, others the rhetoric of cost, others moralism. No matter how independent they may become, policy styles and other cultural meanings issue from concrete institutional arrangements. At the extreme, educational institutions are designed precisely to shape cultural meanings.

d. *Mental grids especially influence decisions concerning technological development, since policies are often based on assumptions about the future rather than on concrete experience.* Sound cost data for nuclear reactors were not available until the late 1970s, long after important policy choices had been made. Information about safety and operations also increased after decisions had been taken. Reactors were ordered about which very little was known, since none like them had ever operated. Policymakers had to guess about future developments, and to do this they relied on their intuitive beliefs about how the world works.

e. *We need to identify variables that are not reducible to political structure in order to judge when structures tighten or loosen, as well as when they change.* The discretion open to policymakers after the oil crisis is clear from the divergent choices they made from a similar set of possibilities and on the basis of similarly inconclusive information about nuclear energy. The tightening of policy paths in the United States in the late 1970s is clear given the inability of a new government in 1981 to implement its nuclear goals. Similarly, even if the French Socialists had wanted a new nuclear policy in 1981, the sunk costs were too great to allow much change.

3. *New arenas for policy debate change not only the actors involved but also the terms of the debate.* Nuclear energy was seen as a different policy issue depending on who was debating it—or even on whether it was be-

ing debated. Mental grids of participants influenced not only their policy positions but what they thought the nuclear issue was about. The dominant partisan cleavages differed in our three countries, but all three frames were different from the framing that occurred before energy policy became a public issue and from the framing that the general public applied. Before we can understand why a political leader or bureaucrat makes the decision she does, we have to know what she believes the alternatives are.

a. *When policy decisions were made by bureaucracies, the issues were typically seen as technical feasibility and economic costs, although some bureaucracies emphasized costs and others feasibility.* EDF and the CEA fought over reactor lines, with one favoring cost criteria and the other national technological independence. After the oil crisis all three countries had debates between cost-benefiters, typically finance ministries or independent commissions, and technological enthusiasts, typically industry ministries or industry representatives.

b. *When the public became involved in policy debates, it raised questions of public health and safety, democracy, and right and wrong.* The antinuclear movements in each country had strong strands of moralist politics and often ignored technical and economic questions. Moral frames involved the protection and rights of future generations, the protection of existing ways of life, and democratic processes. Antinuclear movements were not restricted to moralist rhetoric, but they used it far more than bureaucrats did.

c. *When political parties debated nuclear energy, they twisted the issue to reflect the dominant cleavages between those parties.* In the United States politicians typically saw nuclear energy as a question of free markets versus government intervention; in France it represented private profits for industry; and in Sweden it was associated with the Social Democrats' long rule and urban, pro-growth economic policy. The French and American cleavages were ideological, and the Swedish cleavages more organizational (the Social Democrats, governing for so many years, versus the other parties excluded from government until 1976). But it was the Swedish debate that concerned whether nuclear reactors should be used at all, whereas ideological framing in the other countries twisted the question. In all three countries nuclear policy debates shifted so that they could be used by each political bloc to attack the other.

d. *Economic rationality influenced policy only when it was allowed to.* Sometimes economic analysis was inconclusive, so that it had to be supplemented with additional assumptions—for example, those of technological enthusiasm. Sometimes it was simply resisted or rejected, usually because alternative goals were judged more important.

The rhetoric and precise tools of microeconomics were only one set of criteria among others.

4. *Policy trajectories themselves can become powerful causal variables, reshaping political and economic variables so that they support the chosen trajectories.* This is another way of pointing out the extent to which political and economic structures can be reshaped by actual choices made by policymakers. Part Four showed how the costs of nuclear plants, operating safety, public opinion and antinuclear movements, electricity demand, and even the availability of energy resources were all shaped by state policies toward nuclear energy. The factors that should provide a basis for making policy decisions can be heavily determined by those decisions. This reversal allows elites to rewrite history and to claim that their policy path had been inevitable all along—a revisionism that is almost complete in France. They create what Roberto Unger (1987b) calls "false necessity." Researchers must listen to elites skeptically, since they almost always want to conceal their own power.

Practical Conclusions

In addition to conclusions about how to study policymaking, this study suggests some ways to better practice politics and develop policies. These conclusions are made even more important by the theoretical finding that administrative and legal structures sometimes easily bend to the will of political elites. What state arrangements lead to sound technological choices? Are there ways to avoid the emotional, often vicious conflict that accompanied nuclear policy debates?

1. *Veblen was right that there are penalties for taking the lead in many technological innovations and merits to borrowing from other countries.* The United States lacked the regulatory structure to reshape nuclear energy and recover from the shocks of leaping on the nuclear bandwagon before the technology was fully developed. Likewise France benefited from borrowing light water technology almost intact from the United States. By the time of the Messmer Plan, light water technology was developed enough so that standardization was a reasonable option. Whether a country adopts a technology early or late influences its costs, even though the ultimate effect of costs on policies depends in turn on organizational arrangements.
2. *Technological controversies have many benefits.* They lead to a fuller consideration of technologies, and different points of view are capable of uncovering different kinds of data. Antinuclear organizations uncovered new information, especially by eliciting disgruntled engineers who went public with what they knew. The *delays* that conflicts cause can also be beneficial. The dispute between the CEA and EDF delayed France's adoption of nuclear

energy, but the nation was then able to standardize on a more advanced design. Delays after the oil crisis allowed the United States and Sweden to institute more rational comparisons among different forms of energy. They avoided the overbuilding of electric generating capacity that now plagues France.

3. *Relatively cheap and relatively safe nuclear energy depends above all on good utility management, but regulators can use their powers of prohibition to screen out poorly managed utilities.* There were no cases of creating good management out of bad, although EDF did have to learn to apply its good managerial practices to nuclear energy. But it could do this precisely because it was well managed. Regulators in all three countries were better able to prohibit bad practices than promote good ones. American regulators proved particularly poor at instilling good management practices. Their solution should instead have been to carefully select utilities that could manage the new technology, since there were many such firms. Regulators could have used their negative powers to screen out the utilities (and there were many of these too) that should not have built or operated nuclear reactors.

4. *No state was altogether successful at persuading electricity producers to adopt nuclear programs they did not want. Instead policymakers had to adopt goals and plans already being developed.* Whether the electric utilities were public or private enterprises made little difference. Subsidies and limited insurance liability were important but insufficient. In the United States it was manufacturers rather than the AEC that finally sold utilities on nuclear energy. In France EDF resisted nuclear energy until the state reversed itself and favored the LWR. The image of a strong French state setting out clear goals and then pursuing them gives way to an image of competing bureaucracies where EDF imposed its goals on the rest of the state. The Messmer Plan was successful because it simply removed the constraints from EDF.

5. *In the case of expensive technologies with long construction lead times, policies are often adopted without much information about the full effects and costs of the technology; in fact, it is only by adopting the technology that the full costs and benefits become clear.* Because nuclear energy involves such a large and complicated industrial structure, this infrastructure's effects on costs, public opinion, and society cannot be judged in advance. This is especially true when the technology is still developing rapidly. In all three countries costs changed (that is, increased) constantly, and planners could only pretend to make accurate forecasts. The result of this lack of information in policy choice is that worldviews play an important role in decisions by providing assumptions to fill in the information gaps.

Because accidents are inevitable—in LWRs and any other reactor design—I would not recommend a revival of the nuclear industry in the United States. If political and industrial elites could invent a nuclear program thirty years ago, they could certainly do the same for conservation and renewable sources in the future. Growing awareness of our environmental crisis could foster this kind of political will. But this sense of urgency is also leading to calls for a revival of nuclear energy on the basis of newer, safer designs. The new designs are much safer than the complex LWR, but the arrogance that labels them "inherently safe" is all too familiar. If a revival nevertheless occurs, utilities should be screened carefully before participating; performance must be monitored extensively, and bad performances (or performers) eliminated; and new designs must be deployed gradually so that we can learn from operating experience. Regulators and the industry should try to outdo each other in demanding high standards; but perhaps they should cooperate with each other in developing those standards. Public debate over these options should be encouraged, not stifled.

Bringing Politics Back In

Political and economic elites make and remake the world that the rest of us inhabit. Top bureaucrats were excited by the promise of nuclear energy, so they constructed a system for planting reactors throughout their country. After the oil crisis, political leaders had to decide whether to build new plants or to conserve energy and develop alternative energy resources. Elites change political, regulatory, and economic structures to attain their goals every bit as much as they react to existing structures. At the same time they deny they have any such power. They claim they are simply responding rationally to available data. Sometimes this is right, sometimes the structures do constrain what they can do. But the degree of constraint or flexibility should be an empirical question, not an assumption; it varies over time. We have to watch elites carefully, so that they cannot rewrite history to absolve themselves of responsibility and hide their own power.

List of Informants

IN GATHERING the information for this book, I had formal interviews and informal conversations with 101 people. I only list 98 here, since in three cases anonymous quotes in the text could easily be traced to their sources.

UNITED STATES

JOHN AHEARNE, Resources for the Future; member of Schlesinger energy team, 1977; NRC commissioner 1978–1983; series of conversations, September 1985 to December 1986. DAVE BERRICK, Environmental Policy Institute; telephone, 4 November 1985. SUE BROWN, Pacific Gas & Electric Company; public relations; ninety minutes, 8 October 1984. WILL COLLETTE, Citizens' Clearing House on Hazardous Waste; telephone, 12 November 1985. FRANK COSTAGLIOLA, Navy captain; JCAE staff in the 1960s; AEC commissioner in 1969; two hours, 1 May 1986. NANCY CULVER, Mothers for Peace; involved in media work; two hours, 8 October 1984. DIANE CURRAN, Harmon and Weiss law firm; telephone, 4 November, 1985. MORE DOWNING, Abalone Alliance; two one-hour interviews, 12 and 16 September 1984. RAYE FLEMING, Diablo Project Office; staff of Diablo Project Office and People Generating Energy; two hours, 8 October 1984, and telephone, 10 November, 1985. CHARLES FRANKLIN, Baltimore Gas and Electric; telephone, 22 November 1988. MARTIN F. GITTEN, Consolidated Edison Company of New York; telephone, 29 November 1988. DAVID HARTSOUGH, American Friends Service Committee; helped found Abalone Alliance; thirty minutes, 4 October 1984. JAMES HEWLETT, Department of Energy, Energy Information Agency; telephone, 12 November 1987. MARK JACOBS, Florida Power Corporation; telephone, 5 June 1989. RACHEL JOHNSON, Abalone Alliance; two hours, 10 September 1984, and several other conversations. JIM KILPATRICK, Pacific Gas and Electric Company; telephone, 30 October 1985. PAULA KLINE, Center for Law in the Public Interest; telephone, 11 November 1985. IAN MCMILLAN, Shandon, California; early critic of Diablo Canyon; two hours, 7 October 1984. GEOFF MEREDITH, Abalone Alliance; three hours, 12 September 1984, and several other conversations. PAM METCALFE, Diablo Project Office; active member of DPO group; one hour, 8 October 1984. GENE I. ROCHLIN, University of California; series of conversations, 1983–1985. PETER SANDMAN, Rutgers University; observer and adviser to the Kemeny Commission; three hours; 24 September 1987. DONNA SAN GIMINO, Atomic Industrial Forum; thirty minutes, 22 September 1986. GEORGE SARKISIAN, Pacific Gas and Electric Company; one hour, October 1984, and telephone, 6 November 1986. LEE SCHIPPER, Lawrence Berkeley Laboratory; University of California; energy analyst concentrating in energy efficiency; several conversations, 1984–1985. SANDY SILVER, Mothers for Peace; did legal work and gave testimony for Mothers; two hours, 9 October 1984. RON WEINBERG, Pacific Gas and Electric Company; telephone, 6 November 1985. FRED WOOCHER, Center for Law in the Public Interest; telephone, 11 November 1985.

SWEDEN

MICKI AGERBERG, *Ny Teknik*; editor of engineers' weekly; one hour, 17 June 1985. HARRY ALBINSSON, Sveriges Industriförbund; energy representative; two hours, 18 June 1985. OLLE ALSÉN, *Dagens Nyheter*; editorialist on energy and the environment; one hour, 3 June 1985. SVEN ANÉR, former member of Miljöpartiet, author of books on ecological issues; series of conversations, spring 1985. HADAR CARS, former Liberal Party MP and one of its main energy spokespersons; two hours, 30 May 1985. BIRGITTA DAHL, Industridepartmentet; Social Democrats' Minister of Energy and energy spokesperson; ninety minutes, 2 July 1985. LENNART DALÉUS, helped found Swedish Friends of the Earth; directed campaign for Line 3 in nuclear referendum; Center Party candidate for Riksdag; two hours, 28 May 1985. PER GAHRTON, leader of the Miljöpartiet, former Liberal MP; ninety minutes, 17 June 1985. TOR RAGNAR GERHOLM, University of Stockholm; professor of nuclear physics; actively pronuclear; three hours, 14 May 1985. BRIGITTA HAMBRAEUS, Riksdag; Center Party member since 1971; played crucial role in turning the party antinuclear; three hours, 23 May 1985. OLOF HÖRMANDER, Statenskärnkraftinspection; head of the Nuclear Inspectorate; one hour, 2 July 1985. OLOF JOHANSSON, Riksdag; Center Party MP; energy minister 1976–1978; two hours, 29 May 1985. PER KÅGESON, Sveriges Författarförbund; prominent ecology activist, formerly with the Communist Party; three hours, 25 June 1985. LARS G. LARSSON, Sveriges Industriförbund; has worked for Westinghouse, the Swedish Embassy in the United States, and the Swedish nuclear inspectorate; two hours, 18 June 1985. LARS LILJEGREN, Social Democrats' staff energy expert in 1970s and early 1980s; ninety minutes, 3 July 1985. STEFAN LINDSTRÖM, Department of Political Science; University of Stockholm; political scientist and policy analyst who has studied high technology; two hours, 27 July 1985. MÅNS LÖNNROTH, Secretariat for Future Studies; long-time energy analyst; head of "Energy and Society" project, 1974–1976; two hours, 25 May 1985. OLOF PETERSSON, Uppsala Universitet; Statsvetenskapliga Institutionen; analyst of voting patterns; has done work on nuclear energy issue; one hour, 20 May 1985. INGE PIERRE, Swedish Embassy to the United States; several telephone conversations, 1985–1986. RUNE PREMFORS, Department of Political Science; University of Stockholm; policy analyst; one hour, 27 July 1985. OVE RAINER, formerly a high civil servant and justice minister; chaired the Energy Commission of 1976–1978; two hours, 23 May 1985. JAN RANDERS, managing director of Kraftsam, organization of Sweden's main electricity producers; ninety minutes, 28 May 1985. BIRGIT RENBE, data librarian for SIFO, Sweden's most prominent polling organization; several conversations, June 1985. PETER STEEN, Försvarets Forskningsanstalt (National Defense Research Institute); long-time analyst of Sweden's energy system; series of conversations and lunches in Berkeley 1983 and Stockholm 1985. CARL THAM, former Liberal MP and energy minister in 1978; later head of new Energy Research Administration; one hour, 25 June 1985. PER UNCKEL, Riksdag; Moderate Party MP; directed the campaign for the most pronuclear line of 1980 referendum; ninety minutes, 22 May 1985. EVERT VEDUNG, Uppsala Universitet; Statsvetenskapliga Institutionen; political analyst who has concentrated on the nuclear energy debate; several conversations in Paris and Stockholm, 1985. CARL-ERIK WIKDAHL, assistant vice president for OKG, and active in energy debates; two hours, 28 June 1985.

FRANCE

DOMINIQUE ALLAN MICHAUD, sociologist working for ARTE, a private firm often under contract with the Ministry of the Environment; three hours, 26 March 1985. PIERRE AMOUYEL, Commissariat Général du Plan; in charge of energy and tertiary activities at the CGP; two hours, 15 February 1985. DANIEL BOY, Ecole des Hautes Etudes; political scientist who has studied relationship between national ecology groups and local communities; two hours, 28 March 1985. BERNARD CAZES, Commissariat Général du Plan; in charge of long-term planning and adviser to the commissaire; ninety minutes, 15 February 1985, and telephone, 19 March 1985. DANIEL CHAVARDÈS, French nuclear attaché to the U.S., formerly to Japan; thirty minutes, 18 October 1984. BERNARD CLAVEL, chef du Département Concessions-Contentieux, EDF; lawyer who has done work on land acquisition and administrative procedures; four hours, 14 March 1985. MICHEL CROZIER, sociologist of organizations and of French Grands Corps; ninety minutes, 11 February 1985. JÉROME DAUZET, assistant to nuclear attaché to U.S.; one hour, 18 October 1984. DENIS DUCLOS, sociologist at CNRS who has studied French ecology movement; series of conversations, November 1984 to April 1985. JEAN-LOUIS FABIANI, sociologist at Ecole Normale Supérieure who has studied ecology movement in France; one hour, 28 March, 1985. JEAN-MICHEL FAUVE, chef du Service de l'Information et des Rélations Publiques, EDF; in charge of public relations, formerly site chief at Plogoff; one hour, 6 March 1985. DENIS FOUQUET, chef de la Division "Economie Externe," Services des Etudes Economiques Générales, EDF; working on model of energy demand; two hours, 11 March 1985. ROBERT FRIANT, chef des services administratif et financier, Région d'Equipement Paris, EDF; one hour, 25 March 1985. CLAUDE GABOREAU, EDF; one hour, 6 March 1985. PATRICK GIROD, chef du service administration, Région d'Equipement Paris, EDF; two hours, 25 March 1985, and telephone conversation, 4 April 1985. MICHEL GRAS, Direction des Relations Internationales du CEA; presents French activities and reports to Framatome; 8 and 25 February 1985, ninety minutes each. HENRI HAOND, adjoint au chef du Département CPPI, Direction de l'Equipement, EDF; one hour, 27 March 1985. ZSUZSA HEGEDÜS, sociologist and member of Alain Touraine's team, familiar with ecology movement; series of conversations and seminars, 1984–1985. ANDREAS KITOFF, Lawrence Berkeley Laboratories; University of California; Italian, familiar with antinuclear movements in Europe; August 1984. MICHEL LABROUSSE, chef du Service Evaluation, Planification, Etudes et Recherches Economiques, Agence Française pour la Maîtrise de l'Energie; formerly at CEA, active in CFDT; one hour, 24 April 1985. ROLAND LAGARDE, Agence Française pour la Maîtrise de l'Energie; formerly at CEA and active in CFDT; one hour, 9 May 1985. YVES LENOIR, Groupe Energie et Développement; antinuclear activist; four hours, 5

March 1985. HÉLÈNE MEYNAUD, Groupe de Recherche—Energie, Technologie et Société, EDF; worked on public attitudes to nuclear energy for EDF; series of conversations, September 1982 to winter 1986. ERIC MIGNOT, Service des Etudes Economiques Générales; EDF; engineer working on economic calculations; two hours, 11 March 1985. JEAN-PAUL MOATTI, Institut National de la Santé et de la Recherche Médicale; economist studying technological risk; has written on nuclear debates in France; two hours, 1 March 1985, and a series of conversations in January 1986. ERIC MONNIER, Centre Scientifique et Technique du Bâtiment sociologist doing work on energy efficiency in buildings; one hour, 10 August 1984. GEORGES MORLAT, statistician, Groupe de Recherche—Energie, Technologie et Société, EDF; one hour, 21 February 1985. GUY PEDEN, chef du division IPC, EDF; one hour, 11 April 1985. MICHAEL POLLAK, Institut d'Histoire du Temps Présent; has written about the antinuclear movement; ninety minutes, 10 March 1985. LOUIS PUISEUX, economist, Centre International de Recherche sur l'Environnement et le Développement; formerly in charge of demand forecasting at EDF; author of *La Babel nucléaire* among other works on the politics of nuclear energy; two hours, 20 March 1985. PHILIPPE ROQUEPLO, former EDF researcher, one of the founders of the antinuclear *Gazette nucléaire*; several conversations, January 1986. MME. ROUISSON, Région d'Equipement Paris, EDF; one hour, 25 March 1985. PIERRE SAMUEL, Amis de la Terre; mathematician at Orsay; a founding father of French ecology, long-time member of Amis de la Terre; one hour, 13 March 1985. JEAN-PAUL SCHAPIRA, nuclear physicist at CNRS laboratory at Orsay; member of GSIEN; member of Castaing Commission; several conversations in Berkeley in May 1983, and in Paris 1985. MONIQUE SENÉ, nuclear physicist and guiding force behind GSIEN; two hours, 8 March 1985. TERRY SHINN, historian of the Ecole Polytechnique and engineers in France; currently working on scientists' worldviews, especially in physics; one hour, 15 March 1985. JOCELYNE SMADJA, sociologist at EDF studying public opinion, especially on energy, nuclear opinion; series of conversations, 1984 and 1985. ERIC STEMMELEN, statistician working for CEA, but with an office at GRETS to work with it on AESOP; ninety minutes, 1 March 1985. JEAN TASSART, CFDT; works on energy issues; two hours, 12 April 1985. MARCEL TESSERAUD, chef du Département Budget-Programme, Direction de l'Equipement, EDF; one hour, 26 March 1985. ALAIN TOURAINE, sociologist of social movements; series of conversations, Berkeley, Spring 1983, and Paris, 1984–1985. BÉNÉDICTE VALLET, sociologist who has written on nuclear safety in France; several conversations, February 1985. JEAN-CLAUDE ZERBIB, works in radioprotection at the CEA; member of CFDT, and one of their most active researchers in energy; two hours, 3 April 1985.

Bibliography

NOTE that the Swedish characters ä and å have been alphabetized as if they were English a's, and ö as if it were o.

Aberbach, Joel D., Robert D. Putnam, and Bert A. Rockman. 1981. *Bureaucrats and Politicians in Western Democracies*. Cambridge: Harvard University Press.
Abrahamson, Dean. 1979. "Governments Fall as Consensus Gives Way to Debate." *Bulletin of the Atomic Scientists* 35, no. 9 (November):30–37.
Abrams, Nancy E. 1979. "Nuclear Politics in Sweden." *Environment* 21, no. 4 (May):6–11, 39–40.
Adelman, M. A. 1972. *The World Petroleum Market*. Baltimore: Johns Hopkins University Press for Resources for the Future.
Agnew, Harold M., and Thomas A. Johnson. 1986. "Chernobyl: The Future of Nuclear Power." *Issues in Science and Technology* 3, no. 1:36–39.
Ahearne, John F. 1983. "Prospects for the U.S. Nuclear Reactor Industry." *Annual Review of Energy* 8:355–84.
———. 1986. "Nuclear Power after Chernobyl." Public talk, Resources for the Future, Washington, D.C., 1 October.
Alford, Robert, and Roger Friedland. 1985. *Powers of Theory*. New York: Cambridge University Press.
Alfvén, Hannes. 1972. "Science, Technology and the Politico-Economic Power." *Impact of Science on Society* 22 (January–June):58–83.
Allan Michaud, Dominique. 1979. "Le Discours Ecologique." Doctorat d'Etat diss., University of Bordeaux.
Allison, Graham, and Albert Carnesale. 1983. "The Utility Director's Dilemma: The Governance of Nuclear Power." In Dorothy S. Zinberg, ed., *Power: The Struggle for a National Energy Policy*. New York: Pergamon Press.
Almond, Gabriel A. 1988. "The Return to the State." *American Political Science Review* 82, no. 3:853–74.
Ames, Mary E. 1978. *Outcome Uncertain: Science and the Political Process*. Washington, D.C.: Communications Press.
Les Amis de la Terre (France). 1978. *L'Escroquerie nucléaire*. Paris: Stock.
———. 1980. "Un Réseau." *Journal des Amis de la Terre* 54 (June).
Anastassopoulos, Jean-Pierre. 1980. *La Stratégie des entreprises publiques*. Paris: Dalloz.
———. 1981. "The French Experience: Conflict with Government." In R. Vernon and Y. Aharom, eds., *State-Owned Enterprises in Western Economies*. London: Croom Helm.
———. 1986. "State-owned Enterprises between Autonomy and Dependency." *Journal of Public Policy* 5, no. 4:521–39.

Anderson, Clinton P. 1965. "Atoms for Peace: The Dream, the Reality." *New York Times Magazine*, 1 August, 10–14.

Anderson, Douglas D. 1981. *Regulatory Politics and Electric Utilities: A Case Study in Political Economy*. Boston: Auburn House.

Anderson, Stephen. 1974. "Economics of Nuclear Power Plant Location with Emphasis on the Coastal Zone." Ph.D. diss., University of California, Berkeley.

Anér, Sven. 1982. *Se upp för Sigyn!* Stockholm: Sven Anér Förlag.

Anger, Didier. 1977. *Chronique d'une lutte: Le Combat antinucléaire à Flamanville et dans La Hague*. Paris: Jean-Claude Simoën.

Askin, A. Bradley, and John Kraft. 1976. *Econometric Dimensions of Energy and Supply*. Lexington, Mass.: Lexington Books.

Aviel, S. David. 1982. *The Politics of Nuclear Energy*. Washington, D.C.: University Press of America.

Baecque, Francis de, and Jean-Louis Quermonne. 1981. *Administration et politique sous la cinquième république*. Paris: Presses de la Fondation Nationale des Sciences Politiques.

Barry, Brian. 1978. *Sociologists, Economists, and Democracy*. Chicago: University of Chicago Press, 1978. First published 1970.

Baughman, Martin L., Paul L. Joskow, and Dilip P. Kamat. 1979. *Electric Power in the United States: Models and Policy Analysis*. Cambridge: MIT Press.

Baum, Robert J., ed. 1980. *Ethical Problems in Engineering*. Vol. 2, *Cases*. 2d ed. Troy, N.Y.: Rensselaer Polytechnic, Center for the Study of the Human Dimensions of Science and Technology.

Beaton, Leonard, and John Maddox. 1962. *The Spread of Nuclear Weapons*. New York: Praeger.

Bell, D. S., and Byron Criddle. 1984. *The French Socialist Party. Resurgence and Victory*. Oxford: Clarendon Press.

Bell, D. S., and Eric Shaw. 1983. "The Socialists." In D. S. Bell and Eric Shaw, eds., *The Left in France*. Nottingham: Spokesman.

Benat, Jean. 1983. "The French Nuclear Program: Benefits of Standardization." *Nuclear Europe*, 14–17.

Berman, Larry. 1986. "Sweden's Energy Politics after Chernobyl." *Human Environment in Sweden*, no. 27. Swedish Information Service.

Bernstein, Basil. 1975. *Class, Codes and Control*. New York: Schocken.

Bernstein, Marver H. 1955. *Regulating Business by Independent Commission*. Princeton: Princeton University Press.

Bernstein, Richard J. 1976. *The Restructuring of Social and Political Theory*. Philadelphia: University of Pennsylvania Press.

Bieber, J. 1973. "Calvert Cliffs' Coordinating Committee vs. AEC: The AEC Learns the True Meaning of the National Environmental Policy Act of 1969." *Environmental Law* 3:316–33.

Biquard, Pierre. 1966. *Joliot-Curie*. New York: Paul S. Eriksson.

Bjurulf, Bo, and Urban Swahn. 1980. "Health Policy Proposals and What Happened to Them: Sampling the Twentieth-Century Record." In Arnold J. Heidenheimer and Nils Elvander, eds., *The Shaping of the Swedish Health System*. New York: St. Martin's Press.

Blackwood, Caroline. 1984. *On the Perimeter*. London: William Heinemann Ltd.

Bodde, David L. 1976. "Regulation and Technical Evolution: A Study of the Nuclear Steam Supply System and Commercial Jet Engine." Ph.D. diss., Harvard University.

Bodiguel, Jean-Luc. 1978. *Les Anciens élèves de l'E.N.A.* Presses Universitaires de la Fondation Nationale des Sciences Politiques.

Bodiguel, Jean-Luc and Jean-Louis Quermonne. 1983. *La Haute fonction publique*. Paris: Presses Universitaires de France.

Boffey, Philip M. 1969. "Ernest J. Sternglass: Controversial Prophet of Doom." *Science* 166:195ff.

Bormann, Ernest G. 1972. "Fantasy and Rhetorical Vision: The Rhetorical Criticism of Social Reality." *Quarterly Journal of Speech* 58:396–407.

Bouthillier, Guy. 1969. *La Nationalisation du gaz et l'électricité en France: Acteurs et Processus*. Thesis. University of Paris at Nanterre.

Bradford, Peter A. 1982a. "Remarks before the Groton School," U.S. Nuclear Regulatory Commission Office of Public Affairs, Washington, D.C., 15 January.

———. 1982b. "The Man/Machine Interface." Remarks to the Public Citizen Forum, NRC Office of Public Affairs, Washington, D.C., 8 March.

———. 1982c. "Nuclear Hearings, Nuclear Regulation, and Public Safety: A Reflection on the NRC's Indian Point Hearings." Speech to the Environmental Defense Fund Associates, New York, 7 October.

Bradshaw, Ted K., Todd R. La Porte, Gene I. Rochlin, and P. Brett Hammond. 1984. *Assessment of Alternate Energy Technologies: An Institutional Analysis*. Berkeley, Calif.: Institute of Governmental Studies.

Brady, Rosemary. 1984. "The Sorcerer's Apprentice." *Forbes*, 24 September, 106–10.

Brennan, G., and James M. Buchanan. 1977. "Toward a Tax Constitution for Leviathan." *Journal of Public Economics* 8:255–73.

Breyer, Stephen. 1978. "*Vermont Yankee* and the Courts' Role in the Nuclear Energy Controversy." *Harvard Law Review* 91:1833–45.

Bronstein, Daniel A. 1971. "The AEC Decision-Making Process and the Environment: A Case Study of the Calvert Cliffs Nuclear Power Plant." *Ecology Law Quarterly* 1, no. 4:689–726.

Budnitz, Robert J. 1981. "The Response of the Nuclear Regulatory Commission to the Accident at Three Mile Island." In *The Three Mile Island Accident*. See Moss and Sills 1981.

Bupp, Irvin C. 1979. "The Nuclear Stalemate." In *Energy Future*. See Stobaugh and Yergin 1979.

———. 1981. Preface to *Power Plant Cost Escalation*. See Komanoff 1981.

Bupp, Irvin C., and Jean-Claude Derian. 1978. *Light Water: How the Nuclear Dream Dissolved*. New York: Basic Books.

———. 1981. "Introduction to the Paperback Edition" of *Light Water* (see Bupp and Derian 1978), entitled *The Failed Promise of Nuclear Power*. New York: Basic Books.

Bupp, Irvin C., Jean-Claude Derian, Marie-Paul Donsimoni, and Robert Treitel. 1975. "The Economics of Nuclear Power." *Technology Review* 77, no. 4:15–25.

Bupp, Irvin C., and Charles Komanoff. 1983. "Prometheus Bound: Nuclear Power at the Turning Point." Report to Cambridge Energy Research Associates, Cambridge, Mass.

Burke, Kenneth. 1941. "The Rhetoric of Hitler's Battle." *The Philosophy of Literary Form*. Baton Rouge: Louisiana State University Press.

———. 1950. *The Rhetoric of Motives*. Berkeley: University of California Press.

———. 1954. *Permanence and Change*. 2d ed. Berkeley: University of California Press.

———. 1969. *A Grammar of Motives*. Berkeley: University of California Press.

Burtheret and de Cormis. 1980. "Le régime administratif des centrales nucléaires." *Revue générale nucléaire*, no. 3 (May–June):249–58.

Byse, Clarke. 1978. "*Vermont Yankee* and the Evolution of Administrative Procedure: A Somewhat Different View." *Harvard Law Review* 91:1823–32.

Camilleri, Robert. 1984. *Nuclear Power and the State*. Seattle: University of Washington Press.

Campbell, John. 1986. "The State, Capital Formation, and Industrial Planning: Financing Nuclear Energy in the United States and France." *Social Science Quarterly* 67:707–21.

———. 1988. *Collapse of an Industry: Nuclear Power and the Contradictions of U.S. Policy*. Ithaca, N.Y.: Cornell University Press.

Carson, Rachel. 1962. *Silent Spring*. Greenwich, Conn.: Fawcett Crest Books.

Carter, Luther J. 1976a. "Nuclear Initiative: Impending Vote Stimulates Legislative Action." *Science* 192 (4 June):975–77.

———. 1976b. "Nuclear Initiative: Californians Vote 'No', But Legislature Acts." *Science* 192 (25 June):1317–19.

———. 1987. *Nuclear Imperatives and Public Trust: Dealing with Radioactive Waste*. Washington, D.C.: Resources for the Future.

Catton, William R., Jr., and Riley E. Dunlap. 1980. "A New Ecological Paradigm for Post-Exuberant Sociology." *American Behavioral Scientist* 24:15–47.

Cavers, David F. 1964. "Improving Financial Protection of the Public against the Hazards of Nuclear Power." *Harvard Law Review* 77:644–88.

Centrala Driftledningen (CDL). 1972. *Sveriges Elförsörjning 1975–1990: 1972 års studie*. Stockholm.

Cerny, Philip G., and Martin A. Schain, eds. 1980. *French Politics and Public Policy*. New York: St. Martin's Press.

Chafer, Tony. 1982. "The Anti-Nuclear Movement and the Rise of Political Ecology." In Philip G. Cerny, ed., *Social Movements and Protest in France*. London: Frances Pinter.

Charlot, Monica. 1975. "The Language of Television Campaigning." In Howard R. Penniman, ed., *France at the Polls*. Washington, D.C.: American Enterprise Institute.

Chaudron, Martine, and Yves Le Pape. 1979. "Le Mouvement écologique dans la lutte anti-nucléaire." In *Nucléopolis*. See Fagnani and Nicolon, 1979.

Chevallier, F. 1979. *Les Entreprises publiques en France*. Paris: Documentation Française.

Chubb, John E. 1983. *Interest Groups and the Bureaucracy. The Politics of Energy*. Stanford: Stanford University Press.

Clarfield, Gerard H., and William M. Wiecek. 1984. *Nuclear America: Military and Civilian Nuclear Power in the United States 1940–1980*. New York: Harper & Row.

Clarke, Lee. 1985. "The Social Origins of Nuclear Power: A Case of Institutional Conflict." *Social Problems* 32, no. 5:474–87.

Clausen, Aage R. 1973. *How Congressmen Decide: A Policy Focus*. New York: St. Martin's Press.

Clausen, Aage R., and Carl E. Van Horn. 1977. "The Congressional Response to a Decade of Change: 1963–1972." *Journal of Politics* 39:624–66.

Cochaud, Jean-François. 1981. "Procédures administratives réglementant la construction et l'exploitation des centrales nucléaires." *Revue générale nucléaire*, no. 2 (March–April):115–19.

Cochrane, James L. 1981. "Carter's Energy Policy and the Ninety-fifth Congress." In Crauford D. Goodwin, ed., *Energy Policy in Perspective*. Washington, D.C.: Brookings Institution.

Cohen, Etahn Micah. 1981. "Ideology, Interest Group Formation, and Protest: The Case of the Anti-Nuclear Power Movement, the Clamshell Alliance, and the New Left." Ph.D. diss., Harvard University.

Cohen, Linda. 1979. "Innovation and Atomic Energy: Nuclear Power Regulation, 1966-Present." *Law and Contemporary Problems* 43, no. 1:67–97.

Collier, Charles S. 1966. "Are the 'No Recourse' Provisions of the Price-Anderson Act Valid or Unconstitutional?" *Houston Law Review* 4:236–74.

Colson, Jean-Philippe. 1977. *Le Nucléaire sans les Français. Qui décide? Qui profite?* Paris: Maspero.

Colton, Roger D. 1986. "Utility Involvement in Energy Management: The Role of a State Power Plant Certification Statute." *Environmental Law* 16:175–206.

Columbia Law Review. 1975. "Preemption Doctrine: Shifting Perspectives on Federalism and the Burger Court." Vol. 75:623–54.

Commissariat à l'Energie Atomique. See French Commissariat à l'Energie Atomique.

Confédération Française Démocratique du Travail (CFDT). 1979. *Energie, nucléaire: Choisir notre avenir*. Paris: Montholon-Services.

———. 1980. *Le Dossier électronucléaire*. Paris: Seuil. Revised version of *L'Electronucléaire en France*, 1975.

———. 1982. *Connaître la CFDT*. Brochure. Paris: Montholon-Services.

Connolly, Thomas J., Ulf Hansen, Wolfgang Jaek, and Karl-Heinz Beckurts. 1982. "World Nuclear Energy Paths." In Ian Smart, ed., *World Nuclear Energy: Toward a Bargain of Confidence*. Baltimore: Johns Hopkins University Press.

Conseil Général de l'Isère. 1977. *Creys-Malville: le dernier mot?* Grenoble: Presses Universitaires de Grenoble.
Cook, Constance Ewing. 1980. *Nuclear Power and Legal Advocacy: The Environmentalists and the Courts.* Lexington, Mass.: Lexington Books, D.C. Heath and Co.
Cook, James. 1985. "Nuclear Follies." *Forbes*, 11 February, 82–100.
Cooney, Robert, and Helen Michalowski, eds. 1977. *The Power of the People: Active Nonviolence in the United States.* Culver City, Calif.: Peace Press.
Cooper, Walter. 1981. "A Nuclear Power Plant in Whose Backyard?" *Technology Review* 83, no. 6:64–66.
Corrigan, Richard, Claude E. Barfield, and James A. Noone. 1973. "Administration, Congress Move Swiftly to Counter Shortages of Fuel Supplies." *National Journal Reports*, 17 November 1973, 1722–30.
Cotgrove, Stephen. 1982. *Catastrophe or Cornucopia: The Environment, Politics and the Future.* Chichester and New York: John Wiley & Sons.
Cotgrove, Stephen, and Andrew Duff. 1980. "Environmentalism, Middle-Class Radicalism and Politics." *Sociological Review*, new series, 28, no. 2 (May): 333–51.
Crain, M., and R. Tollison. 1976. "Campaign Expenditures and Political Competition." *Journal of Law and Economics* 19:177–88.
Daglish, James. 1985. "Nuclear Plant Availability: Real Achievements." *IAEA Bulletin* 27, no. 3 (Autumn).
Dahl, Hans. 1984. "Those Equal Folk." *Daedalus* 113, no. 1 (Winter):93–107.
Dallmayr, Fred R. 1981. *Twilight of Subjectivity: Contributions to a Post-Individualist Theory of Politics.* Amherst: University of Massachusetts Press.
———. 1984. *Language and Politics: Why Does Language Matter to Political Philosophy?* Notre Dame, Ind.: Notre Dame University Press.
Darmstadter, Joel, and Hans H. Landsberg. 1976. "The Economic Background." In *The Oil Crisis.* See Vernon 1976.
Davidson, Roger H. 1977. "Breaking up Those 'Cozy Triangles': An Impossible Dream?" In Susan Welch and John G. Peters, eds., *Legislative Reform and Public Policy*, New York: Praeger.
Davis, Thomas P. 1976. "Citizen's Guide to Intervention in Nuclear Power Plant Siting: A Blueprint for Alice in Nuclear Wonderland." *Environmental Law* 6, no. 3:621–74.
deLeon, Peter. 1979. *Development and Diffusion of the Nuclear Power Reactor: A Comparative Analysis.* Cambridge, Mass.: Ballinger.
Del Sesto, Stephen. 1979. *Science, Politics, and Controversy: Civilian Nuclear Power in the United States, 1946–1974.* Boulder, Colo.: Westview Press.
de Marchi, Neil. 1981a. "Energy Policy under Nixon: Mainly Putting Out Fires." In Crauford D. Goodwin, ed., *Energy Policy in Perspective.* Washington, D.C.: Brookings Institution.
———. 1981b. "The Ford Administration: Energy as a Political Good." In Crauford D. Goodwin, ed., *Energy Policy in Perspective.* Washington, D.C.: Brookings Institution.
Devall, William B. 1970. "The Governing of a Voluntary Organization: Oligarchy

and Democracy in the Sierra Club." Ph.D. diss., University of Oregon, Eugene.

Dodman, Michael J. 1988. "Electric Utilities Face Increased Competition." *Resources* 90:20–22.

Dörfer, Ingemar N. H. 1974. "Science and Technology in Sweden: The Fabians versus Europe," *Research Policy* 3:134–55.

Dorget, François. 1984. *Le Choix nucléaire Français*. Paris: Economica.

Douglas, Mary. 1970. *Natural Symbols. Explorations in Cosmology*. New York: Random House.

———. 1978. "Cultural Bias." London: Royal Anthropological Institute, occasional paper 35. Reprinted in *In the Active Voice*, edited by Mary Douglas. London: Routledge and Kegan Paul, 1984.

———. 1986. *How Institutions Think*. Syracuse: Syracuse University Press.

Douglas, Mary, ed. 1982. *Essays in the Sociology of Perception*. London: Routledge and Kegan Paul.

Douglas, Mary, and Aaron Wildavsky. 1982. *Risk and Culture: An Essay on the Selection of Technological and Environmental Dangers*. Berkeley and Los Angeles: University of California Press.

Downs, Anthony. 1957. *An Economic Theory of Democracy*. New York: Harper & Row.

———. 1960. "Why the Government Budget is Too Small in a Democracy." *World Politics* 12:541–63.

Dunlap, Riley E., and Kent D. Van Liere. 1978. "The 'New Environmental Paradigm': A Proposed Instrument and Preliminary Results." *Journal of Environmental Education* 9, no. 4:10–19.

Durrieu, Yves. 1977. "Nucléaire et indépendance nationale." *Repères* 48.

Dworkin, Ronald. 1985. *A Matter of Principle*. Cambridge: Harvard University Press.

Ebbin, Steven, and Raphael Kasper. 1974. *Citizen Groups and the Nuclear Power Controversy: Uses of Scientific and Technological Information*. Cambridge: MIT Press.

Edelman, Murray. 1964. *The Symbolic Uses of Politics*. Urbana: University of Illinois Press.

———. 1977. *Political Language: Words That Succeed and Policies That Fail*. New York: Academic Press.

Electricité de France. 1982. *Eléments de sûreté et de radioprotection des centrales nucléaires de 1300 mégawatts*. Paris: EDF, Direction de l'Equipement. October.

England, William T. 1971. "Nuclear Insurance and the Price-Anderson Act." *Atomic Energy Law Journal* 13, no. 1:27–49.

Epstein, Leon D. 1986. *Political Parties in the American Mold*. Madison: University of Wisconsin Press.

Escoube, Pierre. 1971. *Les Grands Corps de l'Etat*. Que Sais-Je? series. Paris: Presses Universitaires de France.

Evanoff, Mark. 1982. "Seduced by the Friendly Atom: One Utility's Adventure." Draft manuscript. Friends of the Earth, San Francisco.

Evans, Peter B., Dietrich Rueschemeyer, and Theda Skocpol, eds. 1985. *Bringing the State Back In*. New York and Cambridge: Cambridge University Press.

Fagnani, Francis, and Alexandre Nicolon, eds. 1979. *Nucléopolis*. Grenoble: Presses Universitaires de Grenoble.

Fagnani, Jeanne, and Jean-Paul Moatti. 1982. "Politique energétique du gouvernement Socialiste Français et marginalisation de l'opposition anti-nucléaire." Paper presented at conference "Comparative Research on Nuclear Energy Policy: The State of the Art," Berlin, 14–16 December.

———. 1984a. "De l'Opposition anti-nucléaire au contrôle social de la Technologie: Une nécessaire adaptation pour les acteurs sociaux de l'environnement." Paper presented at colloquium "Les Politiques d'Environnement face à la Crise," Paris, Germes, 10–12 January.

———. 1984b. "The Politics of French Nuclear Development." *Journal of Policy Analysis and Management* 3, no. 2:264–75.

Faivret, Jean-Philippe, Jean-Louis Missika, and Dominique Wolton. 1980. *L'Illusion ecologique*. Paris: Seuil.

Feigenbaum, Harvey B. 1985. *The Politics of Public Enterprise: Oil and the French State*. Princeton: Princeton University Press.

Fenn, Scott. 1980. *The Nuclear Power Debate*. Washington, D.C.: Investor Responsibility Research Center.

Ferguson, Thomas, and Joel Rogers. 1986. *Right Turn: The Decline of the Democrats and the Future of American Politics*. New York: Hill and Wang.

Festinger, Leon, Henry W. Riecken, and Stanley Schachter. 1956. *When Prophecy Fails*. New York: Harper & Row.

Feyerabend, Paul. 1977. *Against Method*. London: New Left Books.

Firth, Raymond. 1951. *Elements of Social Organization*. London: Watts.

Flavin, Christopher. 1983. *Nuclear Power: The Market Test*. Worldwatch Paper 57. Washington, D.C.: Worldwatch Institute, December.

———. 1986. "Reforming the Electric Power Industry." In Lester R. Brown et al. *State of the World 1986*. New York: Norton.

Fleck, Ludwik. 1979. *The Genesis and Development of a Scientific Fact*. Chicago: University of Chicago Press. Originally published 1935.

Fletcher, Harold A. 1959. "The Nationalization Debate in France, 1942–1946." Ph.D. diss., Harvard University.

Ford, Daniel. 1982. *The Cult of the Atom: The Secret Papers of the Atomic Energy Commission*. New York: Simon and Schuster.

Fourgous, J. M., J. F. Picard, and C. Raguenel. 1980. *Les Français et l'energie*. Paris: CNRS and EDF.

Fowlkes, Frank V., and Joel Havemann. 1973. "President Forms Federal Energy Body with Broad Regulation, Price Control Powers." *National Journal Reports*, 8 December, 1830–38.

Frears, J. R. 1981. *France in the Giscard Presidency*. London: George Allen & Unwin.

Freeman, Gary P. 1986. "National Styles and Policy Sectors: Exploring Structured Variation." *Journal of Public Policy* 5, no. 4:467–96.

French Commissariat à l'Energie Atomique (CEA). 1964. *Perspectives techniques* et économiques des centrales électronucléaire uranium naturel: Première partie—Réacteurs Modérés au Graphite. Paris: CEA. June.

———. 1983. Rapport annuel. Paris: CEA.

French Ministère de Redéploiement Industriel et du Commerce Extérieur. 1985. *Les Chiffres clés de l'énergie, Edition 1985.* Paris: Dunod.

French Secrétariat Général du Haut Comité de l'Environnement. 1978. *L'Etat de l'environnement: Rapport annuel 1976–1977.* Paris: Documentation Française.

Freudenburg, William R., and Eugene A. Rosa, eds. 1984. *Public Reaction to Nuclear Power: Are There Critical Masses?* American Academy for the Advancement of Science Selected Symposia Series. Boulder, Colo.: Westview Press.

Frost, Robert L. 1983. "Alternating Currents: Technocratic Power and Workers' Resistance at Electricité de France, 1946–1970." Ph.D. diss., University of Wisconsin.

Fuller, John G. 1975. *We Almost Lost Detroit.* New York: Ballantine.

Gandara, Arturo. 1977. *Electric Utility Decisionmaking and the Nuclear Option.* Santa Monica, Calif.: Rand Study for the National Science Foundation.

Garris, Jerome H. 1972. "Sweden and the Spread of Nuclear Weapons: A Study in Restraint." Ph.D. diss., University of California, Los Angeles.

Geertz, Clifford. 1973. *The Interpretation of Cultures.* New York: Basic Books.

———. 1983. *Local Knowledge.* New York: Basic Books.

Gerondeau, Christian. 1984. *L'Energie à revendre.* Paris: J.–C. Lattès.

Gibbs, Lois. 1982. *The Love Canal: My Story.* Albany: State University of New York Press.

Giddens, Anthony. 1976. *New Rules of Sociological Method.* New York: Basic Books.

———. 1979. *Central Problems in Social Theory.* London: Macmillan Press.

———. 1984. *The Constitution of Society.* Berkeley: University of California Press.

Gilinsky, Victor. 1982. "Streamlining Nuclear Regulation?" Remarks to the Atomic Industrial Forum, Kansas City, 24 May.

Gillette, Robert. 1972. "Nuclear Safety." Four part article. *Science* 177. 1, 8, 15, and 22 September.

Gilpin, Robert. 1968. *France in the Age of the Scientific State.* Princeton: Princeton University Press.

Giry, Robert. 1978. *Le Nucléaire inutile? Panorama des énergies de rechange.* Paris: Editions Entente.

Glaize, G. 1977. "The Reorganisation of the French Commissariat à l'Energie Atomique." *Nuclear Law Bulletin* 19:43–52.

Gleizal, Jean-Jacques. 1980. *Le Droit politique de l'état.* Paris: Presses Universitaires de France.

Gofman, John W., and Arthur Tamplin. 1971. *Poisoned Power.* Emmaus, Pa.: Rodale. Reprinted 1979.

Goguel, François. 1983. *Chroniques électorales: La Cinquième république après de Gaulle.* Paris: Presses de la Fondation Nationale des Sciences Politiques.

Golay, Michael W., Isi I. Saragossi, and Jean-Marc Willefert. 1977. "Comparative Analysis of United States and French Nuclear Power Plant Siting and Construction Regulatory Policies and their Economic Consequences." Report no. MIT-EL 77-044-WP, December. Cambridge: MIT Energy Laboratory.

Goldman, Sheldon, and Thomas P. Jahnige. 1976. *The Federal Courts as a Political System.* 2d ed. New York: Harper & Row.

Goldschmidt, Bertrand. 1982. *The Atomic Complex: A Worldwide Political History of Nuclear Energy.* Translated by Bruce M. Adkins. La Grange Park, Ill.: American Nuclear Society, 1982. Revised and updated from *Le Complexe atomique,* Libraire Arthème Fayard, 1980.

Goldstein, Walter. 1978. "Politics of U.S. Energy Policy." *Energy Policy* 2 (September):180–95.

Goodman, Gordon T., Lars A. Kristoferson, and Jack M. Hollander. 1981. *The European Transition to Oil: Societal Impacts and Constraints on Energy Policy.* London and New York: Academic Press Inc.

Gordon, Richard L. 1982. *Reforming the Regulation of Electric Utilities.* Lexington, Mass.: Lexington Books.

Gormley, William T., Jr. 1983. *The Politics of Public Utility Regulation.* Pittsburgh: University of Pittsburgh Press.

Gorz, André [Michel Bosquet]. 1977. *Ecologie et liberté.* Paris: Editions Galilée.

——. 1980. *Ecology as Politics.* Boston: South End Press. Translation of *Ecologie et Politique,* 1975.

Gravelaine, Francis, and S. O'Dy. 1978. *L'Etat EDF.* Paris: Alain Moreau.

Green, Harold P. 1973. "Nuclear Power: Risk, Liability, and Indemnity." *Michigan Law Review* 71:479–510.

——. 1982. "The Peculiar Politics of Nuclear Power." *The Bulletin of the Atomic Scientists,* December, 59–65.

Grenon, Michel. 1973. *Ce Monde affamé d'énergie.* Paris: Laffont.

Grossman, Richard. 1976. "Being Right Is Not Enough." *Environmental Action* 8, no. 16 (18 December).

Groupement de Scientifiques pour l'Information sur l'Energie Nucléaire (GSIEN). 1977. *Electro-nucléaire: Danger.* Paris: Seuil.

——. 1981. *Plutonium sur rhône: Super-Phénix, insécurité et incertitudes.* Paris: Syros.

Guilhamon, Jean. 1983. "The Conditions of Implementation of the French Nuclear Program." Speech to the American Nuclear Society winter meeting, San Francisco, 31 October.

Gusfield, Joseph R. 1963. *Symbolic Crusade: Status Politics and the American Temperance Movement.* Urbana: University of Illinois Press.

——. 1981. *The Culture of Public Problems.* Chicago: University of Chicago Press.

Gyorgy, Ann, & Friends. 1979. *No Nukes.* Boston: South End Press.

Habermas, Jürgen. 1979. *Communication and the Evolution of Society.* Boston: Beacon.

——. 1984. *The Theory of Communicative Action.* Vol. 1, *Reason and the Rationalization of Society.* Boston: Beacon.

———. 1987. *The Theory of Communicative Action.* Vol. 2, *Lifeworld and System: A Critique of Functional Reason.* Boston: Beacon.
Hacking, Ian. 1981. *Scientific Revolutions.* New York: Oxford University Press.
Hall, John A. 1965. "Atoms for Peace or War." *Foreign Affairs* 43:602–15.
Hall, Stuart, Dorothy Hobson, Andrew Lowe, and Paul Willis, eds. 1980. *Culture, Media, Language.* London: Hutchinson.
Hall, Stuart, and Tony Jefferson. 1976. *Resistance Through Rituals. Youth Subcultures in Post-War Britain.* London: Hutchinson.
Hamon, Hervé, and Patrick Rotman. 1982. *La Deuxième gauche: Histoire intellectuelle et politique de la CFDT.* Paris: Editions Ramsay.
Hannerz, K. 1982. "Towards Intrinsically Safe Light Water Reactors." Institute for Energy Analysis, Oak Ridge, Tennessee. Mimeo.
Hansen, Kent, Dietmar Winje, Eric Beckjord, Elias P. Gyftopoulos, Michael Golay, and Richard Lester. 1989. "Making Nuclear Power Work: Lessons from Around the World." *Technology Review* 92, no. 2:30–40.
Hays, Walter V. 1976. "State Power to Ban Nuclear Power Plants: The California Nuclear Safeguards Initiative as a Case in Point." *Environmental Law* 6, no. 3:729–52.
Helman, H. B. 1967. "Pre-emption: Approaching Federal-State Conflict over Licensing Nuclear Power Plants." *Marquette Law Review* 51:43–67.
Hernu, Charles. 1976. "Ecologie et conservatisme." *Nouvel observateur* 621 (4–10 October).
Herring, E. Pendleton. 1936. *Public Administration and the Public Interest.* New York: McGraw-Hill.
Hertsgaard, Mark. 1983. *Nuclear Inc.* New York: Pantheon Books.
Hesse, Mary. 1974. *The Structure of Scientific Inference.* Berkeley: University of California Press.
———. 1980. *Revolutions and Reconstructions in the Philosophy of Science.* Bloomington: Indiana University Press.
Hewlett, James G. 1984. *Investor Perceptions of Nuclear Power.* Washington, D.C.: Department of Energy, Energy Information Office. DOE/EIA-0446.
Hilgartner, Stephen, Richard C. Bell, and Rory O'Connor. 1982. *Nukespeak: The Selling of Nuclear Technology in America.* Middlesex: Penguin.
Hill, John. 1978. "An Assessment of the Carter Administration's Proposed Energy Program." *Natural Resources Lawyer* 10:615–23.
Hill, Stuart L. 1985. "Political Culture and Evaluation of Nuclear Power: A Comparative Analysis of France and the United States." *Public Policy across Nations: Social Welfare in Industrial Settings.* Public Policy Studies series 8:35–69.
Hinckley, Barbara. 1981. *Congressional Elections.* Washington, D.C.: Congressional Quarterly Press.
Hirsch, K. L. 1972. "Toward a New View of Federal Preemption." *University of Illinois Law Forum,* 515–58.
Hoban, Thomas More, and Richard Oliver Brooks. 1987. *Green Justice. The Environment and the Courts.* Boulder, Colo., and London: Westview Press.

Hogerton, John F. 1968. "The Arrival of Nuclear Power." *Scientific American* 218 (February):21–31.
Holifield, Chet. 1970. "Federal-State Responsibilities in Nuclear Development." *Atomic Energy Law Journal* 12:165–83.
Hollis, Martin, and Steven Lukes, eds. 1982. *Rationality and Relativism*. Cambridge: MIT Press.
Holmberg, Sören, and Kent Asp. 1984. *Kampen Om kärnkraften*. Stockholm: Liber Förlag Publica.
Holmberg, Sören, and Olof Petersson. 1980. *Inom felmarginalen: En Bok om politiska opinionsundersökningar*. Stockholm: Liber Förlag Publica.
Holmberg, Sören, Jörgen Westerståhl, and Karl Branzén. 1977. *Väljarna och kärnkraften*. Stockholm: Liber Förlag.
Hookway, Christopher, and Philip Pettit, eds. 1978. *Action and Interpretation: Studies in the Philosophy of the Social Sciences*. Cambridge: Cambridge University Press.
Hymen, Leonard S. 1985. *America's Electric Utilities: Past, Present and Future*. Arlington, Va.: Public Utilities Reports.
Ikenberry, G. John. 1988. *Reasons of State: Oil Politics and the Capacities of American Government*. Ithaca, N.Y.: Cornell University Press.
Inglehart, Ronald. 1977. *The Silent Revolution*. Princeton: Princeton University Press.
Institut Economique et Juridique de l'Energie (IEJE). 1975. *Alternatives au nucléaire*. Grenoble: Presses Universitaires de Grenoble.
International Energy Agency (IEA). See OECD International Energy Agency.
Isberg, Pelle. 1976. *Svensk kärnkraft? En Kärntekniker kommenterar debatten*. Stockholm: Natur ock Kultur.
Jacobson, Gary C. 1980. *Money in Congressional Elections*. New Haven: Yale University Press.
Jacobsson, M. 1985. "State Financial Cover for Nuclear Accidents." In OECD, *Nuclear Third Party Liability and Insurance*. Paris: OECD and IAEA.
Janda, Kenneth. 1980. "A Comparative Analysis of Party Organizations." In William Crotty, ed., *The Party Symbol*. San Francisco: Freeman.
Jasper, James M. 1985. "Mobilization for Protest: Values and Activities in the Movement against Nuclear Power." Paper presented at the American Sociological Association Annual Meetings, Washington, D.C., September.
———. 1987a. "Two or Twenty Countries: Contrasting Styles of Comparative Research." *Comparative Social Research* 10:205–29.
———. 1987b. "French Lessons: Can They Help the U.S. Nuclear Industry?" *Resources* 89 (Fall):13–16.
———. 1987c. "Moral versus Instrumental Politics: Why Experts and the Public Miscommunicate." Paper presented at the International Society of Political Psychology annual meeting, San Francisco, July.
———. 1987d. "L'Energie nucléaire et les attitudes face au risque: l'Approche culturelle." In Jean-Louis Fabiani and Jacques Theys, eds., *La Société vulnerable: Evaluer et maîtriser les risques*. Paris: Presses de l'Ecole Normale Supérieure.

———. 1988. "The Political Life Cycle of Technological Controversies." *Social Forces* 67, no. 2:357–77.

Johansson, Thomas B., and Peter Steen. 1979. "What To Do with the Radioactive Waste?" *Bulletin of the Atomic Scientists* 35, no. 9:38–42.

———. 1981. *Radioactive Waste from Nuclear Power Plants*. Berkeley: University of California Press.

Johnsrud, Judy Ann Hays. 1977. *A Political Geography of the Nuclear Power Controversy: The Peaceful Atom in Pennsylvania*. Ph.D. diss., Pennsylvania State University.

Joskow, Paul, and Richard Schmalensee. 1983. *Markets for Power: An Analysis of Electric Utility Deregulation*. Cambridge: MIT Press.

Jund, Thierry. 1977. *Le Nucléaire Contre l'Alsace*. Paris: Syros.

Kågeson, Per, and Björn Kjellström. 1984. *Fängslad vid kärnkraften?* Stockholm: LiberFörlag.

Karp, Walter. 1985. "Liberty Under Siege: The Reagan Administration's Taste for Autocracy," *Harper's*, November, 53–67.

Katz, James Everett. 1984. *Congress and National Energy Policy*. New Brunswick, N.J.: Transaction Books.

Katzenstein, Peter, ed. 1978. *Between Power and Plenty: Foreign Economic Policies of Advanced Industrial States*. Madison: University of Wisconsin Press.

———. 1985. *Small States in World Markets. Industrial Policy in Europe*. Ithaca, N.Y.: Cornell University Press.

Kelman, Steven. 1976. "Letter from Stockholm." *New Yorker*, 1 November, 104–27.

———. 1981a. *Regulating America, Regulating Sweden: A Comparative Study of Occupational Safety and Health Policy*. Cambridge: MIT Press.

———. 1981b. *What Price Incentives? Economists and the Environment*. Boston: Auburn House.

Kemeny, John G. 1981. "Political Fallout." *Transaction/Society* 18, no. 5:5–9.

Kemeny, John G., and other members of the President's Commission on the Accident at Three Mile Island. 1979. *The Need for Change: The Legacy of TMI*. Washington, D.C.: Government Printing Office.

Kesselman, Mark, ed., with the assistance of Guy Groux. 1984. *The French Workers' Movement: Economic Crisis and Political Change*. Translated by Edouardo Diaz, Arthur Goldhammer, and Richard Shryock. London: Allen & Unwin.

Kessler, Marie-Christine. 1968. *Le Conseil d'Etat*. Paris: Colin.

———. 1978. *La Politique de la haute fonction publique*. Paris: Presses de la Fondation Nationale des Sciences Politiques.

Kitschelt, Herbert. 1982. "Structures and Sequences of Nuclear Policy-Making: Suggestions for a Comparative Perspective." *Political Power and Social Theory* 3:271–308.

———. 1986. "Political Opportunity Structures and Political Protest: Anti-Nuclear Movements in Four Democracies." *British Journal of Political Science* 16, no. 1:57–85.

Klein, Jeffrey S. 1981. "The Nuclear Regulatory Bureaucracy." *Transaction/Society* 18, no. 5:50–56.
Komanoff, Charles. 1981. *Power Plant Cost Escalation: Nuclear and Coal Capital Costs, Regulation, and Economics.* New York: Van Nostrand Reinhold Company.
Kosciusko-Morizet, Jacques A. 1973. *La "Mafia" polytechnicienne.* Paris: Editions du Seuil.
Krasner, Stephen D. 1978. *Defending the National Interest: Raw Materials Investments and U.S. Foreign Policy.* Princeton: Princeton University Press.
———. 1984. "Approaches to the State: Alternative Conceptions and Historical Dynamics." *Comparative Politics* 16:223–46.
Kuhn, Thomas. 1962. *The Structure of Scientific Revolutions.* Chicago: University of Chicago Press.
———. 1977. *The Essential Tension.* Chicago: University of Chicago Press.
Kuisel, Richard F. 1981. *Capitalism and the State in Modern France.* Cambridge: Cambridge University Press.
Ladd, Anthony E., Thomas C. Hood, and Kent D. Van Liere. 1983. "Ideological Themes in the Antinuclear Movement: Consensus and Diversity." *Sociological Inquiry* 53:252–72.
Lakatos, Imre, and Alan Musgrave, eds. 1970. *Criticism and the Growth of Knowledge.* London: Cambridge University Press.
Landau, N. J. 1972. "Postscript to Calvert Cliffs." *Boston College Industry and Commerce Law Review* 13:705–17.
Larsson, Christer. 1985. Series of articles. *Ny Teknik.* 25 April and 2, 9, and 16 May.
Laudan, Larry. 1977. *Progress and Its Problems.* Berkeley: University of California Press.
Laurent, Philippe. 1978. *L'Aventure nucléaire.* Paris: Aubier-Montaigne.
Lave, Lester B. 1981. *The Strategy of Social Regulation.* Washington, D.C.: Brookings Institution.
Leigland, James, and Robert Lamb. 1986. *WPPSS: Who Is To Blame for the WPPSS Disaster?* Cambridge, Mass.: Ballinger.
Lenczowski, George. 1976. "The Oil-Producing Countries." In *The Oil Crisis.* See Vernon 1976.
Lentner, Howard H. 1984. "The Concept of the State: A Response to Stephen Krasner." *Comparative Politics* 16:367–77.
Levine, Adeline. 1982. *Love Canal: Science, Politics, and People.* Lexington, Mass.: Lexington Books.
Lewin, Leif. 1984. *Ideologi och strategi: Svensk politik under 100 år.* Stockholm: Norstedt & Söners.
Lewis, Richard S. 1972. *The Nuclear Power Rebellion.* New York: Viking.
Lewis-Beck, Michael. 1984. "France: The Stalled Electorate." In Russell J. Dalton, Scott C. Flanagan, and Paul Allen Beck, eds., *Electoral Change in Advanced Industrial Democracies: Realignment or Dealignment?* Princeton: Princeton University Press.
Lieberman, Jethro K. 1981. *The Litigious Society.* New York: Basic Books.

Lilienthal, David E. 1963. *Change, Hope and the Bomb*. Princeton: Princeton University Press.
Liljegren, Lars. n.d. "Energy Policy for Greater Prosperity." Report from the Social Democratic Party of Sweden. Mimeo.
Lindberg, Leon, ed. 1977. *The Energy Syndrome: Comparing National Responses to the Energy Crisis*. Lexington, Mass.: Lexington Books, D.C. Heath and Company.
Lindblom, Charles E. 1977. *Politics and Markets*. New York: Basic Books.
Link, Ruth. 1976. "Thorböjrn Fälldin." *Sweden Now* 6:12–15, 44–46.
Little, P. Mark. 1982. "The Nuclear Power Issue in Swedish Politics (March 1979–March 1980)." Hull Papers in Politics no. 27. Department of Politics, University of Hull, England.
Lönnroth, Måns. 1977. "Swedish Energy Policy: Technology in the Political Process." In *The Energy Syndrome*. See Lindberg, 1977.
Lönnroth, M., T. B. Johansson, and P. Steen. 1980. "Sweden Beyond Oil: Nuclear Commitments and Solar Options." *Science* 208:557–63.
Lönnroth, Måns, Peter Steen, and Thomas B. Johansson. 1980. *Energy in Transition: A Report on Energy Policy and Future Options*. Berkeley and Los Angeles: University of California Press.
Loomis, Burdett A. 1979. "The Congressional Office as a Small Business: New Members Set Up Shop." *Publius* 9:35–55.
Los Angeles Times. 1983. "Ruling May Have Driven Last Nail in Industry Coffin." 21 April, p. 1.
Löwbeer, Hans, and other members of the state Commission on Reactor Safety. 1979. *Säker kärnkraft?* Stockholm: Statens Offentliga Utredningar, Industridepartementet.
Löwbeer, Hans, and the other members of the 1981 Energy Committee. 1984. *I stället för kärnkraft*. Stockholm: Statens Offentliga Utredningar, Industridepartementet.
Lowi, Theodore J. 1979. *The End of Liberalism*. 2d ed. New York: Norton.
Lucas, N. J. D. 1979. *Energy in France: Planning, Politics and Policy*. London: Europa Publications.
Luker, Kristin. 1984. *Abortion and the Politics of Motherhood*. Berkeley and Los Angeles: University of California Press.
Lundqvist, Lennart J. 1980. *The Hare and the Tortoise: Clean Air Policies in the United States and Sweden*. Ann Arbor: University of Michigan Press.
McCloskey, Donald N. 1986. *The Rhetoric of Economics*. Madison: University of Wisconsin Press.
McConnell, Grant. 1966. *Private Power and American Democracy*. New York: Vintage Books.
McKim, Vaughn. 1977. "Social and Environmental Values in Power Plant Licensing: A Study in the Regulation of Nuclear Power." In Kenneth Sayre, ed., *Values in the Electric Power Industry*. Notre Dame, Ind.: University of Notre Dame Press.
McPhee, John. 1972. *Encounters with the Archdruid*. New York: Bantam Books.
Maher, Ellen. 1977. "The Dynamics of Growth in the U.S. Electric Power In-

dustry." In Kenneth Sayre, ed., *Values in the Electric Power Industry*. Notre Dame, Ind.: University of Notre Dame Press.

Mandrin, Jacques. 1967. *L'Enarchie ou les mandarins de la société bourgeoise*. Paris: La Table Ronde.

Marrett, Cora Bagley. 1981. "Accident Analysis." *Transaction/Society* 18, no. 5:66–72.

Marsh, David. 1984a. "France Gambles on the 1990's." *Financial Times*, 24 May.

———. 1984b. "Going from a Sprint to Cruising Speed." *Financial Times*, 6 June.

Marshall, Eliot. 1984. "The Gas Reactor Makes a Comeback." *Science* 224 (18 May):699–701.

Mazur, Allan. 1981. *The Dynamics of Technical Controversy*. Washington, D.C.: Communications Press.

Medvedev, Zhores A. 1979. *Nuclear Disaster in the Urals*. New York: Norton.

Meehan, Richard L. 1984. *The Atom and the Fault: Experts, Earthquakes, and Nuclear Power*. Cambridge: MIT Press.

Meijer, Hans. 1969. "Bureaucracy and Policy Formulation in Sweden." *Scandinavian Political Studies* 4:103–16.

Mendershausen, Horst. 1976. *Coping with the Oil Crisis: The French and German Experiences*. Baltimore and London: Resources for the Future and Johns Hopkins University Press.

Metzger, Peter H. 1972. *The Atomic Establishment*. New York: Simon and Schuster.

Milbrath, Lester W. 1984. *Environmentalists: Vanguard for a New Society*. Albany: State University of New York Press.

Miljöpartiet (Swedish Environmental Party). 1982. *Nu kommer miljöpartiet*. Stockholm: Timo Förlag.

Mills, Mark. 1980. "The Swedish Referendum on Nuclear Power: A Report from Stockholm." Mimeo.

Mitchell, Robert Cameron. 1980. "Public Opinion and Nuclear Power before and after Three Mile Island." *Resources* 64:5–8.

———. 1981. "From Elite Quarrel to Mass Movement." *Transaction/Society* 18, no. 5 (July–August):76–84.

———. 1984. "Public Opinion about the Environment in Thirteen Industrial Nations." Report prepared for the Environmental Directorate, OECD. Washington, D.C.: Resources for the Future.

———. 1985. "From Conservation to Environmental Movement: The Development of the Modern Environmental Lobbies." Discussion paper QE85-12. Washington, D.C.: Resources for the Future.

Mitchell, Robert Cameron, and Dorothy Nelkin. 1982. *The Ethical and Value Implications of the Nuclear Power Debate*. Report to the National Science Foundation (EVIST CSS 7824813), Washington, D.C.

Mitnick, Barry M. 1980. *The Political Economy of Regulation*. New York: Columbia University Press.

Moatti, J.-P., J. Fagnani, and F. Fagnani. 1979. "L'Opposition anti-nucléaire dans le champ des forces politiques en France—ses Interpretations comme mouvement social." Paper presented at conference "Bügerinitiativbewegung und die Kernergiefrage: Ein Internationaler Vergleich," Bielefeld, 1–4 November.

Moatti, Jean-Paul, Paul Maitre, Edith Wenger, Edith Msika, and Thierry Nhunfat. 1981. *Analyse des mouvements de défense de l'environnement et du cadre de vie à partir d'oppositions locales à des projets d'equipements.* Vol. 1, *Le Cas des centrales nucléaires de Cattenom et de Nogent-sur-Seine.* Paris: Centre International de Recherche sur l'Environnement et le Développement.

Le Monde. 1974. *L'Election Présidentielle de Mai 1974.* Paris: Le Monde.

The Morgan Guaranty Survey. 1968. "The New Look in Electric Power Generation." (August):3–9.

———. 1969. "Can 14% Be Financed?" (May):6–11.

———. 1971. "Problems in the Powerhouse." (November):4–9.

Morris, Aldon. 1984. *The Origins of the Civil Rights Movement.* New York: Free Press.

Moss, Thomas H., and David L. Sills, eds. 1981. *The Three Mile Island Accident: Lessons and Implications.* Annals of the New York Academy of the Sciences, volume 365. New York: New York Academy of the Sciences.

Mouriaux, René. 1983. *Les Syndicats dans la société Française.* Paris: Presses de la Fondation Nationale des Sciences Politiques.

Moynet, Georges. 1984. "Evolution du coût de l'éléctricité nucléaire en France au cours des dix dernières années." *Revue générale nucléaire*, March–April.

Munson, Richard. 1985. *The Power Makers.* Emmaus, Pa.: Rodale Press.

Murphy, Arthur W. 1976. "Nuclear Power Plant Regulation." In Arthur W. Murphy, ed., *The Nuclear Power Controversy.* New York: American Assembly, Columbia University, Prentice-Hall.

Murphy, Arthur W., and D. Bruce La Pierre. 1976. "Nuclear 'Moratorium' Legislation in the States and the Supremacy Clause: A Case of Express Preemption." *Columbia Law Review* 76:392–456.

Nader, Ralph, and John Abbotts. 1979. *The Menace of Atomic Energy.* Revised ed. New York: Norton.

Nau, Henry R. 1974. *National Politics and International Technology: Nuclear Reactor Development in Western Europe.* Baltimore and London: Johns Hopkins University Press.

Navarro, Peter. 1986. "The Performance of Utility Commissions." In John C. Moorhouse, ed., *Electric Power: Deregulation and the Public Interest.* San Francisco: Pacific Research Institute for Public Policy.

Nelkin, Dorothy. 1971. *Nuclear Power and Its Critics: The Cayuga Lake Controversy.* Ithaca, N.Y.: Cornell University Press.

———. 1974. "The Role of Experts in a Nuclear Siting Controversy." *Bulletin of the Atomic Scientists* 30, no. 9 (November):29–36.

———. 1979. "Nuclear Power and its Critics: A Siting Dispute." In Dorothy Nelkin, ed., *Controversy: Politics of Technical Decisions.* Beverly Hills, Calif.: Sage.

Nelkin, Dorothy, and Michael Pollak. 1977. "The Politics of Participation and the Nuclear Debate in Sweden, the Netherlands, and Austria." *Public Policy* 25:333–57.

Nelkin, Dorothy, and Michael Pollak. 1980. "Political Parties and the Nuclear Energy Debate in France and Germany." *Comparative Politics*, January.

———. 1981. *The Atom Besieged: Anti-Nuclear Movements in France and Germany*. Cambridge: MIT Press.

New York Times. 1986. "Management Cited at 16 'Problem' Nuclear Plants." (16 July): A11.

Nichols, Elizabeth. 1987. "U.S. Nuclear Power and the Success of the American Anti-Nuclear Movement." *Berkeley Journal of Sociology* 32:167–92.

Nicolon, Alexandre. 1979. "Analyse d'une opposition à un site nucléaire." In *Nucléopolis*. *See* Fagnani and Nicolon 1979.

Nicolon, Alexandre, and Marie-Josèphe Carrieu. 1979. "Les Partis face au nucléaire et la contestation." In *Nucléopolis*. *See* Fagnani and Nicolon 1979.

Niskanen, William A., Jr. 1971. *Bureaucracy and Representative Government*. Aldine-Atherton: Chicago.

Nivola, Pietro. 1980. "Energy Policy and the Congress: The Politics of the Natural Gas Policy Act of 1978." *Public Policy* 28 (Fall):491–543.

———. 1986. *The Politics of Energy Conservation*. Washington, D.C.: Brookings Institution.

Nixon, Richard M. 1973. "Address on the Energy Emergency." 7 November. Reported in the U.S. Congress, Senate Committee on Interior and Insular Affairs, *Executive and Energy Messages*. Washington, D.C.: Government Printing Office, 1975.

Nordlinger, Eric A. 1981. *On the Autonomy of the Democratic State*. Cambridge: Harvard University Press.

Novick, Sheldon. 1976. *The Electric War: The Fight over Nuclear Power*. San Francisco: Sierra Club Books.

Nowotny, Helga, and Helmut Hirsch. 1980 "The Consequences of Dissent: Sociological Reflections of the Controversy of the Low Dose Effects." *Research Policy* 9:278–94.

Nuclear Energy Agency (NEA). *See* Organization for Economic Cooperation and Development Nuclear Energy Agency.

The Observer. 1986. *The Worst Accident in the World: Chernobyl, The End of the Nuclear Dream*. London: William Heinemann.

Offe, Claus. 1984. *Contradictions of the Welfare State*. Cambridge: MIT Press.

Okrent, David. 1981. *Nuclear Reactor Safety: On the History of the Regulatory Process*. Madison: University of Wisconsin Press.

Organization for Economic Cooperation and Development (OECD) International Energy Agency (IEA). 1977. *IEA Reviews of National Energy Programmes*. Paris: OECD.

———. 1978. *Energy Policies and Programmes of IEA Countries 1977 Review*. Paris: OECD.

———. 1984. *Energy Balances of OECD Countries 1970–1982*. Paris: OECD.

OECD Nuclear Energy Agency. 1980. *Description of Licensing Systems and Inspection of Nuclear Installations*. Paris: OECD.

———. 1983a. *The Costs of Generating Electricity in Nuclear and Coal Fired Power Stations: A Report by an Expert Group*. Paris: OECD.

———. 1983b. *Nuclear Legislation: Analytical Study*. Vol. 1. Paris: OECD.
———. 1984. *Nuclear Legislation: Analytical Study*. Vol. 2. Paris: OECD.
OECD Nuclear Energy Agency and International Atomic Energy Agency. 1985. *Nuclear Third Party Liability and Insurance: Status and Prospects*. Paris: OECD.
Owen, Daryl H. 1981. "Waste Not, Want Not: The Role of the State in Nuclear Waste Facility Siting." *Louisiana Law Review* 41:1227–55.
Palda, K. S. 1973. "Does Advertising Influence Votes?" *Canadian Journal of Political Science* 6:638–55.
Parenteau, Patrick A. 1976. "Regulation of Nuclear Power Plants: A Constitutional Dilemma for the States." *Environmental Law* 6, no. 3:675–728.
Parti Socialiste. 1978. *Pour une autre politique nucléaire: Rapport du Comité Nucléaire, Environnement et Société au Parti Socialiste*. Paris: Flammarion. Authors include Louis Puiseux, Alain Touraine, and nine others.
———. 1981. *Energie: l'autre politique*. Paris: Club Socialiste du Livre.
Pearson, Frederic S., and Michael Nyden. 1980. "Energy Crisis and Government Regulations: Swedish and Dutch Responses in 1973." *West European Politics* 3, no. 3:406–20.
Penniman, Howard R., ed. 1975. *France at the Polls: The Presidential Election of 1974*. Washington, D.C.: American Enterprise Institute.
Penrose, Edith. 1976. "The Development of Crisis." In *The Oil Crisis*. See Vernon 1976.
Perrow, Charles. 1984. *Normal Accidents*. New York: Basic Books.
Petersson, Olof. 1978a. "The 1976 Election: New Trends in the Swedish Electorate." *Scandinavian Political Studies*, new series, 1:109–21.
———. 1978b. *Väljarna och valet 1976*. Stockholm: Central Bureau of Statistics.
———. 1979a. "The Government in Crisis in Sweden." *Scandinavian Political Studies*, new series, 2:171–78.
———. 1979b. *Regeringensbildningen 1978*. Stockholm: Rabén and Sjögren.
Pettit, Philip. 1978. "Rational Man Theory." In *Action and Interpretation*. See Hookway and Pettit 1978.
Peyrefitte, Alain. 1981. *The Trouble with France*. New York: Alfred A. Knopf. (Translation of *Le Mal Français*, 1976.)
Picard, Jean-François, Alain Beltran, and Martine Bungener. 1985. *Histoire de l'EDF*. Paris: Dunod.
Pickles, Dorothy. 1972. *Government and Politics of France*. New York: Harper & Row.
Pignero, Jean. 1974. "Nous Allons Tous Crever." Special Issue of *Protection Contre les Rayonnements Ionisants*, nos. 48–49 (April–May).
Pilat, Joseph F., Robert E. Pendley, and Charles K. Ebinger, eds. 1985. *Atoms for Peace: An Analysis after Thirty Years*. Boulder, Colo., and London: Westview Press.
Polach, Jaroslav G. 1969. "Nuclear Power in Europe at the Crossroads." *Bulletin of the Atomic Scientists* 25, no. 8:15–20.
President's Commission on the Accident at Three Mile Island. 1979. See Kemeny, John G.

Price, Jerome. 1982. *The Antinuclear Movement.* Boston: G. K. Hall and Company, Twayne.
Primack, Joel, and Frank von Hippel. 1974. *Advice and Dissent: Scientists in the Public Arena.* New York: Basic Books.
Pringle, Peter, and James Spigelman. 1981. *The Nuclear Barons.* New York: Avon Books.
Prodi, Romano, and Alberto Clô. 1976. "Europe." In *The Oil Crisis.* See Vernon 1976.
Pryor, Larry. 1976. "California's Nuclear War." *Washington Post,* 23 May, C3.
Puiseux, Louis. 1977. *La Babel Nucléaire.* Paris: Editions Galilée.
———. 1978. "EDF et la politique énergétique." *Les Entreprises Publiques,* March–April.
———. 1981. "Post-Scriptum." *La Babel Nucléaire,* 3d ed. Paris: Editions Galilée.
———. 1984. "Les choix de la politique énergétique Française." Mimeo. Centre International de Recherche sur l'Environnement et le Développement.
Quirk, Paul J. 1981. *Industry Influence in Federal Regulatory Agencies.* Princeton: Princeton University Press.
Rainer, Ove, and other members of the Energy Commission. 1978. *Energi: Hälso—miljö—och säkerhetsrisker: Slutbetänkande.* Stockholm: Statens Offentliga Utredningar, Industridepartementet.
Rankin, William L., Barbara D. Melber, Thomas D. Overcast, and Stanley M. Nealey. 1981. "Nuclear Power and the Public: An Update of Collected Survey Research on Nuclear Power." Seattle: Battelle.
Ray, Dixy Lee. 1975. "Irrational Fears of Runaway Nuclear Energy Don't Stand up against Scientific Evidence." *Oregonian* (Portland), 26 October, C1.
Raymond, James F. 1979. "A *Vermont Yankee* in King Burger's Court: Constraints on Judicial Review under NEPA." *Boston College Environmental Affairs Law Review* 7, no. 4:629–64.
Réal, B. 1975. "Le Pari du nucléaire: Nucléaire et dépendance." *Economie et Humanisme* 223:31–47.
Revue Générale Nucléaire. 1975. April–June.
Reynaud, Jean-Daniel. 1975. *Les Syndicats en France.* 2 vols. Paris: Seuil.
Rhenman, Eric. 1958. *Atomarbetet i fjörton länder.* Stockholm: Studieförbundet Näringsliv och Samhälle.
Ribes, Jean-Paul, Brice Lalonde, Serge Moscovici, and René Dumont. 1978. *Pourquoi les écologistes font-ils de la politique?* Paris: Seuil.
Richardson, Jeremy, ed. 1982. *Policy Styles in Western Europe.* London: Allen and Unwin.
Rigaud, Jacques, and Xavier Delcros. 1984. *Les Institutions administratives Françaises: Les Structures.* Paris: Presses de la Fondation Nationale des Sciences Politiques.
Ringleb, Al H. 1986. "Environmental Regulation of Electric Utilities." In John C. Moorhouse, ed., *Electric Power: Deregulation and the Public Interest.* San Francisco: Pacific Research Institute for Public Policy.

Ripley, Randall B., and Grace A. Franklin. 1980. *Congress, the Bureaucracy and Public Policy.* Homewood, Ill. Dorsey Press.

Roberts, Marc J., and Jeremy S. Bluhm. 1981. *The Choices of Power. Utilities Face the Environmental Challenge.* Cambridge: Harvard University Press.

Rochlin, Gene. 1979. *Plutonium, Power, and Politics: International Arrangements for the Disposition of Spent Nuclear Fuel.* Berkeley: University of California Press.

Rogovin, Mitchell. 1980. *Three Mile Island: A Report to the Commissioners and the Public.* Washington, D.C.: NRC.

Rolph, Elizabeth S. 1979. *Nuclear Power and the Public Safety: A Study in Regulation.* Lexington, Mass.: Lexington Books.

Rose, Judah L., Alan C. Weinstein, and Julia M. Wondolleck. 1979. *Nuclear Energy Facilities and Public Conflict: Three Case Studies.* Report to the U.S. Department of Energy. Cambridge: Laboratory of Architecture and Planning, MIT.

Rosenbaum, Walter A. 1981. *Energy, Politics and Public Policy.* Washington, D.C.: Congressional Quarterly Press.

Roucaute, Yves. 1983. *Le Parti Socialiste.* Paris: Bruno Huisman.

Ruddick, Sara. 1989. *Maternal Thinking.* Boston: Beacon Press.

Ruin, Olof. 1982. "Sweden in the 1970's: Policy-Making Becomes More Difficult." In *Policy Styles in Western Europe.* See Richardson 1982.

Rustow, Dankwart A. 1955. *The Politics of Compromise: A Study of Parties and Cabinet Government in Sweden.* Princeton: Princeton University Press.

Ryan, Alan, ed. 1973. *The Philosophy of Social Explanation.* London: Oxford University Press.

Rycroft, Robert W., and Robert D. Brenner. 1981. "Nuclear Energy Facility Siting in the United States: Implications of the International Experience." Research monograph no. 46. Princeton: Woodrow Wilson School of Public and International Affairs, Center of International Studies, Princeton University.

Sahr, Robert C. 1985. *The Politics of Energy Policy Change in Sweden.* Ann Arbor: University of Michigan Press.

Samuel, Laurent. 1978. *Guide pratique de l'écologiste.* Paris: Belfond.

Sandgren, Lennart, and other members of the state Commission on the Consequences of a Nuclear Moratorium. 1979. *Om Vi avvecklar kärnkraften: Konsekvenser för ekonomi, sysselsättning och miljö.* Stockholm: Statens Offentliga Utredningar, Industridepartementet.

Särlvik, Bo. 1977. "Recent Electoral Trends in Sweden." In Karl H. Cerny, ed., *Scandinavia at the Polls.* Washington, D.C.: American Enterprise Institute.

Sartori, Giovanni. 1969. "From the Sociology of Politics to Political Sociology." In Seymour Martin Lipset, ed., *Politics and the Social Sciences.* New York: Oxford University Press.

Saumon, Dominique, and Louis Puiseux. 1977. "Actors and Decisions in French Energy Policy." In *The Energy Syndrome.* See Leon Lindberg, 1977.

Sawhill, John C., and Lester P. Silverman. 1983. "Build Flexibility—Not Power Plants." *Public Utilities Fortnightly* 23 (May):17–21.

Scaminaci, James, III, and Riley E. Dunlap. 1986. "No Nukes! A Comparison of

Participants in Two National Antinuclear Demonstrations." *Sociological Inquiry* 56, no. 2:272–82.
Schain, Martin A. 1985. "Politics and Mass Mobilization: Relations Between the CGT and the CFDT." In Philip G. Cerny and Martin A. Schain, eds., *Socialism, the State and Public Policy in France*. New York: Methuen.
Schattschneider, E. E. 1960. *The Semisovereign People: A Realist's View of Democracy in America*. New York: Holt, Rinehart and Winston.
Scheinman, Lawrence. 1965. *Atomic Energy Policy in France under the Fourth Republic*. Princeton: Princeton University Press.
Schipper, Lee, and A. J. Lichtenberg. 1976. "Efficient Energy Use and Well-Being: The Swedish Example." *Science*, 3 December.
Schlesinger, James R. 1971. "Expectations and Responsibilities of the Nuclear Industry." Remarks to the All-Conference Banquet of the Atomic Industrial Forum–American Nuclear Society Annual Meeting, Bal Harbour, Florida, 20 October.
Schnaiberg, Allan. 1985. "Capital Flight from Environmental Regulation: Nonmetropolitan Industrialization and 'Folk' Resistance." Paper presented at the American Sociology Association annual meeting, Washington, D.C., 30 August.
Schneider, Jerrold E. 1979. *Ideological Coalitions in Congress*. Westport, Conn.: Greenwood Press.
Schonfeld, William R. 1985. *Ethnographie du PS et du RPR: Les Eléphants et l'aveugle*. Paris: Economica.
Schwartz, Julien. 1974. *Rapport sur les sociétés pétrolières opérants en France*. Annex to the verbal proceedings of the French National Assembly Meeting of 6 November, 1974, no. 1280. Paris: Documentation Française.
Schwartz, Laurent. 1983. *Pour Sauver l'Université*. Paris: Seuil.
Schwartz, Michael. 1986. "Bank Hegemony and the Relative Autonomy of the State." Paper presented at the Eastern Sociological Society annual meeting, 4–6 April, New York.
Shaffer, William R. 1980. *Party and Ideology in the United States Congress*. Lanham, Md.: University Press of America.
Sharaf, Alan Barry. 1978. *Local Citizen Opposition to Nuclear Power Plants and Oil Refineries*. Ph.D. diss., Clark University.
Shinn, Terry. 1980. *L'Ecole Polytechnique 1794–1914*. Paris: Presses de la Fondation Nationale des Sciences Politiques.
———. 1984. "Enseignement, épistémologie et stratification." Paper presented at roundtable on "Le personnel de l'enseignement supérieur en France aux XIXème et XXème siècles." Paris, 25–26 June.
Shonfield, Andrew. 1965. *Modern Capitalism*. London: Oxford University Press.
Sillin, L. F., Jr. 1984. Speech to the Nuclear Power Assembly. Washington, D.C., 8 May.
Sills, David L., C. P. Wolf, and Vivien B. Shelanski, eds. 1982. *Accident at Three Mile Island: The Human Dimensions*. Boulder, Colo.: Westview Press.
Simonnot, Philippe. 1978. *Les Nucléocrates*. Grenoble: Presses Universitaires de Grenoble.

Skocpol, Theda. 1979. *States and Social Revolutions: A Comparative Analysis of France, Russia, and China*. Cambridge: Cambridge University Press.

———. 1985. "Bringing the State Back In: Strategies of Analysis in Current Research." In *Bringing the State Back In*. See Evans et al. 1985.

Skowronek, Stephen. 1982. *Building a New American State*. Cambridge: Cambridge University Press.

Smith, V. Kerry, ed. 1984. *Environmental Policy under Reagan's Executive Order*. Chapel Hill: University of North Carolina Press.

Sorin, Francis. 1979. "Le Débat nucléaire au sein du Parti Socialiste." *Revue générale nucléaire* (September–October):532–34.

———. 1981a. "La Politique energétique du Parti Socialiste." *Revue Générale Nucléaire* (January–February):48–50.

———. 1981b. "Les Nouvelles orientations de la politique énergétique." *Revue Générale Nucléaire* (September–October):474–80.

Stepan, Alfred. 1978. *The State and Society: Peru in Comparative Perspective*. Princeton: Princeton University Press.

Sternglass, Ernest G. 1969. "Infant Mortality and Nuclear Tests." *Bulletin of the Atomic Scientists* 25 (April):18–20.

Stever, Donald W., Jr. 1980. *Seabrook and the Nuclear Regulatory Commission: The Licensing of a Nuclear Power Plant*. Hanover, N.H.: University Press of New England.

Stewart, Richard B. 1978. "*Vermont Yankee* and the Evolution of Administrative Procedure." *Harvard Law Review* 91:1805–22.

Stobaugh, Robert B. 1976. "The Oil Companies in Crisis." In *The Oil Crisis*. See Vernon 1976.

———. 1979. "After the Peak: The Threat of Hostile Oil." In *Energy Future*. See Stobaugh and Yergin 1979.

Stobaugh, Robert, and Daniel Yergin, eds. 1979. *Energy Future: Report of the Energy Project at the Harvard Business School*. New York: Random House.

Strauss, Lewis L. 1954. Remarks to the National Association of Science Writers, 16 September. United States Atomic Energy Commission press release.

———. 1955. "My Faith in the Atomic Future." *Reader's Digest*, August, 17–21.

———. 1962. *Men and Decisions*. Garden City, N.J.: Doubleday.

Sturmthal, Adolph. 1952. "The Structure of Nationalized Enterprises in France." *Political Science Quarterly* 67:57–77.

Sugai, Wayne H. 1987. *Nuclear Power and Ratepayer Protest: The Washington Public Power Supply System Crisis*. Boulder, Colo.: Westview Press.

Suleiman, Ezra. 1974. *Politics, Power and Bureaucracy in France*. Princeton: Princeton University Press.

———. 1978. *Elites in French Society*. Princeton: Princeton University Press.

Swedish Atomic Forum. No date. *Nuclear Sweden V*. Stockholm: Swedish Atomic Forum.

Swedish Central Bureau of Statistics. 1978. *Allmänna valen 1976*. Vol. 3, *Specialundersökningar*. Stockholm.

Swedish Government. 1970. *Svensk Atomenergipolitik*. Stockholm: Department of Industry.

Swedish Government. 1975. *Energihushållning m.m. regeringens proposition 1975:30.* Stockholm.

———. 1986. *Efter Tchernobyl.* Stockholm: Department of Industry.

Swedish Institute. 1987. "Fact Sheets on Sweden: The Swedish Political Parties." Stockholm.

Swedish Ministry of Trade. 1975. *Energiberedskap för kristid.* Stockholm.

Swedish Riksdagens Förvaltningskontor. 1984. *Riksdagen 1982–1985: Biografiska uppgifter om ledamöterna.* Stockholm: Axplock.

Swidler, Ann. 1986. "Culture in Action: Symbols and Strategies." *American Sociological Review* 51, no. 2:273–86.

Szalay, Robert A. 1981. "The Reaction of the Nuclear Industry to the Three Mile Island Accident." In *The Three Mile Island Accident. See* Moss and Sills 1981.

Tamplin, Arthur R. 1969a. "A Criticism of the Sternglass Article on Fetal and Infant Mortality," UCID-15506. Livermore, Calif.: Lawrence Livermore Laboratory.

———. 1969b. "Fetal and Infant Mortality and the Environment." *Bulletin of the Atomic Scientists* 25, no. 1 (December):23–29.

Taylor, Charles. 1985. *Philosophy and the Human Sciences.* Cambridge: Cambridge University Press.

Taylor, Serge. 1984. *Making Bureaucracies Think.* Stanford: Stanford University Press.

Temples, James R. 1982. "The Nuclear Regulatory Commission and the Politics of Regulatory Reform: Since Three Mile Island." *Public Administration Review* 42, no. 4 (July–August):355–62.

Tham, Carl. 1981. "The Politics of Adjustment." In *The European Transition from Oil. See* Goodman et al. 1981.

Thiriet, Lucien. 1975. "Le coût de l'énergie nucléaire est-il sous-estimé?" *Revue Générale Nucléaire* (November–December).

Thomas, Steve, and John Surrey. 1981. "What Makes Nuclear Power Plants Break Down?" *Technology Review* 83, no. 6 (May–June):56–63.

Thompson, Theos J., and William R. Bibb. 1970. "Response to Gofman and Tamplin: The AEC Position." *Bulletin of the Atomic Scientists* 26 (September):9–12.

Tietenberg, Thomas H. 1976. *Energy Planning and Policy.* Lexington, Mass.: Lexington Books.

Tilly, Charles. 1978. *From Mobilization to Revolution.* Reading, Mass.: Addison-Wesley.

Touchard, Jean, and Jacques Solé. 1965. "Planification et technocratie," *La Planification comme processus de décision.* Cahiers de la Fondation Nationale des Sciences Politiques, no. 140.

Touraine, Alain. 1971. *The Post-Industrial Society.* New York: Random House.

———. 1977. *The Self-Production of Society.* Chicago: University of Chicago Press.

———. 1981. *The Voice and the Eye: An Analysis of Social Movements.* Cambridge: Cambridge University Press.

Touraine, Alain, Zsuzsa Hegedüs, François Dubet, and Michel Wieviorka. 1982.

Anti-Nuclear Protest. The Opposition to Nuclear Energy in France. Cambridge: Cambridge University Press. Translation of *La Prophétie Anti-nucléaire.* Paris: Editions du Seuil, 1980.

Tribe, Laurence H. 1979. "California Declines the Nuclear Gamble: Is Such a State Choice Preempted?" *Ecology Law Journal* 7, no. 3:679–729.

Tullock, Gordon. 1967. *Toward a Mathematics of Politics.* Ann Arbor: University of Michigan Press.

Turner, Stephen P. 1980. *Sociological Explanation as Translation.* Cambridge: Cambridge University Press.

Turner, Victor. 1974. *Dramas, Fields, and Metaphors.* Ithaca, N.Y.: Cornell University Press.

Unger, Roberto. 1987a. *Social Theory: Its Situation and Its Task.* Cambridge: Cambridge University Press.

———. 1987b. *False Necessity: Anti-Necessitarian Social Theory in the Service of Radical Democracy.* Cambridge: Cambridge University Press.

Union of Concerned Scientists. 1977. *The Risks of Nuclear Power Reactors: A Review of the NRC Reactor Safety Study WASH–1400.* Cambridge, Mass.: Union of Concerned Scientists.

———. 1985. *Safety Second: A Critical Evaluation of the NRC's First Decade.* Washington, D.C., and Cambridge, Mass.: Union of Concerned Scientists.

United States Atomic Energy Commission. 1958. *Theoretical Possibilities and Consequences of Major Accidents in Large Nuclear Power Plants.* Washington, D.C.: Government Printing Office.

———. 1962. *Civilian Nuclear Power: A Report to the President—1962.* Washington, D.C.: Government Printing Office.

United States Comptroller General. 1978. *Federal Attempts to Influence the Outcome of the June 1976 California Nuclear Referendum.* Report to the Senate Committee on Energy and Natural Resources. Washington, D.C.: General Accounting Office.

United States Congress. 1975. *Energy Conservation and Oil Policy.* Hearings before the Subcommittee on Energy and Power of the House Committee on Interstate and Foreign Commerce, March and May 1975. Washington D.C.: Government Printing Office.

United States Congress, House Committee on Government Operations. 1978. *Nuclear Power Costs.* Washington, D.C.: Government Printing Office. 26 April.

———. 1981. *Nuclear Safety: Is NRC Enforcement Working?* Washington, D.C.: Government Printing Office, 14 December.

United States Congress, Joint Committee on Atomic Energy. 1968. *Nuclear Power Economics 1962–1967.* 90th Cong., 2d sess., February.

United States Congressional Budget Office. 1979. "Delays in Nuclear Reactor Licensing and Construction: The Possibilities for Reform." Washington, D.C.: Government Printing Office.

United States Department of Energy. 1980. *Nuclear Power Plant Licensing: Opportunities for Improvement.* Washington, D.C.: Energy Information Administration.

United States Department of Energy. 1983. *Commercial Nuclear Power: Prospects for the U.S. and the World*. Washington, D.C.: Energy Information Administration, Department of Energy. November.

———. 1984. *See* Hewlett 1984.

United States Federal Power Commission, 1971. *1970 Annual Report*. Washington, D.C.: Government Printing Office.

United States Nuclear Regulatory Commission. 1980. *Three Mile Island* (The Rogovin Report). *See* Rogovin 1980.

———. 1983. *The Price-Anderson Act—The Third Decade*. NUREG-0957. Washington D.C.: NRC, December.

———. 1984. *Improving Quality and the Assurance of Quality in the Design and Construction of Nuclear Power Plants*. NUREG-1055. Washington, D.C.: Government Printing Office, May.

United States Office of Technology Assessment (OTA). 1984. *Nuclear Power in an Age of Uncertainty*. Washington, D.C.: Government Printing Office.

United States President's Commission on the Accident at Three Mile Island. 1979. See Kemeny et al. 1979.

Vadrot, Claude-Marie. 1978. *L'Ecologie, histoire d'une subversion* Paris: Syros.

Valen, Henry. 1972. "Local Elections in the Shadow of the Common Market." *Scandinavian Political Studies* 7:212–82.

———. 1973. "Norway: 'No' to EEC." *Scandinavian Political Studies* 8:214–26.

Valen, Henry, and Willy Martinussen. 1977. "Electoral Trends and Foreign Politics in Norway: The 1973 *Storting* Election and the EEC Issue." In Karl H. Cerny, ed., *Scandinavia at the Polls*. Washington, D.C.: American Enterprise Institute.

Vallet, Bénédicte. 1986. "The Nuclear Safety Institution in France: Emergence and Development." Ph.D. diss., New York University.

Varanini, Emilio E., III. 1981. "Reassessing the Nuclear Option: Three Mile Island from a State Perspective." In *The Three Mile Island Accident*. *See* Moss and Sills 1981.

Veblen, Thorstein. 1915. *Imperial Germany and the Industrial Revolution*. Ann Arbor: University of Michigan Press paperback edition, 1966.

Vedung, Evert. 1979. *Kärnkraften och regeringen Fälldins fall*. Uppsala: Rabén och Sjögren.

———. 1982. *Energipolitiska Utvärderingar 1973–1981*. Stockholm: Delegationen för Energiforskning.

———. 1984. "Main Trends in Swedish Energy Policy in Response to the Oil Crisis." Unpublished paper. University of Uppsala.

———. 1988. "The Swedish Five-Party Syndrome and the Environmentalists." In Kay Lawson and Peter H. Merkl, eds., *When Parties Fail: Emerging Alternative Organizations*. Princeton: Princeton University Press.

Veitor, Richard H. K. 1984. *Energy Policy in America since 1945*. Cambridge: Cambridge University Press.

Vernon, Raymond, ed. 1976. *The Oil Crisis*. New York: Norton.

Vilain, Michel. 1970. "Politique de l'énergie: Les Grands orientations." *Revue politique et Parlementaire* 72, no. 815 (November).

Vogel, David. 1986. *National Styles of Regulation: Environmental Policy in Great Britain and the United States*. Ithaca, N.Y.: Cornell University Press.

Weart, Spencer R. 1979. *Scientists in Power*. Cambridge: Harvard University Press.

Weber, Max. 1958. "Politics as a Vocation." In *From Max Weber*. New York: Oxford University Press.

Webster, Frank. 1986. "The Politics of New Technology." In Ralph Milibrand, John Saville, Marcel Liebman, and Leo Panitch, eds., *Socialist Register 1985/1986*. London: Merlin.

Weinberg, Alvin M. 1971. "State of the Laboratory, 1970." *Oak Ridge National Laboratory Review*, Winter.

Weinberg, Alvin M., Irving Spiewak, Jack N. Barkenbus, Robert S. Livingston, and Doan L. Phung. 1984. "The Second Nuclear Era." Research memorandum, Institute for Energy Analysis, Oak Ridge, Tennessee.

Weinberg, Alvin, and Gale Young. 1967. "The Nuclear Energy Revolution." *Proceedings of the National Academy of Sciences* 57 (January):1–15.

Weinberg, Philip. 1980. "Power Plant Siting in New York: High Tension Issue." *New York Law School Law Review* 25:569–94.

Wenner, Lettie M. 1981. "Energy-Environmental Trade-offs in the Courts: Nuclear and Fossil Fuels." In Regina S. Axelrod, ed., *Environment, Energy, Public Policy: Toward a Rational Future*. Lexington, Mass.: D.C. Heath.

———. 1982. *The Environmental Decade in Court*. Bloomington: Indiana University Press.

Westinghouse Electric Company. 1967. *Infinite Energy*. 48-page brochure.

Wildavsky, Aaron. 1962. *Dixon-Yates: A Study in Power Politics*. New Haven, Conn.: Yale University Press.

———. 1987. "Choosing Preferences by Constructing Institutions: A Cultural Theory of Preference Formation." *American Political Science Review* 81, no. 1:3–21.

———. 1988. *Searching for Safety*. New Brunswick, N.J.: Transaction Publishers.

Wilensky, Harold L. 1967. *Organizational Intelligence: Knowledge and Policy in Government and Industry*. New York: Basic Books.

Wilson, Bryan R., ed. 1970. *Rationality*. New York: Harper & Row.

Wilson, Caroll L. 1979. "Nuclear Energy: What Went Wrong." *Bulletin of the Atomic Scientists*, June, 13–17.

Wilson, Richard. 1986. "Chernobyl: Assessing the Accident." *Issues in Science and Technology* 3, no. 1:21–29.

Winch, Peter. 1958. *The Idea of a Social Science and its Relation to Philosophy*. London: Routledge and Kegan Paul.

———. 1964. "Understanding a Primitive Society." *American Philosophical Quarterly* 1:307–24.

Wittrock, Björn, and Stefan Lindström. 1982. "Policy-Making and Policy-Breaking." Report no. 23, University of Stockholm, Group for the Study of Higher Education and Research Policy.

———. 1984. *De Stora programmens tid*. Stockholm: Academilitteratur.

Wolfe, Bertram. 1982. "The Hidden Agenda." In Michio Kaku and Jennifer Trainer, eds. *Nuclear Power: Both Sides*. New York: Norton.

Wood, M. Sandra, and Suzanne M. Shultz, eds. 1988. *Three Mile Island: A Selectively Annotated Bibliography*. Westport, Conn.: Greenwood.

Wood, William C. 1982. *Insuring Nuclear Power: Liability, Safety and Economic Efficiency*. Greenwich, Conn.: JAI Press, Inc.

Wooley, Barbara B. 1973. "Nixon Asks New Federal Energy Agency; House Panel Votes Deep Water Port Bill." *National Journal Reports* 8 (December):1844–45.

Woychik, Eric Charles. 1984. "California's Nuclear Disposal Law Confronts the Nuclear Waste Management Dilemma: State Power to Regulate Reactors." *Environmental Law* 14, no. 2:359–463.

Wuthnow, Robert. 1987. *Meaning and Moral Order: Explorations in Cultural Analysis*. Berkeley: University of California Press.

Yellin, Joel. 1981. "High Technology and the Courts: Nuclear Power and the Need for Institutional Reform." *Harvard Law Review* 94, no. 3:489–560.

Zetterberg, Hans. 1980. *The Swedish Public and Nuclear Energy: The Referendum 1980*. Vällingby, Sweden: Svenska Institutet för Opinionsundersökningar.

Zimmerman, J. J. 1973. "Alternatives to Proposed Actions under NEPA: The AEC Response after Calvert Cliffs." *Atomic Energy Law Journal* 14:265–313.

Zimmerman, Martin B. 1987. "The Evolution of Civilian Nuclear Power." In Richard L. Gordon, Henry D. Jacoby, and Martin B. Zimmerman, eds., *Energy: Markets and Regulation*. Cambridge: MIT Press.

Zurcher, Louis A., Jr., and R. George Kirkpatrick. 1976. *Citizens for Decency: Antipornography Crusades as Status Defense*. Austin and London: University of Texas Press.

Zysman, John. 1983. *Governments, Markets, and Growth*. Ithaca, N.Y.: Cornell University Press.

Index

AB Atomenergi, 65, 66, 67, 68, 69, 99
Aberbach, Joel D., 23
Abrahamson, Dean, 221–22
accidents, 18, 43, 99, 169, 187, 196, 213n, 277; and AEC, 52, 53, 125; and public opinion, 121, 262; in U.S. versus France, 260. *See also* liability; risks; safety; and specific accidents
acid rain, 20, 217
actions, interpretation of, 10–13
Act on Nuclear Activities (1984) (Sweden), 71
Adams, Ansel, 109
advanced gas reactors, 51
Advisory Committee on Reactor Safeguards (ACRS), 52, 53, 55, 58, 58n, 59
Agence Français pour la Maîtrise de l'Energie, 249
Agence pour les Economies d'Energie, 172, 242
Agesta plant, 66, 67
agrarians, 132, 146. *See also* farmers; rural regions
Ahearne, John F., 189n
Alford, Robert, 6n
Alfvén, Hannes, 73n, 130, 131, 230
Algeria, 155
Allan Michaud, Dominique, 149n
Almond, Gabriel A., 6n
Alsace, 164, 171
Alsén, Olle, 130
Alsthom-Atlantique, 91, 99, 248, 248n
alternative energy sources, 20, 101, 193, 259, 260–61, 262, 264, 268
Althusser, 10
Amis de la Terre, Les, 150, 151, 163, 166, 169, 170, 238, 244, 244n, 246
Anastassopoulos, Jean-Pierre, 82n
Anderson, Clinton, 44
anthropology, 10, 11n
antinuclear movement(s), xi, xii, 3, 16, 17, 21, 24, 87, 275–76; compared, 102, 136, 150, 178–81, 184, 256, 258, 259, 268, 269; courts and, 55, 87, 205–6; in France, 18, 32, 148–54, 163–71, 176–77; in France, repression of, 237–41, 242–43, 244–46, 248, 249n, 255, 262, 268; influenced by policy, 259, 262, 264; moralist policy style of, 32–35, 274; and police surveillance, 13–14; and policy trajectory, 18; and public policy debate, 19, 88; and regulatory changes, 124, 125, 197; and state, 7–8; in Sweden, 71, 129–38, 218–36, 258, 262; symbolism and, 12; and TMI, 209, 212–14, 216; in U.S., 32, 58, 88, 107–12, 112–14, 120–24, 125, 127–28, 137–38, 187, 194, 202–3, 204–5, 206–9, 262
antitrust, 58, 87
Apollo project, 29, 115, 158
"Appel des 400," 165–66, 250n
APRE, 150
Arabs, 154–56, 159n
Argentina, 76
Army Corps of Engineers, U.S., 58
ASEA, 65, 68, 69, 99
ASEA-ATOM, 41, 69, 232
Asp, Kent, 225n
Asselstine, James, 127n
Atom Committee (Sweden), 64, 65
atomic bomb, 42, 65
Atomic Energy Act (1956) (Sweden), 66
Atomic Energy Act (1954) (U.S.), 42, 42n, 55
Atomic Energy Commission (AEC), 31, 71, 74, 75, 98, 108, 119, 120, 123, 126, 149, 189, 271, 276; abandons LWR, 51–52, 86, 98, 102; and antinuclear movement, 109, 110–11, 121, 124, 202–3, 205, 206; and centralization of power, 66; versus European counterparts, 98; and oil crisis, 115; promotes light water reactors, 41–45, 49, 63; reform of, 54–57, 131, 165; and states, 59–60; tightens standards, 124–25, 196–97; and utilities, 58–59, 59n, 62, 127n, 196–97, 196n, 198
Atomic Safety and Licensing Board, 58, 58n
"Atoms for Peace," 31, 42, 42n, 65, 66
Austin, Texas, 208
autonomes, 170

Babcock and Wilcox, 47
Babel nucléaire, La (Puiseaux), 168
Baltimore Gas and Electric, 60n, 61n
bandwagon market, 47–51, 54, 196, 198
banks, 5, 96, 117, 203n
Barre, Prime Minister, 243
Barsebäck reactors, 70, 134, 219–20
Baughman, Martin L., 50n
Belgium, 64n, 75n, 251
Bell, D. S., 248n
Belleville, 240n
Bernstein, Richard J., 11n
Bevill, Tom, 192, 193
Beznau, Switzerland, 211n
Biquard, Pierre, 74n
Bjurulf, Bo, 221n
Bluhm, 50n
Bodega Head site, 109, 125
Bodiguel, Jean-Luc, 95n, 96n
boiling water reactors, 41, 45, 67, 68, 91, 122, 234
Boiteaux, Marcel, 78n, 91, 168
Bombard, Alain, 248n
Bonneville Power, 48n
Bormann, Ernest G., 21n
Bouchardeau, 246
Bourdieu, 10
Bourjol, Maurice, 247, 247n
Bouthillier, Guy, 80n
Bradford, Peter, 48, 63, 127n, 199n
Brady, Rosemary, 251
breeder reactor, 31, 46, 51, 53, 62, 65, 115, 170, 190, 245, 248, 250, 254
Brenner, Robert D., 200n
Brennilis, France, reactor, 75n
Bretons, 171
Bringing the State Back In (Evans et al.), 6
Britain, 16, 18, 64n, 65, 67, 251
Brittany, 164
Brookhaven report, 43, 43n
Brower, David, 109, 150
Brown's Ferry accident, 121
Buchanan, James, 6n
Bugey reactor, 75, 150, 151
Building Act (1947) (Sweden), 71
Bupp, Irvin C., 8n, 49, 49n, 67–68, 77n, 125n, 177n, 196, 197, 205n, 238n, 240n
bureaucracies, 5, 9, 10, 19, 81, 181, 270, 271, 274
bureaucrats, 6, 6n, 269, 277
Burke, Kenneth, 21n

business, xi, 13, 23n, 33–34, 57, 72
Byron 1 plant, 125

California, 121; Assembly Bill No. 1579, 206–7; Energy Commission, 207; referendum, 144, 206–7
Calvert Cliffs Coordinating Committee v. AEC, 54–55, 56, 124, 200n, 205
Campbell, John, 60n, 61n, 198n, 203n, 205n, 240n
Canada, 16, 64n, 65, 68
CANDU, 68, 78
capitalism, 121, 178
capital-labor issue, 17, 24, 163, 239–40
capture, 59, 126, 127n, 212
carcinogens, 35
Carlsson, Ingvar, 232
Carrieu, 161n
Carson, Rachel, 129
Carter, Jimmy, 187, 188–94, 210, 215, 243, 257, 264
Carter, Luther, 61, 235
Castaing Commission, 249–50
Cattenom reactor, 246n
Center Party (Sweden), 24, 129, 131–33, 135–39, 141, 144–47, 179, 219–22, 223n, 225, 226, 228, 233, 235, 257, 258, 269
centralization, 59, 66, 75, 80, 82, 98, 100, 271
Centre National de la Recherche Scientifique (CNRS), 165, 165n
CFTC, 81. *See also* Confédération Française Démocratique du Travail
Chalmers Technical School, 64, 65, 72
Charlie-Hebdo, 150
Charlot, Monica, 24n
Chaudron, Martine, 149n
Chernobyl, 18, 187, 213–14, 218, 232
Chevallier, F., 82n
Chinon reactor, 75n
Chooz reactor, 75n, 246n, 251
Civaux reactor, 246n
civil disobedience, 122
civil liberties, 164
civil rights movement, 32, 55n, 123n
Clarfield, Gerard H., 45
Clausen, Aage R., 23
Clean Air Act (1970) (U.S.), 35
Clinch River breeder project, 189, 190, 191

coal, 44, 46n, 183, 195–96, 201, 247, 251, 268
Cochrane, James L., 191n
Cogema, 86n, 254
Cohen, Linda, 199
College de France, 165
Collier's, 31
Colson, Jean-Philippe, 162
Comanche Peak plant, 126n, 209
Combustion Engineering, 47
Commissariat à l'Energie Atomique (CEA), 74–75, 76, 85–87, 102, 103, 157, 163, 172n, 176, 244, 248n, 255n; conflict with EDF, 77–79, 81, 82, 96, 244, 270, 271, 274, 275; promotes nuclear energy, 90, 98; and regulation, 86–87, 88, 262; workers, 152, 153
Commission on Energy (France), 90
Committee on Economic Defense (Sweden), 139
Common Cause, 126n
Commonwealth Edison of Chicago, 42n, 50, 195
Communist Party (Sweden), 146, 221, 225, 228, 233
Communists (PCF) (France), 80, 80n, 81, 161, 162, 171, 177, 239n, 249
Compagnie Générale d'Electricité (CGE), 91, 248, 248n
Concorde, 79, 158
Confédération Française Démocratique du Travail (CFDT) (formerly CFTC), 81, 135, 151–54, 164, 166, 168, 169, 170, 178, 238, 240, 244, 245, 246, 249, 273
Confédération Général du Travail (CGT), 80, 153, 244n
Connolly, Thomas J., 101n
Conseil de l'Information Electronucléaire, 241, 244
consensus, 12, 143, 143n, 158, 220n–21n
Consequences Commission, 227n
conservation, 140, 141, 143, 158, 159, 182, 183, 189, 193, 202, 209, 218, 231, 237, 242, 247, 249, 257, 261, 268, 277
Conservative Party (Sweden), 144, 145, 146, 219–20, 222, 223, 225, 227–28
Conservatives (France), 250, 264
Consolidated Edison of New York, 42, 60n, 61n
cooling systems, 54
corporations, 8, 52, 59

Corps des Mines, 78, 80, 83, 84, 85, 87, 172n
Corps des Ponts et Chaussées, 78, 80, 83, 84, 85
cost-benefiters, 19, 20, 21, 22, 101, 180–82, 184, 256, 276; and antinuclear movement, 112–13, 178–79, 179; compared, 178, 267–68; defined, 25, 26–28; in France, 28, 76–78, 90, 93–97, 142, 148, 157–60, 167–69, 171–72, 175–77, 242, 245, 250–54; and growth, 34; and moralists, 35–36; and non-economists, 27–28, 37; and policy in 1980s, 263–64, 267–68; in Sweden, 68–69, 73, 129, 141–42, 159–60, 233–34, 236, 257, 271; and technological enthusiasts, 100, 103, 256–57, 267, 274; in U.S., 28, 44–45, 49–50, 51, 55–56, 61–62, 73, 107–28, 142, 158, 159–60, 187, 188–91, 193–94, 202, 215–17, 257, 268, 271
costs, 20, 44, 99–100, 101, 125, 275; antinuclear movement and, 108; cleanup and decommissioning, 214–15; development, 196–98, 252–53; estimates, in France, 76–77, 93–96, 167–69, 171–72, 172–73; and financing, 183; in France, 237, 241; high, in U.S., 187, 188, 194–204, 208, 209, 216–17, 259–60, 264, 268, 269; influenced by policy, 259–60, 262, 264; low, in France, 251–54, 259–60, 268; overruns, in Sweden, 67–68; overruns, U.S., 45–48, 49–50, 61–62; social, 168; in Sweden, 233, 234, 259; and turn-key contracts, 45–46, 51, 76; who will pay, 214–15, 269. *See also* financing
Cotgrove, Stephen, 33
Council of Economic Advisors (U.S.), 117, 119, 120, 128, 142, 180, 181, 191, 257, 270, 271
Council of Ministers (France), 92, 156, 242
Council of State (France), 83, 87–88, 96n, 240n, 242
Council on Environmental Quality (U.S.), 200
counterculture, 121, 122, 134, 150, 151, 154
Cour des Comptes, 83, 96n
courts, 7, 57, 87, 123, 124; in France, 96n, 240, 242; in France versus U.S., 87–88, 240; U.S, 205–6, 205n, 214–15, 240n
Creative Initiative, 121n

314 · Index

Crépeau, 246
Creusot-Loire, 91, 248n
Creys-Malville reactor, 170, 179, 237–39
Criddle, 248n
crisis, 9, 15–16
"Critical Mass" conference, 120
CRS, 237, 238–39, 245
cultural meaning, xi, xii, 4, 7, 8, 10–14, 16, 37–38; and dominant partisan cleavages, 21; and LWR technology, 100, 103; and policy outcomes, 272–73; and policy styles, 19, 21–22, 264, 271; and structure, 14, 22
CWIP (construction work in progress), 158

Dagens Nyheter, 130, 152
Dahl, Birgitta, 231, 232
Dahl, Hans, 137
Daléus, Lennart, 135
Dallmayr, Fred R., 11n
Davis, Besse, 211n
decentralization, 151, 161, 187–88, 191–94, 204–9, 217, 258, 262, 268, 269
decisionmakers, xii, 5, 13, 270–71. *See also* policymakers
Declaration of Public Utility (DUP), 88
Defense Department (Sweden), 67
de Gaulle, Charles, 79, 80, 82, 88n, 103, 160, 270, 271
Delany Clause, 35
delay, 182–83, 195, 198–201, 264, 275–76
Délégation Générale l'Energie (DGE), 172
deLeon, Peter, 74n
Delouvrier, Paul, 78n
Del Sesto, Stephen, 42n
de Marchi, Neil, 117n, 119n
democracy, xii, 4, 13, 28n, 55n, 81, 107, 120, 162, 164, 274
Democratic Party, American, 17, 23, 45, 51, 107, 108, 118, 135, 188
demonstrations, 134, 150, 163–65, 179, 213, 237–39. *See also* site occupations
Derian, Jean-Claude, 8n, 49, 67–68, 77n, 177n, 205n, 238n, 240n
Devall, William B., 109n
Diablo Canyon reactor, 59, 77n, 109, 122, 135, 199, 200
Dingell, John, 192
direct action, 122–23
discretion, 4, 9, 17–18, 20, 217, 263, 269, 271, 273; and biography, 14, 270–71;

compared, 178–84, 269; in France, 243, 255, 255n; reduction of, 17, 18, 20, 193, 217, 269; in Sweden, 233
Division of Reactor Development and Technology, 53
Dixon-Yates proposal, 43n
Dodman, Michael J., 191n
Domestic Policy Staff (U.S.), 189, 190
dominant partisan cleavage, 16, 19, 21, 23–25, 181, 182, 258, 272, 274; and antinuclear compromise in Sweden, 218–36; compared, 256–57; and decentralization in U.S., 188–94; in France, 21, 24–25, 161–64, 239–40, 249, 258, 274; and policy styles, 21–38; in Sweden, 131, 139, 142–43, 226, 258; U.S., 43, 45, 57–58, 107, 118–20, 128, 258; using, in analysis, 37–38
dominant pro-growth paradigm, 29–30, 30n, 33, 34; in Sweden, 132, 274; utilities and, 50, 100. *See also* economic growth
Douglas, Mary, 34–35, 35n
Dresden plant, 42n
Duff, Andrew, 33
Duke Power, 195, 198
Dumont, René, 151, 164
Dunlap, Riley E., 33n
Duquesne Light v. Barasch, 214–15
Durrieu, Yves, 159n
Dutch electricity, cost of, 251
Dworkin, Ronald, 123n

Ebbin, 126
Ecole Nationale d'Administration (ENA), 83, 95–96
Ecole Polytechnique (X), 83–85, 95, 168
Ecologie, 151
econometric analysis, 116, 118
Economic and Social Council (France), 247–48
economic growth, 13, 34, 77; electricity and, 89–90, 100, 116; environmentalists and, 33, 132; in France, 50, 162, 176, 240, 249, 250. *See also* dominant pro-growth paradigm
economic incentives, 23–24
economic rationality, 4–5, 181, 274–75; in France, 93–94, 95; in U.S., 116
economists, 26, 112–13; antinuclear, 148; head AEC, 55; in France, 78n, 79, 81,

95–96, 168–69, 170, 172, 177n; as policymakers, 17, 181–82; policy styles of, 27, 36; in U.S., 50, 55, 116, 189
Edelman, Murray, 14n
Edison, Thomas, 46n
education, 33, 83–85, 95
efficiency, 19, 141, 250, 261
Egypt, 114
Eisenhower, Dwight D., 31, 42
elected officials, 5, 7, 9, 121, 181–82. *See also* political leaders
elections, 8, 9, 24n, 258; in France, 164, 246; in Sweden, 179
Electricité de France (EDF), 74, 75–76, 80–83, 85–86, 87, 99, 103, 142, 148, 156, 157, 159, 170n, 182, 183n, 196, 242, 243, 245, 248n, 249, 250–55, 257, 259, 260, 269, 271, 273, 276; commercial strategy, 88–91, 98, 261; conflict with CEA, 77–79, 82, 244, 262, 270, 274, 275; conflict with Finance Ministry, 92–97, 102–3, 173, 241, 259; financing, 174–75, 183; management of, 252–54; and Messmer Plan, 161, 167n, 168, 169, 172–73, 176–77; and TMI, 243; workers, 152, 153
electricity: costs of, 31, 251; demand for, 18, 19, 20, 50, 56, 100, 116–18, 120, 128, 141, 142, 158–59, 167, 191, 201, 247–48, 250, 256, 257, 261, 275; excess of, in France, 250–51, 261; and growth, 77; marketing of, in France, 88–96; pooling of, 46; rates, in France, 174, 242; rates, in U.S., 60–61, 62, 118, 158, 209, 214–15
elites, 4, 8, 14n, 18, 137, 176, 178–84, 217, 263, 264, 271, 275, 277; technical, in France, 83–85, 95–96, 158, 181
Emergency Core Cooling Systems (ECCS) hearings, 54, 111–12, 197, 245
Empain-Schneider group, 91
Energie: l'autre politique (Quilès), 248
Energy, Department of (DOE) (U.S.), 193, 199n, 201, 214, 215
Energy Bill (1975) (Sweden), 138n, 140, 141, 142–43, 218, 235
Energy Commission (Sweden), 143n, 219, 220–22, 223
Energy Council (Sweden), 139
Energy Independence Act (1975) (U.S.), 118
energy modeling, 5, 201, 267–68

Energy Plan (1974) (France), 3
energy policy, 5, 120, 180, 267, 269
Energy Policy Council (Sweden), 141
Energy Reorganization Act (1974) (U.S.), 56
Energy Research and Development Administration (ERDA), 56, 124
"Energy Syndrome," 180
engineers, 30, 46, 112–13, 272; and antinuclear movement, in U.S., 110–11, 122, 126n, 135; in France, 65, 74, 77, 78, 79, 81, 83–85, 153, 158, 172, 177n, 245, 249n; as policymakers, 17, 158; policy style of, 29, 30–31, 36; and safety, 245, 272–73; in Sweden, 65, 72, 135; and U.S. utilities, 50, 116, 195–96, 216
Environmental Federation (Sweden), 135
environmental impact statement, 54, 87, 124
environmentalists, 24, 32, 33–35, 46n; and AEC, 55; in France, 149–51, 163–65, 169–70, 238, 238n, 240, 246, 248n, 249; in Sweden, 129–36, 222, 262; in U.S., 110, 134, 135. *See also* new environmental paradigm
environmental law firms, 110
Environmental Party (Sweden), 233
Environmental Protection Agency, 58
Euratom, 87, 100n
evacuation plans, 208, 217
Evanoff, Mark, 109n
Exxon, 117n

Fagnani, Jeanne, 249n
Fälldin, Thorbjörn, 131, 133, 135, 137, 144, 146, 164, 179, 218, 219, 221–22, 227, 229, 231, 233n, 235, 236, 257, 264, 268, 269, 272
farmers, 18, 131, 133, 137, 164, 169–70, 171n
Federal Energy Administration (FEA) (U.S.), 115, 119
Federal Energy Regulatory Commission, 61
Federal Power Commission (U.S.), 199
federal systems, 10, 205
Feigenbaum, Harvey, 82, 82n
Fenn, Scott, 126n
Ferguson, Thomas, 23n
Fermi 1, 108, 121

316 · Index

Fessenheim reactor, 76, 150, 151, 166, 170, 245
financing, 8, 10, 17, 20, 182; compared, 182, 183, 258; in France, 92–96, 174–75; and TMI, 210, 212; in U.S., 42, 60–62, 116–18, 125, 127, 128, 158, 175, 191, 195, 199n, 200–203, 208, 209, 214–16, 217, 259, 268
Firth, Raymond, 10n
FKA, 70
Flamanville, 240n
Flavin, Christopher, 194n
Fleck, Ludwik, 21n
Fletcher, Harold A., 81n
flexibility, 256–64, 269–70
Florida Power, 60n, 61n
Fluegge, Ronald, 126n
Food, Drug, and Cosmetic Act (U.S.), 35
Forbes magazine, 195
Force Ouvrière, 153, 245
Ford, Daniel, 43n, 59n
Ford, Gerald R., 56, 118–20, 128, 143, 158, 175, 181, 183, 187, 188, 191, 257, 258, 270, 271, 272
Ford Foundation, 110
Ford Motors, 175
Forsmark reactors, 70, 222, 223, 234
fossil fuels, 89, 96, 101, 117, 156, 159, 183
Foucault, 10
Fouquet, Denis, 94n
Fourgous, J. M., 159
Fournier, Pierre, 150, 151
Framatome, 41, 86n, 87, 89, 91, 99, 157, 234, 248, 250, 252, 253, 259
France: development of policy in, 26, 187; divergence of policies from Sweden and U.S., 267–70; early commercialization, versus Sweden and U.S., 98–103; and elite discretion after oil crisis, versus Sweden and U.S., 178–84; and future of nuclear energy in, 269; low costs in, 251–53, 259–60; and nuclear weapons, 74; overbuilding in, 250–54, 269; pro-technology values in, 30n; repression of antinuclear sentiment in, 237–50; structures and flexibility, versus Sweden and U.S., 256–64; versus Sweden and U.S., xii, 3–7, 8, 16–20, 21, 24–25, 26, 28, 50, 76, 80, 132n, 146, 193, 274; transition to light water in, 41, 49, 61, 65, 66, 70, 73, 74–97, 275; triumph of technological enthusiasm in, after oil crisis, 148–77, 193. *See also* specific agencies, commissions, issues and sites
Freedom of Information Act (U.S.), 43
French Federation of Societies for the Protection of Nature (FFSPN), 149, 150, 169
French Left, 17, 24, 80, 81, 160, 161n, 162, 163, 174, 177, 239n, 258
French Right, 17, 24, 163, 174, 258
Friedland, Roger, 6n
Friends of the Earth, 109, 129, 150, 178
Frost, Robert, 81, 81n
fusion, 31

Gabon, 155
Gallen, Hugh, 209
Gandara, Arturo, 46n, 117, 202
Garfinkel, Harold, 10n
Garris, Jerome, 64n, 66n
gas, 188, 247; coolant, 74
gas-graphite reactor, 74–80, 81, 96, 100n, 153, 158, 173, 260n, 271
Gaspard, Roger, 81
Gaz de France, 80, 172
General Atomic Corporation, 47
General Electric (GE), 41, 42n, 45–46, 47, 51, 99, 103, 122, 245
General Motors, 175
General Public Utilities, 212
Gibbs, Lois, 135n
Giddens, Anthony, 7n, 11n, 102
Gilinsky, Victor, 127n, 195, 201, 207
Gillette, Robert, 53
Giraud, André, 86, 172n, 243
Girod, M. Patrick, 87n
Giscard d'Estaing, Valery, 80, 163, 173–74, 175, 177, 241, 242, 245–46, 248, 249, 257
goals, xi, 9, 12–13, 82, 270; in France, 81, 91
Goffman, Erving, 10n
Gofman, John, 110–11, 124, 149
Golay, Michael W., 200n
Gold, Thomas, 231
Goldman, Sheldon, 55n
Goldschmidt, Bertrand, 74n
Golfech reactor, 170n, 242, 246n
Gordon, Richard L., 50n
Gore-Holifield Bill (1956) (U.S.), 43, 45, 52
Gorz, André, 151

Index · 317

government, 28, 98; agencies, 8, 9, 19; intervention of, 17, 23–24, 43, 51–52, 57–58, 107, 108–9, 118, 119–20, 124, 128, 182, 188–89, 196, 258, 274; and mixed ownership, in Sweden, 65, 69–70; ownership, in U.S., 44–45, 52; role of, 52. *See also* public ownership
Government Accountability Project, 122
graphite moderator, 74
grass roots activists, 189, 229, 248
Gravelaine, Francis, 74n
Gravelines reactor, 244n
Green, Harold, 42
Greenham Common protesters, 123n
greenhouse effect, 217
Greenspan, Alan, 119
Gregory, Bernard, 250
Grossman, Richard, 123
Grothendieck, A., 150
Groupement de Scientifiques pour l'Information sur l'Energie Nucléaire (GSIEN), 166, 168, 169, 170, 244
groups, 4, 7; cultural resources of, 8; formal power of, and outcomes, 270–72; motivations of, and cultural analysis, 12–13; voluntary, 34
Gueule Ouverte, La, 150, 163
Guilhamon, Jean, 255n

Habermas, Jürgen, 10, 11n
Hambraeus, Birgitta, 130, 131, 133, 135
Hart, Gary, 192, 193
Hawaii, 209
heat production, 65n, 66; home, in France, 89, 167, 172, 242
heavy industry, 20, 76, 80, 160, 174
heavy water reactor, 64, 65, 66, 67–69, 77, 100n
Henry, Claude, 168
Hervé, Alain, 150
Hervé, Edmond, 247
Hewlett, James G., 212
Holifield, Chet, 47, 114
Hollis, Martin, 11n
Holmberg, Sören, 225n
Hookway, Christopher, 11n
Hugon Report, 247, 248, 250
hydroelectricity, 92, 101, 174, 195
Hymen, Leonard S., 50n

IAEA, 42n

Idaho laboratory, 53
ideologies, 13, 14, 23–24, 52, 153, 270, 274; defined, 11
Indiana, 209
Indian Point plant, 42n, 126n
individuals, xi, xii, 6n, 7, 8n, 9, 12n, 14–15, 16, 272–73; and LWR, 100, 103; skills, and formal power, 271
industrial and trade policies, 171, 173, 174–75; versus national self-sufficiency, 68–69
industry, 24, 30, 33; in France, 89, 177; NRC and, 126–27; in Sweden, 140, 141. *See also* heavy industry; nuclear industry
"Infinite Energy" (pamphlet), 31
Inglehart, Ronald, 29
insider politics, 19
Inspection Général des Finances, 83, 96
Institut de Protection et de Süreté Nucléaire (IPSN), 87
Institut Economique et Juridique de l'Energie (IEJE), 166–68
Institute of Nuclear Power Operations (INPO), 210, 211
insurance, 276; in Europe, 70, 167, 168; in U.S., 43, 43n–44n, 49, 194, 212
intellectuals, 135, 137, 153, 154
interest groups, 5, 23, 190n, 268
interest rates, 201–3
interests: versus beliefs, 272–73; ideal, 12; versus policy styles, 37
International Energy Agency (IEA), 155
international markets, 80, 100–101, 158, 174
iron triangle, 58–59, 107, 126–27
Israel, 114, 154, 155, 239n

Jackson, Henry "Scoop," 115
Jahnige, Thomas P., 55n
Japan, 7, 16
Jersey Central Power and Light, 45
Johansson, Olof, 219, 221, 222
Johansson, Thomas B., 219n, 223n
Jolio-Curie, 74n

Kaku, Michio, 210n
Karp, Walter, 28n
Kasper, 126
Katz, James E., 188n, 193n
Katzenstein, Peter, 101

Kelman, Steven, 23, 72, 132, 137, 220n–21n
Kemeny Commission, 126, 210, 211, 212, 230
Kessler, Marie-Christine, 95n
Kitschelt, Herbert, 179, 205n
Komanoff, Charles, 125n, 196, 197, 203
Kraftsam, 69–70
Krasner, Steven D., 6n
Kuhn, Thomas, 27
Kuisel, Richard R., 81n

labor leaders, 33, 34
Ladd, Anthony E., 33n
Lagarde, Roland, 153
La Hague, 152–53, 245, 248, 254, 273
Lake Cayuga site, 109, 124
Lalonde, Brice, 246
Lamb, Robert, 213n
Langer, R. M., 31
Larsson, Christer, 64n
Lave, Lester B., 27n, 35
lawyers, 50, 57, 110
Lebreton, Philippe, 32
left, 178. *See also* French Left; Swedish Left
legislation and laws, 262, 264; state, in U.S., 206–8; and structural change, 15
Leigland, James, 213n
Le Monde, 164, 168
Lentner, Howard H., 6n
Le Pape, Yves, 149n
Le Pellerin, site, 246n
Levine, Adeline, 135n
Lewis-Beck, Michael, 24n
L'Express, 164
liability, 49, 70, 100, 194, 231, 276
Liberal Party (Sweden), 144, 145, 146, 218, 219–20, 222–24, 225, 226–27, 236, 264
licensing procedure, 3, 262, 267; in France, 85–88; in Sweden, 70–73; in U.S., 54–60, 62, 125–27, 195, 198–200, 205, 208, 211–12
Lichtenberg, A. J., 141n
Lieberman, Jethro K., 55n
light water reactor (LWR), 3, 16, 18, 19, 31, 101–2, 103, 267, 276; abandoned by AEC, 51–52, 62; costs of, 48, 77–78, 99–100, 168, 252–53; in France, 73, 74–97, 99, 100, 101, 149, 153, 158, 162, 271; in Sweden, 64–73, 99, 100–101; in U.S., 41–54, 61–63, 99, 100, 101–2, 191n, 198, 204, 252–53
Liljegren, Lars, 145
Lindberg, Leon, 180
Lindblom, Charles, 99n
Lindström, Stefan, 68
Lloyd, Marilyn, 192
Lönnroth, Måns, 138n, 234n
Love Canal, 135n
Löwbeer, Hans, 227
Lucas, N.J.D., 74n, 91, 92, 152n, 157, 172, 238, 239, 241n, 257n
Lukes, Steven, 11n
Lundqvist, Lennart J., 182n

McCloskey, Donald N., 26n
MacLeod, Gordon, 230
McMillan, Ian, 135
Maher, Ellen, 50
Maine, 208
Malibu site, 109
management, 6n, 213n, 271, 276; in France, 76, 81, 89, 252–54, 259, 260; in Sweden, 234, 236, 260; in U.S., 128, 187, 188, 194–97, 198, 199n, 200–201, 203, 213n, 215, 216–17, 236, 253, 259, 260, 268
Mandrin, Jacques, 95n
Manhattan project, 29, 52, 115
Marble Hill plant, 201, 209
market(s), 28; competition, 10; environmentalists and, 33; free, 17, 23–24, 26–27, 43, 44, 45, 51, 57–58, 107, 118, 119–20, 124, 128, 161, 181, 191, 268, 271, 272, 274. *See also* international markets
Markey, Edward, 192, 193
Marsh, David, 251n
Marviken plant, 66, 67, 68, 69
Marxism, 10
Maryland, 207–8
Mass, Pierre, 76
Matthews, E., 150
Mauroy, Pierre, 248
media, 5, 88, 107, 110, 111, 112, 120, 127–28, 136, 190, 225, 230, 256, 267
Medvedev, Zhores, 230
Meidner Plan, 144, 145
Mendershausen, Horst, 155n
mental grids, 14, 270, 272, 273–74
Mentré de Loye, Paul, 172n

Messmer, Pierre, 3, 139, 155, 157, 158, 181
Messmer Plan, 143, 153, 154, 156, 159, 160–71, 173, 175, 176–77, 183, 237, 243, 252, 255, 275, 276
Midi-Pyrenées Regional Council, 242
Mignot, Eric, 94n
Milbrath, Lester W., 33
military, 42, 64, 66–67, 214
Ministry of Finance (France), 91–97, 98, 102–3, 141, 142, 148, 154, 157–58, 171–72, 173, 174–75, 176, 180, 181, 182, 241, 259
Ministry of Finance (Sweden), 141
Ministry of Health (France), 87
Ministry of Industry (France), 86, 87, 172n, 176, 241, 244, 247, 257n, 262
Ministry of Industry (Sweden), 142
Ministry of Trade (Sweden), 138n
minority groups, 123
Mitchell, Robert Cameron, 123n
Mitterand, 246, 248n, 257
mixed economies, 99n
Moatti, Jean-Paul, 249n
Modular High Temperature Gas-Cooled Reactor, 217n
Moffett, Toby, 192
moralists (ecological), 19, 22, 26, 27, 28, 55n, 178–79, 256, 274; Carter and, 188–89, 190, 191; defined, 25, 26n, 31–35; in France, 153, 166; and other styles, 35–37; in Sweden, 129, 131, 133, 136, 144–45, 146–47, 218–22, 228, 233, 235, 236, 269, 272; and U.S. antinuclear movement, 107–8, 109, 112–14, 121, 123
moratoriums: in France, 161, 162, 163, 166, 169; in Sweden, 131, 138
Morgan Guaranty Survey, 202
Mothers for Peace, 135
motivations, 14, 18, 273
Mouvement des Radicaux de Gauche, 244, 246
Mouvement Ecologique, 164
Moynet, Georges, 94n
multinational firms, 162, 163, 177
municipal government: in France, 161; in Sweden, 71, 134
Munson, Richard, 50n
Muskie, Edmund, 115

Nader, Ralph, 112, 120

National Assembly (France), 164, 247–48; Finance Committee of, 241
National Energy Production Board, proposed (U.S.), 118
National Environmental Policy Act (1970) (U.S.), 205
National Environmental Protection Act (1969) (U.S.), 54, 110
national independence or self-sufficiency, 68–69, 78–80, 89–90, 158, 159–60, 268
National Institute for Statistics and Economic Studies (INSEE), 172
national interest, 69, 94, 100, 103, 158, 162
nationalization, 80n–81n, 81, 82–83, 84, 161–62, 182, 248, 249; and financing, 92, 96n; of oil companies, 156; and profits, 89
natural resources, 4–5, 64, 275
Nature and Progress, 164
Nau, Henry R., 74n
Netherlands, 7, 70
new environmental paradigm, 29–30, 33–34, 35, 132
New Hampshire, 209
New York State, 208
New York State Electric and Gas, 60n, 61n
New York Times, 213n
Nichols, Elizabeth, 125n
Nicolon, Alexandre, 161, 161n
Niger, 155
Nine Mile Point, 201
Ninth Plan, 172
Nipomo dunes, 109
Nixon, Richard M., 3, 55, 56, 107, 115, 120n, 124, 139, 154, 156, 157, 165, 271
Nogent-sur-Seine site, 238, 246n
nonprofessionals, 36
Nora, Simon, 89
Nora Report of 1966, 82, 89
Nordlinger, Eric, xi, 6–7
North Carolina, 209
Northern Indiana Public Service Company, 50n
Northern States Power Company v. Minnesota, 59
Norway, 64n, 228
Not In My Backyard (NIMBY), 108–9
Nova program, 210n
Nucléaire sans les Français, Les (Colson), 162

320 · Index

Nuclear Disaster in the Urals (Medvedev), 230
Nuclear Electric Insurance Limited (NEIL), 212
nuclear energy: and change, 15–16; comparative analysis of, 16–20; and cultural meanings, 10–14, 37–38; and dominant partisan cleavages, 22–25; early history of, compared, 98–103; early history of, in France, 74–97; early history of, in Sweden, 64–73; early history of, in U.S., 41–63; high costs and decentralization lead to decline of, in U.S., 187–217; and individual biographies, 14–15; and oil crisis and economic perspective, in U.S., 107–28; and oil crisis and elite discretion, compared, 178–84; and oil crisis and party politics, in Sweden, 129–47; and oil crisis and triumph of technological enthusiasm in France, 148–77; paralysis and compromise on, in Sweden, 218–36; and policy styles, 25–37; and political and economic structures, 7–10; and practice of politics and policymaking, 275–76; reasons for divergence of policies on, 267–70; repression and low costs lead to institutionalization of, in France, 237–55; state-centered explanation, 4–7; and structures and flexibility, compared, 256–64; and theory of politics and policymaking, 270–75. *See also* specific agencies, countries and issues
nuclear fuel: enrichment fees, 215; in France, 86n; leasing of, 42, 44; reprocessing plants, 66, 189, 190, 215, 245, 250, 254, 259; spent, in Sweden, 219, 223–24
nuclear industry, 16; access of, 7–8; and EDF, 75–76, 88–96; and iron triangle, 59n; and Messmer Plan, 160–63; in Sweden, 72–73, 218; in U.S., 49, 76. *See also* nuclear reactor manufacturers; utilities; and specific countries
Nuclear Liability Act (1983) (Sweden), 231
nuclear policy: comparative, 7n–8n, 17, 18–20; and cultural analysis, 13–14; divergence of, 267–70; influence of, on variables thought to influence policy, 259–64, 275; and political and economic structures, 7–9, 16–18. *See also* nuclear energy; and specific countries

nuclear reactor design: changes, in U.S., 197–98, 200, 201, 203, 216; new, 217; and safety, 20; and standardization, 252–54; in Sweden, 64, 68, 77. *See also* specific types
nuclear reactor manufacturers, 5, 8, 16, 19; compared, 99–100, 276; in France, 90–91, 252–53; in U.S., 41–42, 45–51, 91, 198, 214
nuclear reactors: construction problems, 195–203, 214–15, 252, 259, 269; early development of, in Sweden, 64–73, 77; early development of, in U.S., 41–63; orders for, France, Sweden and U.S., compared, 98–100; orders for, in France, 89–90, 98, 250–51; orders for, in Sweden, 69, 98; orders for, in U.S., 45–51, 55n, 60n, 98, 128; orders for cancelled in U.S., 116–18, 187, 193, 195, 197–204, 207–9, 213, 214–15, 263n, 268; siting of, in France, 164, 165, 242, 246–47; siting of, in Sweden, 71, 134; siting of, in U.S., 3, 55, 62, 108–9, 110, 118, 123n, 124, 125n, 131, 134, 208n, 209; size of, in U.S., 48; unplanned shutdowns of, 211. *See also* specific issues, sites and types
Nuclear Regulatory Commission (NRC) (U.S.), 28, 56–60, 61, 88, 121, 122, 124–27, 165, 199, 204n, 205, 206, 214, 245, 262; and Congress, 192–93, 193n; evacuation rules, 208, 217; versus Sweden, 71–72; and TMI, 210, 211–12, 211n; and utilities, 58–59, 126, 127n, 196–97, 212, 260
Nuclear Safety Analysis Center (NSAC) (U.S.), 210
nuclear waste, 130, 215, 234–35, 250, 254
nuclear weapons, 18, 64n, 66–67, 74, 86, 98, 130, 170, 189, 215, 230, 237
Nucleonics Week, 195, 203
Nyden, Michael, 138n

Oak Ridge Laboratories, 53, 192, 230
O'Dy, S., 74n
Office of Management and Budget, U.S., 120n
oil companies, 82, 156
oil crisis, xii, 3–4, 5, 9, 16, 56, 57, 62, 82, 101, 178; compared, 180–81, 183–84; and divergence of policies, 267–68; in

France, 96–97, 114, 120, 154–77; and policy debates, 17, 18, 19–20, 24; and reduced orders, 60n, 197; and structural change, 15; and structures and flexibility, 256–64; in Sweden, 129, 138–47; in U.S., 107–28, 146, 188, 201, 202, 204–5
OKG (Oskarshamn Kraftgrupp), 68, 70, 223
Okrent, David, 58n
OPEC, 154, 155n
operators, 212, 234; training of, 210–11
Organization for Economic Cooperation and Development (OECD), 155
organizations: changes in sub-systems, 15–16; formal power of, 270–72; interests of, 9, 94
ORGEL reactor, 78
Ornano, Michel, 165, 169, 171
Oskarshamn reactors, 70, 233, 234
Oyster Creek, N.J., plant, 45, 46n, 51, 78

Pacific Gas and Electric (PG&E), 50, 109, 122
Palme, Olof, 133, 139, 140, 224–25, 226, 227, 233n
Palo Verde plant, 214n
Paribas, 248
Paris, University of, at Orsay, 165
Paris and Brussels Conventions, 70
Parsonian system, 10, 12n
Paul, Marcel, 80
Paul Jacobs and the Nuclear Gang (film), 230
Peach Bottom reactors, 204n, 211n
Pearson, Frederic S., 138n
Pecqueur, 243
PEON (Production de'Electricité d'Origine Nucléaire) Commission, 74–75, 77, 79, 81, 90, 91, 92, 93, 94, 96, 97, 148, 156–57, 172, 173, 241
"People's Campaign," 225
Perrow, Charles, 213n
Petersson, Olof, 145
Pettit, Philip, 11n
Peyrefitte, Alain, 91–92
Philadelphia Electric, 211n, 214n
physicists, 74
Picard, J. F., 74n
Pignero, Jean, 149
Pilgrim plant, 111, 193
PIUS reactor, 232, 234
Planning Commission (France), 172

Plogoff reactor, 245, 246
plutonium, 65, 66, 74, 170, 254
Point Arena site, 109
Polach, Jaroslav G., 100n
police state, 13–14, 162, 239
policy: cultural analysis and definition of, 13; divergence, reasons for, 16–18, 267–70; and individual biography, 15; influence of, 259–64, 275; practical conclusions about, 275–77; symbolic role of, 14n; theoretical conclusions about, 270–75. *See also* nuclear policy; policymakers; policymaking; policy styles
policymakers, 9, 17–18, 20, 100, 217, 269, 273; change in structures, 9. *See also* political leaders
policymaking: compared, 181–84, 267–70; by nonelected officials, 81–82; organizations, subsystems of, 15–16; state-centered explanation of, 4–7; structures changed by policy, 259; theoretical conclusions about, 270–75
policy styles, 13, 18, 22, 271, 272, 273; defined, 11, 19; and goals, 270; interactions of, with other styles, 35–37; national, 26n; partisan cleavages and, 21–24; symbolism and, 36; types of, for technical policymaking, 25–37; using, in analysis, 37–38. *See also* cost-benefiters; moralists; technological enthusiasts
policy trajectories, 17–18, 19, 268, 275
political leaders, 9, 14, 15, 158, 182, 257–58, 270–71, 272, 277
political movements, 11
political parties, 9, 10, 11, 14, 178; compared, 256–57; in France, 160–62; in Sweden, 66; in U.S., 191–92. *See also* dominant partisan cleavages; and specific parties
political sociology, 6
political systems, differences in, 17, 178, 181–82
politicians, 8, 179–80, 181, 184, 267; and antinuclear movement, 112, 113–14, 124, 188, 194; culture of, 14; French, local, 176; infighting among, 5
Pollard, Robert, 126
pollution, 167; air, 35; thermal, 124
Pompidou, Georges, 79–80, 82, 85, 100, 103, 157, 158, 160, 164, 174, 175, 270, 271

Poperen, Jean, 162
Poulantzas, 10
power, 5, 12; formal, 8, 270–72; informal, 8, 9, 10; "of preconceptions," 49n; state, 184
Power Reactor Demonstration Program, 42
premature deployment, 30, 62, 128, 215, 275
pressurized water reactor, 41, 69, 91, 154, 234
Price-Anderson Act (1957) (U.S), 43, 43n–44n, 49, 70, 194
prices, 27, 28, 116, 118, 120, 188, 189, 190, 216, 250, 267
Pringle, Peter, 42n
"Private Ownership of Special Nuclear Fuels Act," 44
private sector, 42, 45, 47–48; compared, 99–100, 181; and government, in France, 76, 84, 86, 162–63; and government, in U.S., 45, 51–52, 57, 108, 119, 126–27, 193, 199; and mixed ownership, in Sweden, 65, 69–70; versus public ownership, 271
Process Inherent Ultimately Safe Reactor, 217n
professionals, 9, 14, 32, 36. See also specific professions
profits, 162, 177, 182, 248–49, 274; of state-owned enterprises, 81, 82, 89
Project Independence, 3, 115–16, 139, 154, 156, 157
Project Independence Evaluation System (PIES), 116, 118
pronuclear position, 13, 95, 227, 228
Pryor, Larry, 121n
public, the: choice theory of, 6, 6n; debate by, 19–20, 178, 256, 268, 274; goods of, 28; hearings involving, in U.S., 55, 56, 58, 59, 71, 88, 121–22, 124, 126, 134, 179, 200, 262; inquiries by, in France, 71, 88, 240, 247; participation of, 57, 58, 98, 126, 165, 167, 171, 200n, 247
Public Law 85-256, U.S., 43n
public officials, xi, 7, 8, 9, 10, 258. See also elected officials; political leaders
public opinion, 8, 19, 20, 26n, 101, 101n, 259, 275; effect of policy on, 261–62, 264; in France, 159n, 164, 165, 238, 241, 243–44, 261, 268; in Sweden, 72–73, 137–38, 140, 230, 261–62; and TMI, 212–

13; in U.S., 56, 57–58, 98, 121–22, 124, 261, 262, 268, 269
public ownership, 102, 271. See also government; nationalization
Public Service Company, 209
public utility commissions (PUCs), 60, 62, 117, 118, 182, 187–88, 193, 201, 202, 205, 207–9, 214n, 215, 216, 217, 257, 262
Public Utility Regulatory Policies Act (1978) (U.S.), 191n
Puiseux, Louis, 74n, 76, 91, 93, 157n, 168, 180n, 239

Queens site, 109
Quermonne, Jean-Louis, 96n
Quilès, Paul, 247, 248

radiation: low-dose, 110–11; research, 214; standards, 124, 149
radioactive discharges, 152
Radiology Institute at Karolinska Hospital, 64
Rainer, Ove, 221
Ramey, James, 58, 114
Rancho Seco plant, 208
Rankin, William L., 261
rational actions, 12
Ray, Dixy Lee, 56n, 113, 120
Reactor Development and Technology Division, 56n
Reactor Safety Commission, 227n
Reagan, Ronald, 27, 28, 188, 193, 194, 208, 264
Réal, B., 159n
referenda: in California, 144, 206–7; in Sweden, 67, 218, 225–31, 236; in U.S., 121, 123, 124, 206–9
regions: opposition by, 164; pooling of, power of, 46n; regulation by, 57–58
regulation, 30; compared, 17, 98, 102, 179, 200n, 260, 262, 264, 267, 276; in France, 3, 17, 85–88, 200n, 256, 260; in Sweden, 17, 70–73, 260; in U.S., 3, 35, 43, 52–60, 72, 121–22, 123, 124–27, 187–88, 194–212, 215, 216–17, 256, 259, 260, 275
regulatory agencies, 7, 57, 58–59, 72
regulatory bureaucracy, 13
regulatory structures, 3, 8, 16, 17
regulatory styles, 57, 62, 72, 73

renewable energy resources, 159, 188, 191n, 193, 231–32, 233, 257, 260–61, 268, 277
rentabilité, 81, 89
Republican Party (U.S.), 17, 23, 44, 45, 51, 107, 119, 128, 135, 146, 182, 188, 258
research and development, 64n, 98; compared, 102; in France, 86; in Sweden, 64–65; in U.S., 56
Réseau des Amis de la Terre (RAT), 244n
"Respite" law (Sweden), 227
Reuther, Walter, 108
revolution, 107
revolutionaries, 120–21, 134, 163, 169–71
Revolutionary Communist League (France), 161
rhetoric, 11, 13, 14, 21, 22. *See also* policy styles
Ribicoff, Abraham, 59n
Rickover, Hyman, 41, 51, 52
Riesman, David, xi
Riksdag (Swedish parliament), 66, 72, 129, 130, 134, 136, 137, 139, 146, 223, 227, 229, 231
Ringhals reactors, 70, 220
RI reactor, 65
risks, 28, 35, 99, 125, 152, 167, 168, 197, 204, 210, 260; attitudes toward, 35n, 53
rituals, 11
Roberts, Marc. J., 50n
Roberts, Thomas, 212
Rocard, Michel, 162
Rocky Flats plant, 230
Rogers, Joel, 23n
Rogovin Report, 126, 210n, 211, 212, 230
Rolph, Elizabeth, 53, 53n, 58
Ruddick, Sara, 32n
Ruin, Olof, 226n
rural regions: cooperatives of, 48n; populations of, 132n. *See also* agrarians; farmers
Ryan, Alan, 11n
Rycroft, Robert W., 200n

SAFE, 225
safety, 5, 20, 30, 108, 209, 264, 272, 274, 275, 277; AEC and, 52–54; compared, 260; and cost, 28, 67, 204; in France, 86, 152–53, 169, 234, 245, 253–54; influence of policy on, 260, 262; issue, U.S. scientists and, 110–11, 122, 126; and new reactor designs, 217n; in Sweden, 67, 72, 234, 253; after TMI, 210–12; in U.S., 52–54, 58, 59, 60, 126–27, 199, 268
Sahr, Robert, 70n, 138n, 142, 145
Saint-Laurent reactor, 75, 78
Samuel, Pierre, 150
San Diego Gas and Electric, 207
Sandgren, Lennart, 227n
San Onofre plant, 126n, 214n
Saudi Arabia, 114
Saumon, Dominique, 74n, 76
Sauvage, Le, 150
Sawhill, John C., 193n
Scaminaci, James, III, 33
Schattschneider, E. E., 23
Scheinman, Lawrence, 74n
Schipper, Lee, 141n
Schlesinger, James, 55–56, 120, 124, 127n, 189–91, 257
Schloesing report, 242
Schmidt, Ed, 234n
Schnaiberg, Allan, 135n
Schwartz, Michael, 203n
Schwartz Report, 82n
science, philosophy of, 10
scientists, 30, 42, 46, 54; antinuclear, 110–11, 135, 148, 165–66, 170, 176, 178, 238
Seaborg, Glenn, 53, 120
Seabrook plant, 58n, 123, 179, 208, 209, 217, 263n
secrecy, 165
Secretariat for Future Studies (*Sekretariatet för framtidsstudier*) (Sweden), 142, 180, 232, 271
securities markets, 175n, 183
Sené, Monique, 249n
Servant, Jean, 245, 253
Service Central de Sûreté des Installations Nucléaire (SCSIN), 86, 87
service sector, 33
Seventh Plan, 173
Shaw, Milton, 51, 52, 53
Shinn, Terry, 84
Shippingport reactor, 44
Shonfield, Andrew, 57
Shoreham reactor, 199, 208, 217, 263n
Sierra Club, 109, 121n, 150
SIFO, 230n
sigyn, 234
Silent Spring (Carson), 129
Silverman, Lester P., 193n

Simon, William, 115
Simonnot, Philippe, 74n, 78, 85, 158
site occupations, 122–23, 179, 239
Sixth Plan, 90, 92, 93
SKI. *See* State Nuclear Inspectorate
Skowronek, Stephen, 15
Smith, V. Kerry, 28n
social: psychology, 4, 14; status, 12; structure, 7n; theory, 10–11
Social Democrats (Sweden), 17, 24, 66–67, 129, 131–32, 133, 137–47, 181–82, 218, 219, 221, 222, 223, 232–33, 249, 257, 258, 264, 268, 274; and TMI, 224–31
Socialist Party (PS) (France), 148, 161–62, 169, 170, 171, 174, 176, 177, 237, 239–40, 244, 246–50, 257, 258, 264, 269, 273
socialization, 13
Söder, Karin, 233n
SOGERCA, 91
solar energy, 193, 232
Solé, Jacques, 95n
Southern California Edison, 214n
South Texas Nuclear Project, 201, 208, 209
Soviet Union, 7, 64, 213n, 230
Spiegelman, James, 42n
Sporn, Philip, 49n
standardization, 198, 234, 252–53, 259, 268, 275
state, 20, 121; and antinuclear movement, 179; autonomy perspective, 6; -centered explanation, 4–7; conflict within, 7, 8, 270; and credit, 175n; defined, xi–xii, 6–7; and legal and regulatory styles, 72–73; power, 184; structures, 178; support for nuclear expansion, 16, 17; weak, versus strong, 98
State Department (U.S.), 190
State Nuclear Inspectorate (*Statens Karnkraftinspektion* of SKI) (Sweden), 71–72, 88, 220, 223, 234
State Power Board (*Statens Vattenfalsverk*) (Sweden), 65, 67, 68, 69–70, 220, 231, 234
states (U.S.), 205–9, 217; legislatures, 188, 206–7; and NRC, 59, 208; referenda on nuclear power, 121, 123, 124, 144, 206–9; utilities, 48n; and waste disposal, 215. *See also* public utility commissions
Steen, Peter, 219n
Stello, Victor, 212n

Stemmelen, Eric, 245n
Sternglass, Ernest, 110, 149
Stipulation Act (1977) (Sweden), 71, 219–20, 223–24, 231, 235
Stobaugh, Robert, 156n
Stockholm environmental conference, 130
Stockholm Technical Institute, 230
strategies, 13
Strauss, Lewis, 31, 44, 44n
structural approach: and antinuclear movement, 179; and Carter Energy Plan, 190n; change, crisis and, 15–16; to French energy policy, 82; pitfalls of, 200n; to policy decision, 5, 7n–8n, 8–10; and rational actions, 12–13
structural-functionalism, 37
structuralism, 11
structures, 6–10, 14, 98, 100, 217; changes in, 4, 9, 12n, 15–16, 277; and cultural factors, 14, 103, 273; defined, 7n; elites and, 277; enabling versus constraining, 7n; and flexibility, 256–64; and individual biography, 15; and international markets, 101; limitations of, in explaining policies, xi–xii, 8–10; theoretical considerations, 270–75
Sturmthal, Adolph, 80n–81n
subcultures, 11n
submarine propulsion project, 52
Suleiman, Ezra, 84, 96n
Sundesert reactor, 207
Super-Phoenix breeder reactor, 170, 237–38
Surrey, John, 253n
Survivre, 150
Survivre et Vivre, 150
Swahn, Urban, 221n
Sweden, 32, 49; divergence of policies from France and U.S., 267–70; early commercialization, versus France and U.S., 98–103; early victory for light water reactor in, 41, 61, 64–73, 77; elite discretion in, versus France and U.S., 178–84; versus France and U.S., xii, 3–7, 8, 16–20, 21, 24–25, 26, 28, 50, 76, 80, 132n, 146, 193, 274; future of nuclear energy in, 269–70; and nuclear waste, 254; and nuclear weapons, 74n; and oil crisis, 114–15, 120, 156, 180; oil crisis and party politics in, 21, 24–25, 129–47, 274; paralysis and antinuclear

compromise in, 218–36; phaseout of nuclear energy in, 232–33; policy styles in, 26; protechnology values in, 30n; referendum on nuclear energy, 67, 218, 225–31, 236; structures and flexibility, versus France and U.S., 256–64. *See also* specific agencies, commissions, issues, laws and sites
Swedish Academy of Science, 134–36
Swedish Institute, 132
Swedish Left, 17
Swedish Right, 17
Swidler, Anne, 22
Sydkraft, 70, 234
symbolic: interactionist tradition, 10n; lightning rod, 13, 24, 120–21
symbols, 11–12, 13, 14n, 36
synthetic fuels, 191n

Tamplin, Arthur, 110–11, 124, 149, 164–65
Tanguy, Pierre, 243n
Taylor, Charles, 11n
technicians, antinuclear, 170, 176, 178
technocracy: in France, 83–85, 95–96, 95n, 158, 162, 179, 181; in Sweden, 137; in U.S., 107
technological change or development, 13, 21, 174; antinuclear movement and, 121, 133; business leaders and, 33–34; and costs versus risks, 28; environmentalists and, 33–35; as solution to political problems, 31; symbolism of, 36
technological choices, 275–77
technological enthusiasts, 19, 22, 26, 27, 28, 180–82, 184, 256; allied with cost-benefiters, 100, 103, 267; antinuclear, 134–35, 148, 178; versus antinuclear movement, 113–14, 123; Carter and, 188–89, 190, 191n; compared, 178; cost-benefiters versus, 56, 68, 73, 107–8, 115, 157–58, 215–17; defined, 25, 26n, 29–31; in France, 75, 76–78, 79, 84–85, 93–95, 96–97, 148–77, 249, 254–55; and growth, 33, 34; nonprofessional versus professional, 30–31, 36; and other styles, 36; versus realists, 43, 46, 52–54, 84; in Sweden, 67, 68, 73, 158, 220; in U.S., 41–63, 99, 158, 188, 192, 196, 202, 259
technological realists, 31, 43, 46, 84; and safety, 52–54, 111

Teller, Edward, 230
Temples, James R., 212n
Tennessee Valley Authority (TVA), 43n, 47, 48n, 212n
Texas, 209
Tham, Carl, 223, 226
Third World, 155
Thomas, Steve, 253n
Thomson, Meldrim, 209
thought style, 21n
Three Mile Island (TMI), 18, 187, 193, 201, 203, 204, 208, 209–14, 217, 256; effect of, in France, 243–45, 253, 264; effect of, in Sweden, 218, 224–31, 264; and policy flexibility, 264
Touchard, Jean, 95n
Touraine, Alain, 7n, 32, 123, 149n, 244
trade balances, 159, 168, 180n
trade unions, 135, 144, 152–54, 227. *See also* labor leaders
Treasury (France), 92
Tricastin reactor, 244n
Turner, Stephen, 11n
Turner, Victor, 23
turnkey plants, 45–46, 51, 67–68, 76, 78, 99, 100, 103

Udall, Morris, 192, 193
UDC-Que Choisir, 244
Ullsten, Ola, 223
Unckel, Per, 223n, 234
Unger, Roberto, 7n, 275
Unified Socialist Party (PSU) (France), 161, 162, 169, 170, 171, 174, 176, 238, 244, 246
Union of Concerned Scientists, 54, 111, 126, 133–34, 166, 230, 262
Unitary Systems, 10
United Auto Workers, 108
United States, 142; and collapse of nuclear energy, 215–17; commercialization, versus France and Sweden, 98–103; development costs in, 252–53; divergence of policies from France and Sweden, 267–70, 273; and dominant partisan cleavages, 21, 23–24, 51–52, 274; early triumph of technological enthusiasm in, 41–63; and elite discretion, versus France and Sweden, 17–18, 178–84, 217; versus France and Sweden, xii, 3–7, 8,

326 · Index

United States (cont.)
16–20, 21, 24–25, 26, 28, 50, 76, 80, 132n, 146, 193, 274; future of nuclear energy in, 269, 277; hidden weaknesses, 61–63, 76, 128, 194, 210, 215, 259, 268–69; high costs and decentralization of control in, 187–217; and nuclear weapons, 74n; and oil crisis, 154, 180, 180n; protechnology values in, 30n; reassertion of economic perspective in, 107–28; structures and flexibility, versus France and Sweden, 256–64; technology used too early in, 30, 62, 128, 215, 275. See also specific agencies, commissions, issues, laws and sites

U.S. Congress, 18, 23, 24, 43n, 56, 118, 119, 124, 126, 136, 187, 188, 190, 191–94, 197, 200, 205, 212, 213, 214, 258

U.S. Executive Order 12291 (1981), 28

U.S. House of Representatives, 190, 190n, 192–93; Committee on Government Operations, 195, 241n; Energy and Commerce Committee, 192n; Science, Space, and Technology Committee, 192; Subcommittee on Energy and the Environment, 192; Subcommittee on Energy Research and Production, 192

U.S. Senate, 190, 192–93

U.S. Senate-House Joint Committee on Atomic Energy (JCAE), 42–43, 44, 51, 52, 55, 58, 59n, 63, 75, 108, 112, 123, 193; dissolved, 187, 191, 194, 205; versus European counterparts, 98; power reduced, 124, 127n; and safety, 53, 111

U.S. Supreme Court, 43n–44n, 55, 59, 205–7, 208n, 214–15

universities, 14, 189

uranium, 3, 16, 44, 68; enriched, 78, 79, 119; in France, 74, 75; in Sweden, 64, 65, 67–68, 71n, 134; in U.S., 180n

urban voters, 133, 164

Utah, 209

utilitarian approach, 6, 9n, 12

utilities, 5, 8, 19, 99, 193, 272; and antinuclear movement, 109–10; compared, 99–100, 102, 103, 184; cooperation between government and, 98, 182–83, 276; in France, 74, 75, 80–83, 86–87, 157, 158; in Sweden, 68–69; in U.S., 41–51, 57–58, 59, 60–62, 116–20, 127, 128, 131, 142, 157, 158, 175, 187, 188, 191n, 193–204, 208–9, 210–12, 213, 214–17, 257, 259, 268, 271; U.S. state and municipal, 48n. See also financing; management; public utility commissions; and specific issues and companies

Utility Finance Corporation, proposed (U.S.), 119–20

Vallecitos plant, 42n
values, 10, 12n, 37
Vandellos, Spain, reactor, 75
Vänner, Jordens, 130, 135
Vanoise national park, 149
Veblen, Thorstein, 101, 102n, 275
Vedung, Evert, 138n, 222n
Veil, Simone, 241
Vermont Yankee v. Natural Resources Defense Council, 206
veto, municipal, 71, 134
violence, 169–70, 239n, 245
Vogel, David, 57
voters and voting, 5, 6n, 24n, 132, 140n. See also elections

Wååg, Nils-Erik, 233n
Washington Public Power Supply System, 208, 213
Washington State, 208
"WASH-740," 43
Waste Isolation Pilot Plant (WIPP), 215
Watergate scandal, 115
Weart, Spencer R., 74n
Weber, Max, 33
Webster, Frank, 161n
Weinberg, Alvin, 47, 217
Wenner, Lettie, 205
West Germany, 7, 16, 18, 67, 70, 122, 179, 251
Westinghouse, George, 46n
Westinghouse, 31, 41, 45–46, 47, 51, 69, 91, 99, 103, 154, 162
whistleblowers, 122, 197. See also engineers, and U.S. antinuclear movement; scientists, antinuclear; technicians, antinuclear
Whyle, West Germany, 179
Wiecek, William M., 45
Wikdahl, Carl-Erik, 232, 233
Wildavsky, Aaron, 12n, 34–35, 253n
Wilson, Byran R., 11n
Winch, Peter, 11n

Wirth, Tim, 217
Wittrock, Björn, 68
women, 32, 189, 222; in environmental movement, 33, 131, 135
"world time," 102
worldviews, 11, 13, 14, 15, 21. *See also* policy styles
Wright, J. Skelly, 54, 55n

"yellow energy bill," 223n
young people, 33, 131, 132, 189, 222
Young Socialists (Sweden), 225
Yucca Mountain site, 215

Zetterberg, Hans, 145, 229n, 230n
Zimmer plant, 126n
Zysman, John, 175n